Walter's Vegetation of the Earth

M000311336

# Springer

*Berlin*
*Heidelberg*
*New York*
*Barcelona*
*Hong Kong*
*London*
*Milan*
*Paris*
*Tokyo*

**Fig. 1.** *Euryops walterorum,* a bushy Asteracea, named in honour of Heinrich and Erna Walter. It is an endemic plant of the Gamsberg region in Namibia (Photo: U. KULL, February 1993)

Siegmar-Walter Breckle

# Walter's Vegetation of the Earth

## The Ecological Systems of the Geo-Biosphere

4th, Completely
Revised and
Enlarged Edition

Translated from the 7th, Completely
Revised and Enlarged German Edition
by Gudrun and David Lawlor

With 300 Figures

 Springer

Heinrich Walter †

**Siegmar-Walter Breckle**
University of Bielefeld
Department of Ecology
Universitätsstr. 25
33615 Bielefeld
Germany

*Translators*
Gudrun und David Lawlor
9 Burywick
Harpenden
Hertfordshire, AL5 2AQ
UK

Translated from the 7th German Edition "Vegetation und Klimazonen"
© 1999 by Eugen Ulmer GmbH & Co., Stuttgart, Germany

| | | | |
|---|---|---|---|
| First edition | 1973 | $2^{nd}$ printing | 1983 |
| $2^{nd}$ printing | 1975 | Third edition | 1985 |
| $3^{rd}$ printing | 1977 | $2^{nd}$ printing | 1994 |
| $4^{th}$ printing | 1978 | Fourth edition | 2002 |
| Second edition | 1979 | | |

# ISBN 3-540-43315-5 Springer-Verlag Berlin Heidelberg New York

Library of Congress Cataloging-in-Publication Data

Walter, Heinrich, 1898-
    [Vegetationszonen und Klima. English]
    Walter's Vegetation of the earth: the ecological systems of the geo-biosphere/
Siegmar-Walter Breckle; translated from the 7th, completely revised and enlarged
German edition by Gudrun and David Lawlor. – 4th, completely rev. and enl. ed.
    p. cm.
    Includes bibliographical references (p. ) and index.
    ISBN 3540433155 (Softcover: alk. paper)
    1. Plant ecology. 2. Vegetation and climate. 3. Phytogeography. 4. Plant phy-
siology. 5. Life zones. I. Breckle, Sigmar-W. II. Title.
    QK901.W2613 2002
    581.7–dc21

Springer-Verlag Berlin Heidelberg New York
a member of BertelsmannSpringer Science+Business Media GmbH
http://www.springer.de

© Springer-Verlag Berlin Heidelberg 2002
Printed in Germany

Cover design: Design & Production GmbH, Heidelberg
Typesetting: K + V Fotosatz GmbH, Beerfelden
SPIN 10794300        31/3130-5  4  3  2  1  0 – Printed on acid-free paper

# Preface to the Fourth English Edition

The German edition of this book, the seventh edition, was published 9 years after the death of the original author of this comprehensive book, Prof. Heinrich Walter. He had actually thought that the translation of the fifth German edition was likely to be the last translation – it was done when he was 85 years old.

Over the last 10 years, ecological research and new worldwide studies have brought many new insights and detailed knowledge. Therefore, the seventh German edition had to be intensively revised and amended. Our insights and priorities have changed. In ecology, more and more diverse fields of science have developed. Also in politics and socio-economics, ecology becomes more important, however, often in a very imprecise way. Nowadays, the prefix 'eco-' is very often combined with other terms which have almost nothing to do with any ecological background, but which appear to represent some kind of hope or even faith. And often, political declarations and real measures contrast or even contradict each other strongly. Scientists have to be careful, but have to raise their voices.

The conference of the UN in Rio 1992 was remarkable. The consequences of political decisions, however, are still very few. The three conventions – climate change, conservation of biodiversity and combating desertification – are important for all nations. They are even binding laws. The bureaucracy of international bodies makes the processes of innovation and improvement very slow.

Sound ecological knowledge and, in some cases, enormous research efforts are needed. However, this does not yet seem to have reached the political level in the necessary way. This new edition is a short, comprehensive textbook. The four volumes of *Ökologie der Erde* (second edition, three volumes in English: *Ecological Systems of the Geobiosphere*) cover the various zonobiomes of the globe in a more detailed way. This is possible because of the author's own experience of most floristic realms and climatic zones which leads to comparable material on a global scale.

Our aim was to include many graphs and figures in order to draw a clear and illustrative picture of the various results and regions. However, ecology cannot only be learned and understood from books, in the laboratory or in the library, but rather according to the following principle: "The ecologist's laboratory is God's nature and the whole earth is his field". Therefore, it is important for students studying ecology to participate in many excursions. This seems to have been forgotten in many universities.

I would like to thank the publishers, my wife, Uta, and my daughter, Margit, for their unstinting help. Also, the technical help of Anja Scheffer is gratefully acknowledged.

Bielefeld, April 2002                                   S.-W. BRECKLE

# Foreword to the First English Edition

Ecology is current, exciting, relevant, and offers guides to action and even some hope of harmony and order, as well as congenial environment for mankind in an overpopulating world. Plant ecology is basic to general, animal systems, paleo-, and human ecology. Plants are the primary producers. They dominate the flow and cycling of energy, water, and mineral nutrients within ecosystems. The structure of the vegetation determines much of the character of the landscapes in which other organisms live and prosper, including men and women.

Plants are immediately at hand for study. They are evident, mobile only in certain stages, familiar, easily identified, and related to a rich literature on the properties of various kinds of plants. If we know why plants grow where they do, we know a good deal about why organisms other than plants live where *they* do.

Plant ecologists need a general botanical background. Professor WALTER has that background. His multivolume textbook series, "Introduction to Plant Science," includes books on general botany, systematics, ecology in a strict sense (2nd ed., 1960), plant geography (2nd ed., with H. STRAKA, 1970), and on vegetation. ELLENBERG covers the last topic with books on principles of vegetation organization (1956) and the vegetation of central Europe (1963). These texts are in German, published by Ulmer (Stuttgart), and the reader can be referred to reviews of these (Ecology 38(4):666–668, 1957; 43(2):346, 1962; 47(1):167–168; 1966; and J. Ecology 55(1):234–245, 1967).

It is not enough for the plant ecologist to be well-grounded in principles. Our so-called ecological principles need continual reexamination, questioning, and testing. Principles must be drawn from, and applied to the specific ecological relationships of plants in particular ecosystems. Concrete ecosystems are the testing ground for prinicples as well as their source. So the ecologist needs an overview of the earth's plants and its vegetation. What are the possibilities and actualities of plant growth and vegetation organization?

Professor WALTER's two volumes in German entitled "Vegetation of the Earth Considered Ecophysiologically"

(1964 and 1968) are an admirable summary of much of this basic plant ecology. They are separated by a language barrier from many persons who need to use them. They have been justly praised by A. Löve (Ecology 50(6):1105–1106, 1969) and by GRUBB (J. Ecology 58(1)315–316, 1970[1]. They record WALTER's very extensive firsthand knowledge of much of the earth's vegetation. They add the progress made in our ecophysiological understanding of how and why plants grow where they do throughout the world since SCHIMPER's founding work, "Plant Geography on a Physiological Basis" (1898, 1903 in English, 2nd German edition by VON FABER in 1935). WALTER's books are indispensable to every practicing plant ecologist, as Löve says. Their 1593 pages are a rich feast. They also stand apart, separated by their awesome scholarship, bulk, cost, and rich detail. Students who lack a good geographical schooling or have only a rudimentary plant taxomonic background may find them difficult to digest. Further, WALTER's books are traditionally scientific, international in intent and scope, but this tradition contrasts with a widespread general retreat into parochialism which sometimes seems to be a corollary of the potentially wide intellectual scope of ecology. Or perhaps we all just have too much to do.

In 1970, Professor WALTER published a small volume on the ecology of zonal kinds of vegetation, in relation to climate, viewed causally, and covering all the continents. The English edition of that book you have in your hand. The German edition was reviewed in Ecology (52(5):949, 1971). The book is short but replete with facts. It places these facts in a consistent frame of reference. It suggests where more factual data are needed. It is also neat, precise, very readable, an excellent summary of two larger books, and current.

In recent years ecophysiology has attracted much interest from very skilled botanists. Technical progress in instrumentation has made prossible some accurate measurements of photosynthesis and transpiration in the field, rapid chemical analyses of soils and plants, the sensing of elements of the heat balances of organisms, measurements of water potential in both soils and plants, computer modeling of the photosynthetic and respiratory processes in stands of vegetation, summarization of masses of data which would have formerly overwhelmed the investigator, etc. No synthesis of the principles of ecophysiology has appeared. But WALTER's short book does something else

---

[1] English edition of Vol. 1 cf. J. Ecology 60(3):940–941 (1972)

that is very valuable. It provides a framework of ecological descriptions of kinds of zonal vegetation into which ecophysiological data must fit.

Plant ecophysiology can be used variously. Plant success, adaptations to environment whether ecotypic or plastic, should be definable in ecophysiological terms. For others the goal may be understanding the physiology of vegetation, those organizations of individual plants which many botanists ignore. Progress has been made in studying the physiology of crop stands. WALTER records in this book new data on biomasses which summarize the results of the physiological processes in various kinds of vegetation and express their structure.

Under the International Biological Program many kinds of ecosystems are being studied. In the U.S. these include grasslands, deserts, deciduous, coniferous, and tropical forests, and tundra. A project on "Origin and Structure of Ecosystems" says, "The fundamental biological question that this program is asking is whether two very similar physical environments acting on phylogenetically dissimilar organisms in different parts of the world will produce structurally and functionally similar ecosystems. If the answer is no, there cannot be any predictive science of ecology. In fact, knowledge acquired from studying a given ecosystem cannot be applied to an analogous ecosystem, unless similar physical environment indeed means similar ecosystem" (U.S. National Committee for the International Biological Program. National Academy of Sciences, Washington, D.C., Report 4:46). The answer to the question is obviously no in general, but the conclusion suggested does not follow. True, there are many similarities between the plants and vegetations in different parts of the world that have similar climates, soil parent materials, topographies, fire histories, and plant successional and soil developmental histories even though their biotas differ, and WALTER mentions examples repeatedly. But he also mentions exceptions. Ecology is not a simple matter of adaptations to environment. It must consider that biotic diversity over the earth is an ecological factor, and WALTER outlines some of the classical conclusions on floristic diversity in the first pages of his text. The evidence is already clear that the partial effects of differing genetic inputs into environmentally closely similar ecosystems persist ecologically in the form of specific structures and functions. The C-4 photosynthetic pathway is a splendid, so far almost ecologically meaningless, example.

Further, structure and function should be considered separately. They have no necessary and invariant connection. Premature correlation, such as Schimper's bog xero-

morphosis, is a mistake. WALTER's generalization of "peinomorphosis" represents real progress. While the apparent congruence of structure and function are striking,
study of them in detail and the evident exceptions have
led to solution of many ecological puzzles.

Approaching the problem from another direction, ecological principles are applied widely in such kinds of land
management as agriculture, forestry, range and pasture
management, pollution control, park and wilderness management, and natural area maintenance. Ecological principles evidently exist and do allow predictions. WALTER's
little book is a good corrective to the kind of hurried,
mistaken generalization that was quoted.

Of course evolution is a dimension of ecology. Unfortunately the concept of adaptation is only used by evolution-
ists and not often measured. In fact, there is often a shifting from form to function and vice versa in discussions of
adaptation, and substitution of two unknowns for one.
WALTER does not make this mistake either. He does make
clear how much more research needs to be done to describe plant and vegetation structure and function, their
correlations, and their relationships to environment. And
his book records a fine start.

Professor WALTER is a plant ecologist of very wide and
long experience. He and his wife are knowledgeable, en-
thusiastic, indefatigable, and helpful field companions. His
investigations, and those of his students, have suggested,
checked, and documented his ecological ideas derived
from extensive travel and residence in all the continents.
His teaching has clarified their presentation.

Many have questioned whether plant ecology in its century of development since HAECKEL coined the parent
term in 1866 has developed any principles. This book
provides an affirmative answer.

University of California, Davis                    JACK MAJOR
February 8, 1972

# Contents

## General Section

## Special Section

# Physical Quantities, Units and Conversion Factors

## Basic Units

| | | |
|---|---|---|
| Length | Metre | m |
| Mass | Kilogram | kg |
| Time | Second | s |
| Temperature | Kelvin | K |
| Luminous intensity | Candela | cd |
| Amount of substance | Mole | mol |

## Derived Units

Force      Newton   N
$$1\ N = 1\ kg \cdot m \cdot s^{-2} = 0.102\ kp$$

Pressure      Pascal   Pa
$$1\ Pa = 10^5\ Pa = 0.9869\ atm =$$
$$750\ Torr = 750\ mm\ Hg$$

Energy      Joule   J
$$1\ J = 1\ N \cdot m = 10^7\ erg$$

Heat units
$$1\ kcal = 4.187\ kJ = 1.163\ Wh$$
$$1\ J = 0.102\ kp \cdot m = 2.29 \cdot 10^4\ kcal$$
$$= 2.78 \cdot 10^{-7}\ kWh$$

Power      Watt   W
$$1\ W = 1\ J \cdot s^{-1} = 1\ N \cdot m \cdot s^{-1}$$
$$= 0.102\ kp \cdot m \cdot s^{-1} = 0.236\ cal \cdot s^{-1}$$
$$= 0.86\ kcal \cdot h^{-1}$$

Illuminance, light flux      Lux   lx
$$1\ lx = 1\ lm \cdot m^{-2} = ca.\ 10^{-2}\ W \cdot m^{-2}$$

Luminous flow      Lumen   lm
Light intensity      $cd\ m^{-2}$
$$1\ lx\ (red\ light) = ca.\ 4 \cdot 10^{-3}\ W \cdot m^{-2}$$
$$1\ lx\ (blue\ light = white\ light)$$
$$= ca.\ 10^{-2}\ W \cdot m^{-2}$$
$$1\ W \cdot m^{-2}\ (PhAR) \approx 3\text{--}5\ \mu\ Einstein\ m^{-2} \cdot s^{-1}$$
$$1\ Einstein = 1\ mol\ photons = 75\ kcal$$
$$(blue) = 3 \cdot 10^5\ J$$

## Further Conversions

- $1 \text{ g DW} \cdot \text{m}^{-2} = 10^{-2} \text{ t} \cdot \text{ha}^{-1}$
- 1 g organic matter $\approx 0.45$ g C $\approx 1.5$ g $CO_2$

## Transformation Energy for Changes in the State of Water

- solid $\leftrightarrow$ fluid (melting; freezing):
  $0.3337 \text{ MJ} \cdot \text{kg}^{-1}$ ($79.5 \text{ cal} \cdot \text{g}^{-1}$)
- fluid $\leftrightarrow$ gaseous (evaporation, condensation):
  $2.26 \text{ MJ} \cdot \text{kg}^{-1}$ ($539 \text{ cal} \cdot \text{g}^{-1}$)
- gaseous $\leftrightarrow$ solid (sublimation):
  $2.86 \text{ MJ} \cdot \text{kg}^{-2}$ ($684 \text{ cal} \cdot \text{g}^{-1}$)

## Internationally Agreed Prefixes for Units and Appropriate Factors

| | | | | | | |
|---|---|---|---|---|---|---|
| $10^1$ | Ten | deca | da | $10^{-1}$ | deci | d (tenth) |
| $10^2$ | Hundred | hecto | h | $10^{-2}$ | centi | c (hundreth) |
| $10^3$ | Thousand | kilo | k | $10^{-3}$ | milli | m (thousandth) |
| $10^6$ | Million | mega | M | $10^{-6}$ | micro | μ (millionth) |
| $10^9$ | Billion | giga | G | $10^{-9}$ | nano | n |
| $10^{12}$ | Trillion | tera | T | $10^{-12}$ | pico | p |
| $10^{15}$ | Quadrillion | peta | P | $10^{-15}$ | femto | f |
| $10^{18}$ | Quintillion | exa | E | $10^{-18}$ | atto | a |

# Abbreviations and Symbols

| | |
|---|---|
| a | year |
| A | flux per unit area of soil surface |
| A | A-horizon in soils (with predominantly organic material) |
| B | B-horizon in soils (transition horizon between organic cover and weathered parent rock) |
| BHD | breast height diameter of tree stems in centimetre |
| C | C-horizon in soils (subsoil: weathered parent rock in soil profile) |
| °C | degree Celsius |
| cal | calorie |
| CAM | Crassulacean Acid Metabolism |
| CEC | Cation Exchange Capacity |
| D | day length |
| DI | diversity index |
| DW | dry weight |
| E | Einstein (light quantum) |
| E | east |
| Ea | actual evaporation |
| Ep | potential evaporation |
| ET | evapotranspiration |
| FW | fresh weight |
| g | gram |
| G | G-horizon in soils (stagnant, oxygen-deficient gley horizon) |
| GPP | gross primary production |
| h | hour |
| ha | hectare ($10^4$ $m^2$) |
| I | interception |
| J | Joule |
| K | Kelvin |
| kg | kilogram |
| kW | kilowatt |
| l | litre |
| LAI | leaf area index |
| lx | lux |
| m | metre |
| M | mass |

| | |
|---|---|
| mg | milligram |
| min | minute |
| ml | millilitre |
| mm | millimetre |
| mol | mole |
| μm | micrometre |
| N | newton |
| NPP | net primary production |
| OB | orobiome |
| p | vapour pressure |
| $p_o$ | vapour pressure of pure water |
| P | turgor pressure |
| P | precipitation |
| Pa | Pascal (1 Pa $= 10^{-5}$ bar) |
| PB | pediome |
| pH | negative logarithm to the base 10 of the hydrogen ion concentration (strength of acid) |
| Ph | photosynthesis |
| PhAR | photosynthetically active radiation |
| ppb | parts per billion |
| ppm | parts per million |
| $\pi^*$ | osmotic potential |
| R | respiration |
| RH | relative humidity |
| RQ | respiration quotient (carbohydrates $= 1$, fats $= 0.7$) |
| s | second |
| S | south |
| sZB | subzonobiome |
| t | time |
| t | tonne ($10^3$ kg) |
| T | transpiration |
| Torr | $=$ mm Hg, previous pressure measure (750 Torr $= 10^5$ Pa) |
| UV | ultraviolet light |
| W | west |
| WC | water content |
| ZB | zonobiome |
| ZE | zonoecotone |
| Φ | water potential |

# Introduction and Remarks

## 1 Aims of Ecological Science

From our present knowledge, life occurs only on Earth within the solar system, and depends on processes involving open cycles of fluxes of matter coupled with energy fluxes, i.e. the synthesis of materials, driven by solar energy, and the degradation of the material with the accumulated energy usually released as heat.

The smallest autonomous unit of life is the cell. Its structure and function are investigated by the sciences of molecular biology, biochemistry and physiology, which include analysing ultrastructures with the latest techniques as well as the understanding and manipulation of genetic material.

Green plants, algae and bacteria, which photosynthesise, are autotrophic and their metabolism leads to the accumulation of matter and energy in the biosphere. In contrast to all other organisms (including non-green plants, fungi, bacteria and others), which are heterotrophic and their metabolism leads to the loss of matter and energy from the atmosphere. Single cell micro-organisms are the primary targets of the science of microbiology; fungi are now usually classified as a separate group of organisms and are the subject of the discipline of mycology. The next level of organisation is the multi-cellular organism with its tissues and organs.

Ecology, as a biological science, deals with the interactions of organisms and their environment. Ecological factors affect the different levels of complexity, of course, even at the molecular level (Table 1) and cause certain effects and interactions. At the level of the individual, adaptation results from modification, mutation and selection, the topic of investigation of **autoecology.** At the level of ecosystems, these adaptations are expressed as continuously changing structures of populations and continuously changing dynamics of metabolic cycles and energy flows. Populations are investigated in the field of **population ecology.** Whilst **synecology** considers communities and their composition (static analysis), **ecosystem biology** researches the dy-

**Table 1.** Different levels of complexity and examples of what affects them

| Level of complexity | Example of responses, possible effects of environmental conditions |
|---|---|
| Interactions and effects in biomes within the biosphere (large scale ecosystems) | Cycles of materials, balances of matter, energy fluxes, sedimentation, accumulation of eroded material in basins, geomorphic long-term processes |
| Interactions and effects in ecosystems | Acquisition, cycles, balances and accumulation of materials, species composition (frequency and dominance) |
| Effects on populations | Reproduction, age distribution, competitive ability, selection |
| Interactions with intact, whole plants, individuals | Mineral metabolism, fitness and vigour, water balance, adaptation of growth, developmental stages, hormone balances |
| Interactions with cells | Developmental effects, changes in differentiation, accelerated senescence |
| Interaction with tissues | Effects of stresses (e.g. osmotic and ionic concentration) on development, including damage |
| Effects on cell organelles | Respiration, photosynthesis, biosynthesis of secondary plant metabolites |
| Effects on biomembranes | Permeability, changes in potential |
| Biological effects on macromolecules | Gene regulation, enzyme activity, changes to DNA |

namics of communities and, thus, the characteristics which cause energy flows and metabolic cycles.

The highest units of life are the communities of organisms, plants and animals which, together with abiotic environmental factors (climate and soil, see p. 19, 33), form ecosystems. These are characterised by a continuous cycling of material and flow of energy. Ecology, in the widest sense of the word, is the science of these ecosystems, from the very smallest to the global level – the biosphere.

This book is intended to serve as a brief comprehensive introduction to this global ecological system.

Walter, the initial author of this book, expressed the relation between human beings and biosphere as follows:

"The biosphere comprises the natural world in which man has been placed and which, thanks to his mental capacities, he is able to regard objectively, thus raising himself above it. On the one hand, he is a child of this external, apparent world, and dependent upon nature, but on the other hand, through the world within himself, has access to the divine.

Only an awareness of both sides of his nature enables man to develop into a wise and harmonious being with the hope of divine fulfilment upon death. It is not the sole calling of man to use nature to his own ends. He also bears the responsibility for maintaining the earth's ecological equilibrium, of tending and preserving it to the best of his ability."

If man is to fulfil this task and not exploit the environment in a way which, in the long run, jeopardises his own existence, he has to recognise the ecological laws of nature and act upon them, even if, at times, there are people who believe that they can do away with nature and rely completely on technology.

We shall limit our observations to the conditions in natural ecosystems, since it would be beyond the scope of this book to embark upon the consideration of secondary, man-made ecosystems and to consider the various degradation stages in detail. Ecological laws of nature can be better understood in natural ecosystems, which are also in a dynamic equilibrium. Natural ecosystems are the model and the reference point for sustainability. They have become optimised over millions of years of evolution.

## Significance of Present-Day Ecology as a Philosophy

Ecology is a part of biology, namely the science of the management of nature and – as formulated by HAECKEL (1866) – the science of the relation of organisms to their surrounding environment.

Scientific ecology is now defined comprehensively as the "science of the interactions of organisms with each other and with their environment".

The population, as a whole, has only become aware of the importance of ecology during the last two decades. However, ecology has changed completely with the advance of the "green movement". The prefix "eco" has been added to terms which are not at all related to the science of ecology and has, in part, become embellished as a type of saviour religion. In ecology there are no value judgements: a natural catastrophe is, in the ecological sense, nothing bad. A hurricane, a tsunami, hail, invasion of savannas by woody vegetation, fires in the steppe, all these are natural processes. They are part of the natural dynamics, which, however, man can now significantly modify and thus affect nature beyond the local scale via global mechanisms.

The term 'ecological' is commonly used in advertising to imply healthy and wholesome (organic) foods and ways of production and living. It is frequently only make-believe or deception for the general public. Where is the rational thinking against consumerism and indoctrination by advertising?

These processes are immediately awarded value judgements if they negatively change the conditions of life and the environment of humans. To understand the laws of ecology it is important to first consider situations in the abstract and only in a second or third step evaluate the "ecological effect on humans".

Today, it appears even more important to work with clearly defined terms and concepts which, if at all possible, should be used by all biologists and natural scientists, otherwise it is impossible to avoid misunderstandings. The handbook of ecology (KUTTLER 1995) is an important step forward in this area.

## Human Impact on Ecosystems

The human impact on ecosystems, world-wide, is now very substantial and rapid with the loss of habitats for organisms which have adapted to particular conditions over very long periods, This has resulted in reductions in the extent of ecosystems and a huge decrease in the number of individual organisms, loss of biodiversity in ecosystems and extinction of species. This trend is exemplified by the impact of humans on tropical ecosystems and particularly forests.

No other community on our continents is so colourful and varied with in such an unbelievable richness of species and interlinked processes as the tropical rain forest. Indeed, tropical ecology is of great importance for research and teaching. The twentieth century is characterised by the accelerated destruction of tropical forests: monotonous fields, pastures for grazing, banana or coffee plantations and settlements have eaten into the forests. The exploitation is well documented. Biodiversity is being lost at an increasing rate. Many species will be lost forever without ever having been recognised and researched or they have already become extinct. The conviction that tropical forests must be maintained because of their enormous genetic potential and their global importance for climate and soil is only slowly gaining ground.

A fraction of the research spent world-wide for nuclear physics, gene technology or research in the Antarctic would greatly improve the situation of tropical ecology in developing countries.

The discipline of ecology was largely developed in temperate climates, initially in Europe. The importance and fascination of tropical forests were recognised early; for example, ALEXANDER VON HUMBOLDT paved the way for modern developments on this topic (SCHALLER 1993). However, current interest and research in tropical ecology are still relatively small, given the magnitude of the problem. A large scientific effort is required to understand the often much greater biodiversity and the enormously varied interactions associated with tropical ecosystems. Tropical ecology should be significantly strengthened through its own university departments and centres of research. Without a broad basis in tropical ecological research there will be no well-founded teaching relating to this vital area of global ecology.

Ecological principles can, of course, also be understood in temperate climates, for example in Europe, an area which has only been sparsely colonised by plants since the last ice age. However, this understanding helps students little in the tropics with the variety of species and the complex interactions, particularly in tropical forests.

Ecological principles should now be part of our general education and tropical ecology is an important part of this discipline, as it touches the basis of human life. A corresponding system of education is required to enable objective and comprehensive actions.

Ecology, including tropical ecology, should be a core subject in universities, not only for natural scientists.

## Significance of Systematics and Taxonomy for Biology

The destruction of tropical ecosystems not only increases degraded areas and renders them infertile because of erosion, much more important is the loss of biodiversity (BOERBOOM & WIERSUM 1983). This destruction leads to an over-proportional loss of plant and animal species on our globe and the corresponding finely tuned communities of organisms. The rate of loss of species through the destruction of jungles is proceeding much faster than the extinction of the dinosaurs or the changes during the ice ages.

At present, about 1.5 million animal and plant species have been described, i.e. scientifically documented. This is probably only a fraction of all the species living on the globe. Using different methods of determination, the diversity of particular areas can be established through extrapolation leading to values of 5 to 10 million species. Other attempts, such as fractal geometry, result in a value of 30 million. It is difficult to evaluate real numbers, but the material from every new expedition to the tropics leads to a wealth of new species. Scientific evaluation of the material often limps years behind. The number of specialists with detailed knowledge of animal and plant groups is so small that they cannot cope with processing the material, and most of it remains unanalysed. Systematic grouping, exact taxonomic descriptions and especially the phylogenetic connections are known only roughly for many animal groups. The scientific assessment is significantly better for higher plants and this is because there are fewer species. However, many more new species are to

Systems and taxonomy are the essential basis of understanding between biological disciplines. Systematics brings order into diversity. Systematics must provide a fixed framework for the understanding, but also a flexible framework for progress in phylogenetics. Without well-funded methods in systematics and taxonomy not only ecology but all biology drifts without a point of reference.

be expected for algae, and even more so for fungi, so it is important that teaching and research into systematics should be considerably enhanced and re-introduced where lost. Although all biologists work with organisms, some researchers give the impression that they do not know with which organism they are actually working, and the importance of phylogenetic connections.

As GAMS states: "All knowledge of different sub-disciplines of biology, and where possible all characteristics, should, in the end, be used to arrive at a better understanding of how organisms function in the natural world" (personal communication).

Systematics is the biological science of the future. However, it remains questionable, whether world-wide there is sufficient investment in the topic. Perhaps the international umbrella organisation DIVERSITAS will help to reduce some of the gaps.

## Importance of Scientific Documentation

Museums have the responsibility to collect and mark material and, above all, to document it and describe relationships and processes scientifically and present the importance of it to the public.

Documentation is very important for the systematic taxonomic analysis of biodiversity. The material upon which the description of a species is based, the type of specimen, must be kept as an important basis for documentation in museums or large herbaria.

Catalogues and taxonomic listings, keys for identification, aerial maps of distribution etc. may be stored electronically and provide ready access via the internet or other means. The speed and extent of these developments will depend on the number of scientists engaged in this area and thus on the priorities and the political will.

There are still many amateur scientists who deal with certain groups of organisms in their leisure time. Many of these private collectors possess valuable small collections which would enhance existing museum collections. However, museums must be able to accept these, either as donations, or even purchase them; this is often not possible because of financial constraints or the lack of personnel and space resources. Much valuable material, which is perhaps no longer obtainable, is thrown away.

## Importance of Excursions for Young Scientists

Students can only really understand the biology of organisms if they are given the opportunity to study organisms in their environment. Many universities no longer include excursions as part of the syllabus. Many biologists do not have the opportunity to participate in excursions and benefit from this intensive way of learning, which not only brings biological but also general scientific knowledge. It is insufficient to only **watch** how something is presented in a film on television. It is vital to **see, grasp** and **recognise** connections, for example the geological and the geomorphological situation, the possibility of agriculture and forestry in the area, fauna and flora and their interdependence, the time and space dynamics of producers, consumers and the degradation processes, phenology of the historic basis of the development of the landscape, the possibility of sustainability – all this can be explained much better to students standing on a hill. However, do faculties (or ministries) still want this nowadays? Biology without a significant part of environmental biology experienced first hand during excursions is an amputated biology. During excursions, participants stand in the midst of the events. Only then can students understand the ecology and even confront perceived dangers and take suitable preventative measures without experiencing frightened hysteria (for example against ticks). Only then do students learn to move in nature according to nature's laws.

Excursions are the most intensive form of learning. Students learn, by analysis using all the senses, to synthesise and understand the links and connections between the organism and the environment from which they develop, the concepts, vision and understanding of the whole ecosystem.

**Questions**

1. *Why is basic knowledge of biological systematics indispensable for all biologists?*
2. *How many species of animals and plants exist?*
3. *What is the value of high biodiversity?*
4. *At which level of complexity in biology do ecologists work?*
5. *What is an ecological catastrophe?*
6. *What is the objective of museums of natural science?*
7. *Why are excursions a 'must' in the training of biologists?*
8. *Is it possible to classify plants and animals on the computer?*
9. *What are the differences between phylogenetics, systematics, taxonomy and nomenclature?*
10. *What are the most important differences in teaching biology in the laboratory and in the natural environment?*

# General Section

# Ecological Fundamentals

## Historical perspective

The present geo-biosphere is intimately linked with the history of the earth. This is the result of a long process of development in the plant and animal kingdom as well as the long geotectonic history of the solid earth surface. Therefore, ecology cannot afford to neglect the historical perspective.

Continents have not always had their present shape and previously occupied different positions with respect to the poles and the equator. WEGENER's hypothesis of continental drift has developed into the theory of **plate tectonics** which

Plate tectonics: the present position of the plates is, from a geotectonic point of view, only a moment in time. In order to understand the present distribution of organisms, it is essential to know the early position of the plates relative to each other and during the course of evolution.

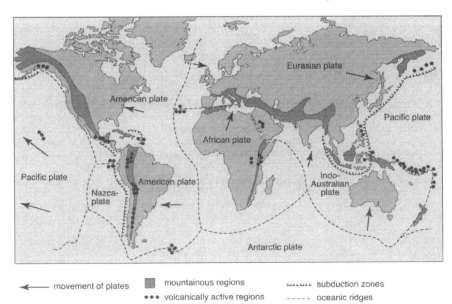

**Fig. 2.** Outline of the most important tectonic plates of the earth's crust. The direction of plate movements, the formation of mountains, subduction zones and particularly active volcanic zones are shown (after SCHÖNWIESE 1994)

explains the movements of land masses by large-scale tectonics and convection currents in the earth's mantle. The
movement of the plates by a few centimetres a year causes
very slow changes in their relative position. The present position of the plates is shown in Fig. 2. Because of the areas of
rising magma (for example opening and widening of the
Atlantic), 'submergence' of plate material at other sites,
called **subduction zones**, must follow. Particularly active
volcanic areas are usually found there and are important
for the evolution of the flora and fauna.

Atmospheric wind movements with associated climatic
zones appears to be a very stable system, in contrast to
the drifting continental plates, and go back in their current expression, in comparable form, to the Mesozoic. The
climatic system is thus the more stable part of the total
system of the biosphere with the continents, as the lithosphere, swimming beneath it and being more variable
(KRUTZSCH 1992).

All life began in water, and the earliest fossils of terrestrial plants originate from the transitional period between
the Silurian and the Devonian. Since cormophytes do not
require NaCl, the chief salt component of seawater, and
since NaCl is toxic for all plants except for halophytes, it
seems that the ancestors of terrestrial plants were freshwater algae, probably living in coastal lagoons in a wet-
tropical climate region. Halophytic angiosperms are recent
secondary adaptations to salty soils in coastal areas or salt
deserts.

The conquest of land was made possible by the presence of large cell vacuoles, together constituting the vacu-
ome, which provided an internal aqueous medium for the
cytoplasm. Around the plasma the cell wall forms a water-
saturated, spongy external medium which enclosed the
cell. As protection from the external medium, land plants
developed a cuticle which protected them against desiccation. Evolution of stomata made the uptake of $CO_2$ for
photosynthesis possible, root and transport systems compensated for transpiration losses (WALTER 1967) and
served, at the same time, as a transport system for mineral nutrients.

Because of the increasing isolation of continents which
took place in the waning Mesozoic, after the development
of angiosperms, the flora of the earth and angiosperms in
particular evolved along different routes which led to the
formation of six floristic realms (Fig. 3). They correspond
essentially to the faunal realms. However, at times they
are given different terms.

The complexity of evolution is shown by the phylogenetically relatively old group of conifers with the family Po-

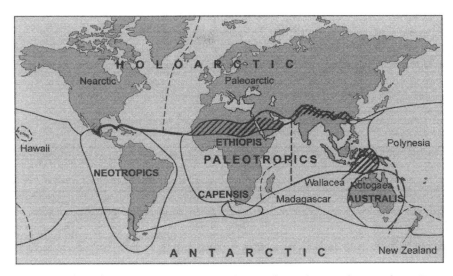

**Fig. 3.** Floristic realms of the earth *black uppercase letters* and animal regions *lowercase letters*. Transition areas between animal regions are *hatched*. There is no differentiation of a Capensis region for fauna. New Zealand and Tasmania contain Antarctic as well as Paleotropic and Australian floristic elements (after WALTER & BRECKLE 1990)

docarpaceae, especially the genus *Araucaria*, which occurs only in the southern hemisphere, whereas the large family Pinaceae and nearly all Taxodiaceae occur in the northern hemisphere. The Cupressaceae, however, are scattered over all continents.

The distribution of flowering plants, or angiosperms, the youngest branch of the plant kingdom, is much more sharply differentiated. Original forms, partly relicts, are found most of all in South-East Asia. The oldest families of this plant group are known to have existed in the Lower Cretaceous, but their main development took place in the Tertiary period, after the Gondwana land mass had already split up into different continents. In the northern hemisphere, however, this held true only to a limited extent, since it was not until the Pleistocene epoch that North America and Greenland finally separated from Eurasia. As a result, floristic differences in this area are so small that these continents can be considered as one floristic realm, the **Holoartic**. Much larger differences exist, however, between the tropical floras of the so-called Old and New Worlds, so that two floristic realms must be distinguished, the **Neotropics** and the **Paleotropics**. Floristically, the southern-most parts of South America and Africa, and Australia and New Zealand in their extreme isolation, have still less in common. Therefore, three floristic realms have been distinguished: the **Antarctic** comprising the southern tip of South America and the subantarctic islands, the **Australis**, geographically identical with the continent of Australia, and the **Capensis** the smallest floristic realm, but one especially rich in species, in the outermost south-western corner of Africa (Fig. 3.)

These six realms are not sharply delineated, and elements from one can be found well into the next. In New Zealand, both Paleotropic-Melanesian elements and Antarctic elements are to be found, often in mosaic-like distribution. Thus, the allocation of these islands to one or the other floristic realm is a question of informed judgement.

The animal regions of the zoologist correspond, broadly speaking, to the floristic realms; only the Capensis has no typical fauna.

Plant species are the building blocks of the plant communities that together constitute the vegetation of the different regions. Even if the species are not the same, extreme environmental conditions can lead to similar life forms; these are termed **convergences**, but are, nevertheless, rather exceptional. A well-known example is provided by the stem-succulent plants, which in the arid areas of the Americas belong to the family Cactaceae (cacti) and in Africa to the genus *Euphorbia* (spurges). In climatically similar arid regions of Australia, on the other hand, there are no stem-succulent plants whatsoever, although Australia is especially rich in other kinds of convergences which have not developed on other continents. In the temperate climate of New Zealand, there are none of the deciduous forests that are widespread in the Holoarctic realm. Since the genetic stock of the individual floristic realms is limited, the same life-form could not necessarily develop everywhere. This is particularly marked in the Australian realm. The vegetation of this floristic realm is physiognomically very different from that of other continents, and even its mammalian fauna is unique.

The Pleistocene, with its many ice ages, left a very distinct mark, especially on the northern hemisphere. The flora of Europe was significantly impoverished and many genera became extinct, although they are still found in North America and East Asia today where it was easier to find a north-south escape route. In Europe, however, the east-west running Alpine range blocked escape to the south and a later return.

The ice ages affected the Sahara by bringing rain, i.e. pluvial periods. The tropics, on the other hand, were subject to dry periods.

The historical factor, therefore, has to be taken into consideration in dealing with zonobiomes that extend over several floristic realms. Zonobiome IV with winter rain is a very good example, since it covers parts of the Holoartic, Neotropic, Australian and Capensic realms. Thus, it is convenient to subdivide zonobiome IV into five historically determined biome groups (Mediterranean, Californian, Central Chilean, Australian, and Capensic) which are distinguished by the flora, despite similar life forms.

Isolation of islands has also led to a considerable degree of **endemism**, i.e. the occurrence of species found nowhere else. The figures in the margin indicate the percentage of flora accounted for by endemic species in islands or groups of islands. The further an island is situated from the mainland and the longer it has been isolated, the larger the proportion of endemic species, although oceanic currents also play a role.

Endemism on islands (percentage of endemic species): Hawaii 97.5%, New Zealand 72%, Fiji Islands 70%, Juan Fernandez 68%, Madagascar, 66%, Galapagos Islands (in the dry belt) 64% (only 8–27% in the humid mountain belt and 12% in the coastal region), New Caledonia 60%, Canary Islands 50–55%, islands near coast 0–12%.

## 2. Co-Evolution and Symbiosis

It is impossible to understand the expression of different ecosystems without discussing the processes of co-evolution in the course of historical development. In many ecosystems mutual interdependence between certain plants and animals is so tight that it must be considered as an obligate relationship. Very often this is expressed as a linkage between pollinators, herbivores and certain plant species which changes during the course of the year and can only be maintained in a population over a large area. In the course of evolution such tight dependencies developed because of mutual stimulation. This also applies to many relations between different organisms. Such a tight **interlinkage** is particularly varied in ecosystems with a long development phase (in and since the Tertiary period). Tight functional linking of organisms makes it much more difficult to clearly distinguish functional compartments in an ecological analysis.

This is clearly seen in different **symbioses** where two partners live almost as mutual "parasites": on balance each one supplies the other with something essential for life. Symbioses which occur everywhere, for example the different forms of mycorrhizae, will be explained later (see p. 101). Nitrogen-binding symbionts are an example, occurring not only in the form of nodules with *Rhizobium* in legumes, but also in several other species (for example *Frankia* with alder) thus improving the competitiveness of species. Symbiosis may even make it possible for plants to

**Fig. 4.** Individual
components for the
regeneration of a plant
species and maintain-
ing a population at a
particular site
(after BURROWS 1990)

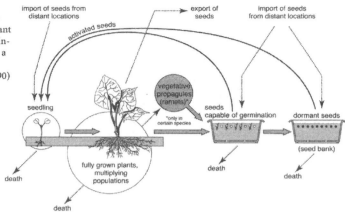

establish in certain, hostile areas; this is the case for lichens which have become the dominant primary producing organism in the Antarctic and in the permanent snow
regions of mountains.

The particularly tight interdependence of very many
different organisms leads, during long periods of evolution, to an unbelievably varied network of relations and to
a network of functions such as the tropical rain forest,
which is very stable under the steady climatic conditions
in the equatorial regions. After destruction of such forests,
however, this feed-back-coupled network cannot be regen-
erated in the foreseeable future. Secondary woods, which
are much poorer in species, have a much wider, looser
functional network.

## Biodiversity and Ecology of Populations

Biodiversity:
$a$-Diversity: the number of
species within the individual
groups of organisms or the
diversity of the taxa.
$\beta$-diversity: the difference in
diversity between biotopes
and ecosystems.

Multiplicity of relationships is a typical qualitative characteristic of an ecosystem. The number of species of individual groups of organisms can only give a rough indication.

Biodiversity is often a good measure for an original
and undisturbed ecosystem. Under extreme ecological
conditions it is, however, possible that a completely virgin
ecosystem is poor in species, particularly if only highly
adapted specialists are able to survive. In situations of
high diversity the question of regeneration of the many
species can often not be answered. Fluctuations in sizes of
populations of the many participating species (seed bank

→ seedling → young plant → adult plant, or egg → larva → pupa → imago etc.) can usually not be measured. Birth and death rates with their appropriate time spans are only known for very few species, even less is known about the effect of their influence in regulating the size of populations. Other effects are variability in time and space of the occurrence and dispersion of seeds (or propagules); in particular ecosystems some species may have large seed banks which can quickly be reactivated, even after years, if conditions change (for example if a meadow becomes disturbed waste land). This principle is shown in Fig. 4 for the reproduction of plants. It is usually not possible to consider seed banks in determinations of biodiversity, which in itself is very difficult to determine. Values can only be given for certain groups of organisms. Many different procedures, indices etc. are used (HUMPHRIES et al. 1995).

Very often certain decisive events lead to new directions in the development of species. The continuous development of populations is not so much determined by periodic events, but is again and again disrupted and given a renewed stimulus by episodic damaging events (fire, storm, floods).

**Periodic** events are predictable and occur regularly (→ winter in ZB VII; tides on the coast etc.)
**Episodic** events cannot be predicted, they occur irregularly and infrequently.
→ thunderstorms in ZB III, El Niño, frosts in the coffee plantations of Brazil etc.)

This is particularly striking in evergreen tropical rain forests where the structure of the habitat is very heterogeneous and where, because of falling branches or trees **gaps** of different sizes are formed (see Fig. 78, p. 127). These are quickly filled again by fast growing species and the type of habitat is thus rejuvenated. Such episodic events are probably the precondition in many more ecosystems than previously thought and secure their long-term maintenance through the successive renewal of their structures. This then leads to a very varied and long cyclic renewal which is predominantly stochastic (dependent on accidents) and less deterministic; individual parts of a biome are younger, others older; the mosaic character and periodic dynamics of natural ecosystems was characterised decades ago by AUBREVILLE (1938). It is an important principle for the maintenance of many species in a dynamic interaction next to each other and with each other.

The biodiversity of large areas in regions of the earth was shown by BARTHLOTT et al. (1996) as a map with diversity indices. Examples for regions which are especially rich in species are the following six diversity centres (DI=10, above 5000 plant species per 10,000 km$^2$):

1. Chocó-Costa Rica centre (southern Central America, ZB I)
2. Tropical eastern Andes centre (Columbia and Ecuador, ZB I)

**Table 2.** The 25 countries with the highest biodiversity with respect to the
number of species of flowering plants, as well as some European countries

| | Country | Approx. number of species | Endemism (%) |
|---|---|---|---|
| 1 | Brazil | 55,000 | ??[a] |
| 2 | Columbia | 35,000 | 4.3 |
| 3 | China | 30,000 | 55.9 |
| 4 | Mexico | 20–30,000 | 13.9 |
| 5 | USSR[b] | 22,000 | ?? |
| 6 | Indonesia | 20,000 | 66.7 |
| 7 | Venezuela | 15–25,000 | 38.0 |
| 8 | Ecuador | 17–20,000 | 20.7 |
| 9 | USA | 18,956 | 20.7 |
| 10 | Bolivia | 15–18,000 | ?? |
| 11 | Australia | 15,000 | ca. 80 |
| 12 | India | 15,000 | 31.3 |
| 13 | Peru | 13,000 | ?? |
| 14 | South Africa | 13,000 | ca. 70–80 |
| 14 | Costa Rica | 10–12,500 | 15.0 |
| 15 | Malaysia | 12,000 | ?? |
| 16 | Thailand | 12,000 | ?? |
| 17 | Zaire | 11,000 | 29.1 |
| 18 | Papua/New Guinea | 10,000 | ca. 55 |
| 19 | Tanzania | 10,000 | 11.2 |
| 20 | Argentina | 9000 | ca. 25–30 |
| 21 | Madagascar | 8–10,000 | 68.4 |
| 22 | Panama | 9000 | 12.7 |
| 23 | Turkey | 8472 | 30.9 |
| 24 | Cameroon | 8000 | 1.9 |
| 25 | Guatemala | 8000 | 13.5 |
| | Italy | 5463 | 12.7 |
| | Yugoslavia[b] | 5250 | 2.6 |
| | Spain | 4916 | 16.8 |
| | Greece | 4900 | 14.9 |
| | Austria | 2850–3050 | 1.2 |
| | Switzerland | 2927 | 0.1 |
| | Germany | 2600 | 0.2 |
| | Finland | 1040 | 0.0 |
| | Andorra | 980 | 0.0 |

In part after Groombridge (1982)
[a] ?? No data known, [b] Countries within their former boundaries

3. Atlantic Brazil centre (ZB II, V)
4. Eastern Himalayas Yunnan centre (ZB II)
5. Northern Borneo centre (ZB I)
6. New Guinea centre (ZB I)

Almost as rich in species (DI = 9, 4500 to 5000 species per 10,000 km$^2$) are also: the Cape region in South Africa (ZB IV), Cameroon-Gabon (ZB I/II), South Mexico/Chiapas (ZB II) followed by Cuba, East Africa, Madagascar, South India, Southwest and Northeast Australia and Venezuela/Guyana.

Regions which are especially poor in species (DI = 1, below 100 vascular plant species per 10,000 km$^2$) are: Libyan desert (ZB III), Arabian desert (ZB III), Gobi desert (ZB VII), parts of Central Australia (ZB III), Atacama (ZB III), Arctic tundra in northern Canada and northern Siberia (ZB IX), Antarctic (ZB IX).

GROOMBRIDGE (1992) provides a comprehensive overview of the number of species for individual countries of the earth. Countries with the most species of flowering plants are listed in Table 2 as well as the degree of endemism, i.e. of species that are limited to that country.

In recent years many national and international initiatives were started to maintain biodiversity (GROOMBRIDGE 1992; HEYWOOD 1995; WILSON 1992). It has been recognised that the current rate of extinction of organisms has increased by much more than the natural "geological rate of extinction". The Convention of Rio in 1992 (Biodiversity) tried for the first time to work against this trend.

One of the well documented laws of ecology states that the number of species per area increases strongly from the poles to the equator. This applies for almost all areas, habitats and groups of organisms.

## Climate and its Presentation (Homoclimates as well as Climate Diagram Maps)

The classification of the geo-biosphere into zonobiomes is based upon climate (see p. 78 f.). Spatial and temporal scales will be discussed later (compare p. 84 f.). It is often not clear how the integration from weather (actual meteorological conditions) to climate (meteorological condition over many years) is made, via the annual average condition. For ecologists they are very important, however, remote formulae and indices are not suitable. One method is the graphic presentation of climate, which must also include changes in conditions during the year.

This kind of information is provided by the **ecological climate diagram** where the total climate near the earth surface is shown pictorially. The figure must be clear and comprehensible at a glance and contain only the most essential data from the ecologist's point of view, namely the

Climate is the average course of weather for a year, averaged over as many years as possible. Meteorological conditions (large-scale weather) are the interaction of different meteorological factors over shorter periods (weeks, months). Weather is the actual meteorological condition at a distinct day.

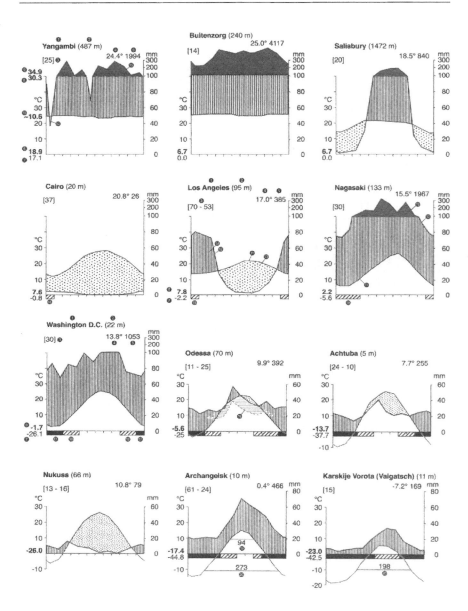

◄ **Fig. 5.** Key to the climate diagrams with typical examples which are also examples for the different zonobiomes (see below).

*Abscissa (horizontal axis):* northern hemisphere, January to December, southern hemisphere, July to June (warm season thus always in the centre of the diagram).

*Ordinate (vertical axis):* temperature in °C, precipitation in mm. One division = 10 °C or 20 mm precipitation (figures are normally omitted).

*Letters* and *numbers* on the diagrams indicate the following: *1* station; *2* height above sea level, *3* number of years of observation (where two figures are given the first indicates temperature and the second precipitation); *4* mean annual temperature; *5* mean annual precipitation; *6* mean daily temperature minimum of the coldest month; *7* absolute minimum temperature (lowest recorded); *8* mean daily maximum temperature of the warmest month; *9* absolute maximum temperature (highest recorded); *10* mean daily temperature fluctuation. *8, 9* and *10* are only indicated for tropical stations with a diurnal climate. *11* curve of mean monthly temperature; *12* curve of mean monthly precipitation (1 scale unit = 20 mm, thus 10 °C=20 mm); *13* = period of relative drought (*dotted*) for the climate region concerned; *14* corresponding relative humid season (*vertical hatching*); *15* mean monthly precipitation >100 (scale reduced to one-tenth, *dark areas* perhumid season; *16* supplementary precipitation curve, reduced to 10 °C=30 mm, horizontal area above = relative dry period (only for steppe stations); *17* months with a mean daily minimum below 0 °C (*black*) = cold season; *18* months with absolute minimum below 0 °C (*diagonally hatched*), i.e. late or early frosts possible; *19* number of days with mean temperature above +10 °C (duration of vegetation period); *20* number of days with mean temperature above –10 °C. Not all of the above are available for every station. Where data are missing, the relevant places in the diagram are left empty.

The diagrams belong to the following zonobiomes:

| | |
|---|---|
| **ZB I** | (humid equatorial day climate): Yangami on the middle Congo; Buitenzorg (Bogor) Java: |
| **ZB II** | (tropical summer climate): Salisbury in Zimbabwe; |
| **ZB III** | (subtropical desert climate): Cairo on the lower Nile; |
| **ZB IV** | (Mediterranean winter rain climate): Los Angeles in southern California; |
| **ZB V** | (warm temperate climate): Nagasaki in Japan; |
| **ZB VI** | (temperate, nemoral climate with short cold season): Washington, DC; |
| **ZB VII** | (temperate semi-arid steppe climate with long dry season and slight drought): Odessa on the Black Sea; |
| **ZB VIIa** | (temperate arid semi-desert climate with long drought): Achtuba on the lower Volga; |
| **ZB VII (rIII)** | (extreme arid desert climate with cold winters): Nukuss in central Asia; |
| **ZB VIII** | (cold temperate climate with a very long winter) Archangelsk in the boreal taiga zone; |
| **ZB IX** | (arctic tundra climate with July mean temperature below +10 °C): Karsije Vorota (island Vaigatsch) |

**Fig. 6.** Examples from mountain stations in the various orobiomes: **OB I:** Páramo de Mucuchies in Venezuela; **OB II:** San Antonio de Los Covres in the Peruvian puna: **OB III:** Calama in the North Chilean desert puna; **OB IV:** Cedres in the Lebanon; **OB V:** Hotham Heights in the Snowy Mountains of Australia; **OB VI:** Zugspitze in the northern Alps; **OB VII:** Pikes Peak in the Rocky Mountains above the Great Plains of North America; **OB VIII:** Aishihik in southern Alaska; **OB IX:** Vostok on the ice cap of the Antarctic

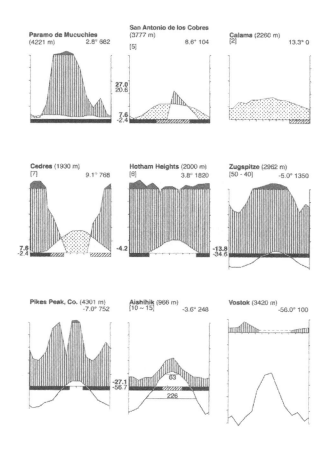

temperature and precipitation throughout the year. The Climate Diagram World Atlas by WALTER and LIETH (1967) contains more than 8000 climate diagrams from stations all over the world.

Explanations of some typical diagrams are given in Fig. 5 where examples of the nine zonobiomes are given from stations at low altitude. Figure 6 shows climate diagrams for orobiomes (OB): OB I with a diurnal type of climate (Páramo) and the remaining orobiomes II–IX. Vostock, in orobiome IX, with a mean annual temperature of –56 °C is perhaps the coldest station of the earth, but still has vegetation, contrary to other similarly cold Antarctic stations.

Climate diagrams show not only temperature and precipitation, but also the duration and intensity of relatively

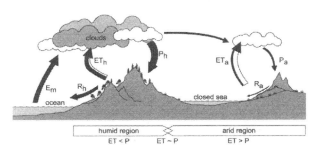

**Fig. 7.** Scheme of a hydrological cycle of the earth in humid and arid regions: *a* arid; *E* evaporation; *ET* evapotranspiration; *h* humid; *m* marine; *P* precipitation; *R* surface and subsurface runoff

humid and relatively arid seasons, as well as the possibility of late and early frosts.

In an **arid** region evaporation predominates in the hydrological water balance (see also p. 55 for equation for the water balance). There is, thus, in contrast to a humid region, no closed permanent river system. Basin landscapes contain only small lakes (see Fig. 7) which were salinated (Dead Sea, Great Salt Lake in Utah, Aral, Lop-Nor), in **humid** regions such basins are "full to the brim" with water (for example Lake Constance in Germany) and have runoff. The internal hydrological cycle (P; ET in Fig. 7) of larger regions is thus very different and depends on whether the region is humid or arid.

With the schematic climate diagrams it is possible to evaluate the climate from an ecological point of view. The aridity and humidity of the different seasons can also be read off the diagrams by using the scale 10 °C, which is approximately equal to 20 mm of precipitation. The temperature curve thus approximately replaces the curve of potential evaporation (values of which are usually not known) and can thus be used to relate the water balance in comparison with the precipitation curve. The height of the dotted area (drought) indicates the intensity of the drought, the width is proportional to its duration. Humidity is indicated in the same manner. The relation 10 °C approximately equals 20 mm of precipitation was found by GAUSSEN (1954) to agree well with the weather conditions of the Mediterranean region. In diagrams for steppes and prairies, a scale of 10 °C that is approximately equal to 30 mm is recommended to show dry periods which are less extreme than droughts.

The arid (drought) season shown in the climate diagram is only arid **relative** to the humid season of the particular type of climate under consideration. This is because the potential evaporation curve and the temperature curve which is used in its place are not identical, but run

**Homoclimates** on different continents are characterised by very similar climatic conditions, i.e. they are characterised by stations with very similar to almost identical climates.

only more or less parallel to one another. The more arid or windier (for example Patagonia) the climate, the larger the quantitative deviation of the temperature curve below the potential evaporation curve. HENNING (1994) and LAUER et al. (1996) have assembled more exactly constructed hydroclimate diagrams. In absolute terms, this means that the arid season in an ecological climate diagram is more arid the drier the climate. An arid season on the climate diagram for a station in the steppe region, for example, is not as extreme as one for a Mediterranean station or for the Sahara. From an ecological point of view this is very favourable since the sensitivity of plants to drought decreases, the drier the climate. For species growing in the tropical rain forest a month with less than 100 mm rain is already relatively dry and the xerophytes of Central Europe would be hydrophytes in a desert region. When discussing vegetational regions the appropriate diagram will be given in each case in order to avoid long tables.

Climate diagrams are particularly useful for tracing homoclimates. This is a tedious process when lengthy climate tables have to be consulted. However, using the Climate Diagram World Atlas a given climate diagram can be easily compared within regions where a homoclimate is suspected. Figure 8 shows the homoclimate of Karachi (Pakistan) from zonobiome III (slight transition to ZB II) and of Bombay (very typical for ZB II) in other parts of the world. Information about homoclimates is essential

**Fig. 8.** Homoclimates of stations Karachi (Pakistan) and Bombay (India) in other continents

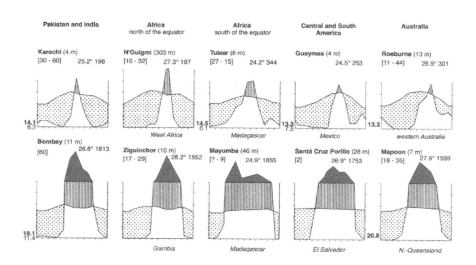

before introducing plants into regions where they have
not previously been cultivated.

A climate diagram map reveals at a glance the type of
climate within a whole continent or a large section of a
continent. Such a map can be constructed by sticking the
climate diagrams from the Climate Diagram World Atlas
onto the appropriate spot on a large wall map. It becomes
still clearer if the areas corresponding to drought are coloured red and those corresponding to wet periods coloured blue. Large black-and-white climate diagram maps
of every continent have been published elsewhere (WAL-
TER et al. 1975). Because of the limited space, only one
small-scale map of Africa showing a few climate diagrams
can be reproduced here (Fig. 9). The World Atlas contains
over 1000 diagrams for Africa.

# 5 Environment and Competition

The climate of a site determines its vegetation. However,
the widespread assumption that the distribution of plant
species is directly dependent upon the physical conditions
prevailing in the habitat is almost never correct. These
conditions are only indirectly important in as much as
they change the competitiveness of species. Only at the
absolute distribution limit, in arid or icy deserts or on the
edge of salt deserts are physical environmental factors
(usually one particular, extreme factor) of direct importance. Apart from such exceptions, plant species are capable of existing far beyond their natural distributional
areas if they are protected from competition by other species. For example, the north-eastern limit of distribution
of the European beech (*Fagus sylvatica*) runs through the
Vistula region of Poland, although beach is found growing
far to the north and southeast of this limit in the botanical gardens in Helsinki and Kiev. The Mediterranean evergreen holm oak (*Quercus ilex*) has its northern limit in
the southern Rhône valley, but cultivated trees survive in
the botanical gardens of Bonn, Copenhagen and Leipzig
and the cork oak (*Quercus suber*) in gardens near Stuttgart and Bielefeld.

The natural limit of distribution of a particular species
is reached, when, as a result of changing physical environmental factors, its ability to compete, or its competitive
power, is so much reduced that it can be ousted by other
species. The range also depends upon the presence of
competitors or of a certain fauna. For beech, these are, at
the eastern limits, hornbeam (*Carpinus betulus*), to the

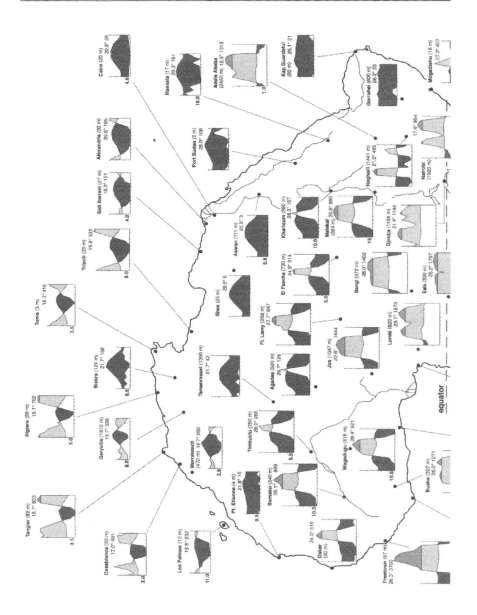

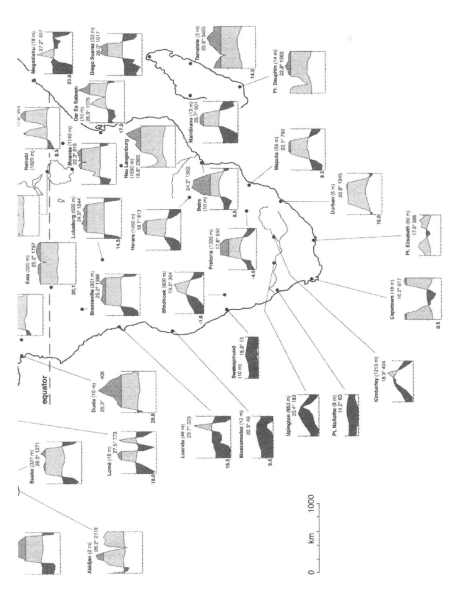

**Fig. 9.** Example of a climate diagram map with only 66 stations. Zonobiomes from north to south: IV–III–II–I–II–III–IV, but north of the equator the east is too dry (monsoon), to the south, in contrast, too humid (SE passat winds)

**Fig. 10.** Growth curve (*grey area*) of one species without (**A**) or with (**B–F**) pressure from competition (*dark area*). *y*-axis: rate of growth and biomass production; *x*-axis: variable habitat factors

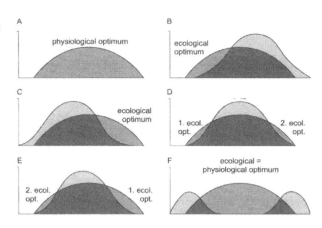

north oak (*Quercus robur*), and in mountainous regions, spruce (*Picea abies*).

The fact that the north-eastern beech limit takes a course similar to the January isotherm for –2 °C and that the northern limits of oak distribution follow the line indicating 4 months of the year above +10 °C, does not necessarily indicate a direct causal relationship. It can, at most, be concluded that, in the case of beech, the increasingly cold winter toward the east and, for oak and spruce, the shorter summer to the north, radically reduce the competitive power of these species.

The conditions under which a species occurs most abundantly in nature may be termed the **ecological optimum,** and the conditions under which it grows best in the laboratory (phytotron or growth chamber) or in individual cultures may be termed the **physiological optimum.** These optima are rarely identical (Fig. 10).

The distribution of a species, therefore, is not an absolute guide to its physiological requirements. For example, the fact that in western Europe the Scotch pine (*Pinus sylvestris*) is found under natural conditions not only on dry calcareous slopes, but also on very dry acid sandy slopes and even on boggy acid soil (see ecogram in Fig. 11) is due to its having been supplanted from a more suitable habitat by a stronger competitor. The information acquired in the phytotron about the physiological needs of a species thus forms an insufficient basis for either predicting or explaining its natural distribution. Whether or not a species colonises a habitat which is physiologically suitable usually depends, apart from the historical factor, on the nature of its competitors. An ecogram shows which

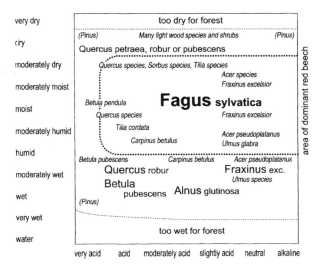

| very dry | | too dry for forest | |
|---|---|---|---|
| dry | *(Pinus)* | *Many light wood species and shrubs* | *(Pinus)* |
| | Quercus petraea, robur or pubescens | | |

**Fig. 11.** Ecogram of the most important submontane woodland-forming tree species in central Europe in a moderate suboceanic climate. The size of the lettering expresses roughly the coverage of trees to be expected as a result of the natural competition (from ELLENBERG 1983)

species are dominant in certain habitats, where their ecological optima are and how they are affected by particular ecological combinations of factors. This is shown in Fig. 11 for central European tree species.

Competition does not imply a direct dependence between species. It can be recognised when a particular plant develops more luxuriantly in isolation than in a plant community. Inhibition in the course of competition results mainly from a reduction in light due to organs above ground, or a lack of water or nutrients due to root competition. It is very difficult to prove, whether apart from these factors, inhibitors excreted by the plant (**allelopathy**) play an important role under natural conditions. There are only a few examples of this which are well substantiated. In other instances there appears to be a mutual advancement, particularly through the metabolic exchange via the hyphae of fungi in the soil which connect the mycorrhizae of different trees with each other and supply young saplings with additional nutrients (system of "wet nursing"). However, in ecosystems the processes of competition are more predominant than those of co-operation.

A distinction must be made between **intraspecific** competition, occurring between individuals of the same species, and **interspecific** competition, taking place between individuals of different species. The former helps to preserve the survival of the species. In interspecific competition species can achieve dominance and supplant others

**Competition** is generally considered to be where growth or development of a species is unfavourably affected by the presence of a different species (without the presence of parasites). Competition operates where several species occur close to each other. Competition is related to the limited resources, water, light, nutrients or space, available to the root or shoot.

or, in a mixed population, a state of equilibrium may be established based on the competitive power of individual partners. In mountainous regions, for example, it can be seen that at the beech-spruce boundary, beech is absolutely dominant on southern slopes and spruce is dominant on the northern slopes, whereas on eastern and western slopes, the two are fairly well balanced and a mixed population is formed. This can also happen if, as seems to be the case in tropical rain forests, the seedlings of a certain species develop better beneath other species than beneath individuals of the same species, perhaps because the pressure from herbivores and parasites or other inhibiting factors are correspondingly weakened.

The competitive power of a species is a highly complex phenomenon and very difficult to define as it can change significantly with the stage of development. It is weakest in seedlings and young plants and increases with age, particularly in trees. In this respect all morphological and physiological properties of a species are of significance. Biennial species are competitively more powerful than annuals because they commence their second year of growth with larger reserves accumulated during the first year. For the same reason, perennial herbs are superior to biennials from the second year onward. Woody species win-out against perennial herbs if the former have not been suppressed during their early years and have succeeded in producing ligneous axial organs to raise them above the herbage layer. As a result of competition, similar combinations of plant species occur repeatedly in similar habitats within a limited area and are termed plant communities (phytocenoses). Examples in central Europe are beech woods on calcareous soil together with their herbaceous flora or flood plain forests, various types of bogs and fens, and so on.

In a stable plant community, the different species are in a state of ecological equilibrium with each other and with the environment. Together with the animal organisms, they constitute a biogeocene. Apart from the influence of animals, the following factors are important for their equilibrium:

1. competition between species
2. dependence of one species (e.g. shade species) on the presence of others
3. occurrence of complementary species adapted to one another either in space or time so that every ecological niche is filled.

The natural community is thus reasonably "saturated" and introduced foreign species can hardly gain a foothold whilst there is a much better opportunity for invasion in a disturbed equilibrium. Therefore, long-range transport of seeds plays a role in the distribution of plants only in areas that have not been colonised, e.g. recent volcanic islands devoid of vegetation.

The equilibrium of a plant community is dynamic rather than static, in as much as some individuals die off while others germinate and grow. At the same time, the individual species are continually exchanging places. A continuously changing **mosaic** of different developmental phases occurs particularly in unspoilt habitats, on moors and even more so in rain forests where these processes are obviously very long term. In larger areas this leads to the presence of all phases. Phases that can be distinguished are dependent on each other and can merge in different ways with certain **cycles** (see Fig. 12).

Quantitatively, too, the composition of species deviates substantially, since external conditions vary from year to year, rainy years follow upon dry ones, and so on. Consequently, sometimes one species will be favoured, sometimes another. If the conditions in the habitat alter con-

**Fig. 12.** Schematic of different phases of original forests and their transitions, derived from data obtained in Rothwald near Lunz am See (Lower Austria, modified after ZUKRIGL et al. 1963). It is possible to classify corresponding phases and mosaics noted in Bialowiecz (eastern Poland) into such a cyclic scheme

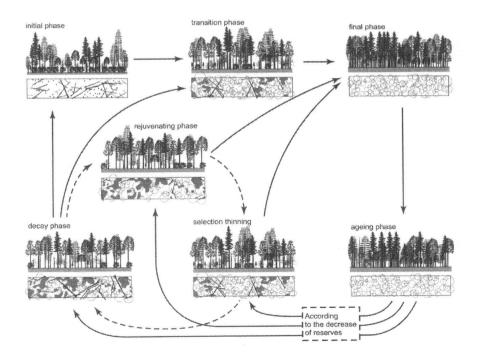

A temporal sequence of plant communities in a certain area is called **succession.** If changes in the habitat arise from natural causes and originate on parent rock, then it is called **primary succession;** this is usually a very slow process because of the lack of propagules (e.g. diaspores or seed banks). More often, a rapidly progressing so-called **secondary succession** results from human interference, e.g. the draining of water meadows, deforestation, abandonment of fallow land, regrowth of meadows or fallow land etc. or catastrophes, e.g. storm damage, floods etc.

tinuously in one direction, e.g. if the groundwater table rises slowly over many years, then the combination of species also changes, some species disappearing and other infiltrating from outside until finally a new plant community arises.

If interference by humans continues over a long period in the same manner, an anthropogenically determined equilibrium occurs resulting in plant communities which are called **cultural formations** if they are intensively utilised or semi-cultural formations if more extensively utilised. They constitute the vegetation in areas densely populated by man. The most important cultural formations are maintained by certain measures with the succession restarted again through a change of utilisation (see Fig. 13). The sequence of succession is usually accidental, dependent on the type of seeds or other propagules arriving first and in large quantities at the area and according to the conditions for germination of seeds and the possibilities for the establishment of seedlings.

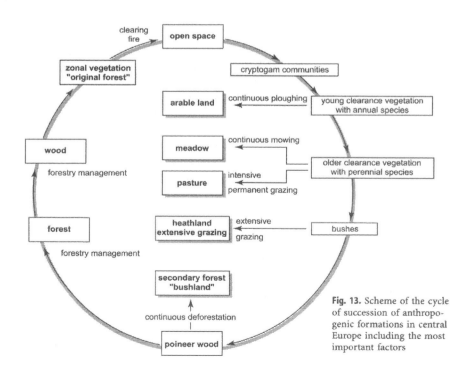

**Fig. 13.** Scheme of the cycle of succession of anthropogenic formations in central Europe including the most important factors

## Ecological Factors

Differentiation of the vegetation in a floristically uniform area is influenced most of all by the climate and the soil; the influence is, however, indirect, as shown above, via the effects of the competitive power of the prevailing species. Single ecological factors supplement each other in their selective effects, often to very different degrees. Climate has a direct effect on vegetation and an indirect effect via the soil. These interactions are shown in Fig. 14.

The type of soil and type of vegetation are determined by the climate. For the former the bedrock, and for the latter the flora (and secondarily the fauna) are also important. Furthermore, there are very tight interactions between soil and vegetation so that one could almost call it a unit. On the other hand, soil and vegetation also affect the climate, however, only directly in the area of the air layer next to the soil, i.e. they affect the micro-climate. All factors influencing plants, or more generally organisms, are called the **environment**, whereby the physico-chemical factors (without competition) are called the **habitat**, whilst the place in which they grow is called the **growing site**, **biotope** or **ecotope**. The main factors responsible for growth and development of the plant can be divided into five groups of primary factors (see marginalia).

It is unimportant for plants whether favourable temperature conditions are determined by the large-scale climate or by the growing site, for example, a warm climate or a protected slope (e.g. southern slope in the northern hemisphere). Similarly, it is of no consequence to the

**Ecological primary factors:**
1. warmth or temperature condition – temperature factor
2. water or hydrological conditions: water factor
3. light intensity and day length: light factor
4. various chemical factors (nutrients and poisons)
5. mechanical factors (wind, fire, herbivory)

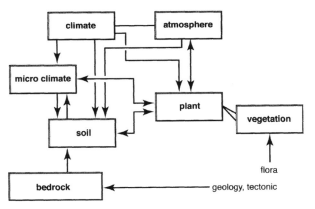

**Fig. 14.** Scheme of interactions between various environmental conditions and plants

**Fig. 15.** Scheme of different ecological factors and their effects on the plant

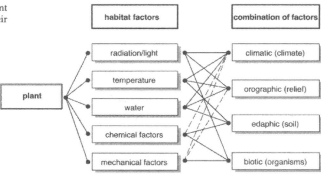

plant, whether the necessary soil moisture is derived from a favourable distribution of rainfall or reduced evaporation on a north-facing slope or even from soil structure and availability of groundwater; the most important factor is that the plant does not experience water deficits.

These five groups of habitat factors and their interaction cause the expression of complex habitat factors (secondary factors, complex factors), i.e. climatic, orographic, edaphic (soil) and biotic factors, as shown in the schematic (see Fig. 15) with some important habitat parameters.

## Radiation, Light and Basic Astronomy Elements

Radiation from the sun falling on the earth is the precondition of almost all living processes (disregarding the special circumstances of the "black smokers" in the deep-sea rift valleys). The energy of the incident solar radiation striking the earth is, for geometric reasons (angle of the rays on the globe of the earth) and astronomical reasons (orbit of the earth, rotation of the earth), dependent on latitude and season (see Fig. 16). Radiation reaches the surface of the earth only indirectly; some wavelengths are absorbed in the atmosphere and scattered and some are reflected back into space.

Maximum radiation in the polar summer (polar day = 24 h) even exceeds values in the tropics; it has a marked annual cycle, which is completely lacking in the tropics. However, over the whole year radiation is highest on the equator and decreases continuously towards the poles. The sum of the summer and winter radiation is highest on the equator. The highest position of the sun (zenith) in summer ($H_S$ in Fig. 17) is in the area of the corresponding

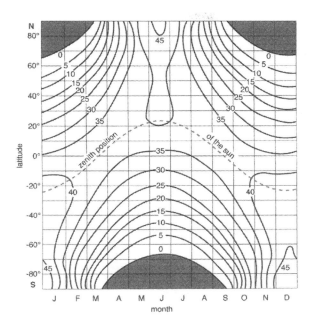

**Fig. 16.** Extra-terrestrial solar radiation (above the earth's atmosphere) during the course of the year and its dependence on geographical latitude (numbers in $10^6$ J m$^{-2}$ day$^{-1}$) (after SCHÖNWIESE 1994)

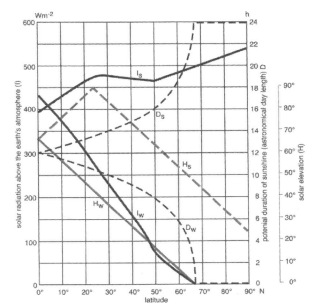

**Fig. 17.** Extra-terrestrial solar radiation ($I$), day length ($D$) and solar elevation ($H$) during the summer ($s$) and winter ($w$) on the northern hemisphere of the earth (after SCHÖNWIESE 1994)

**Fig. 18.** Energy exchange of
solar radiation energy in
the atmosphere and at the
surface of the earth and the
solar constant in relation to
the geographical latitude
(after SCHÖNWIESE 1994)

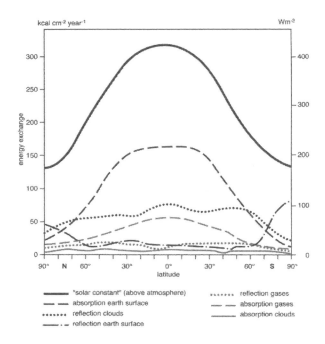

Solar radiation reaching a
particular point on the
earth's surface determines
the season (→ **annual** periodic course of the year)
and also, because of the rotation of the earth, day
length (→ **diurnal** periodic
course of the day). Most
other ecological factors also
show annual and diurnal
rhythms.

tropic (ca. 23.5°, solstice: summer or winter solstice respectively). Conversely, the day length, D, within the area
of the polar circle (ca. 66.5°) and up to the poles, reaches
the maximum value of 24 h, because of the inclination of
the earth's axis towards the orbit of the earth. The atmospheric processes and the radiation angles, dependent on
the latitude, determine the radiation reaching the earth's
surface (Fig. 18). They also explain the sequence of seasons in different latitudes and the difference between the
northern and southern hemispheres. The orbit of the
earth around the sun is not circular, but slightly elliptical
with one point of the earth nearer and the other further
away from the sun. This does not greatly effect the seasons. The sequence of seasons and their development is
explained in Fig. 19.

The course of irradiation during the day leads to continuous changes of the air temperature as a result of various components of irradiance and reflection in the balance of irradiance. The relative turnover of energy reaches
its maximum of energy gain around midday, the energy
losses are particularly large immediately after sunset (Fig.
20).

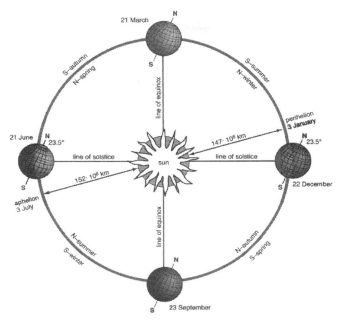

**Fig. 19.** Schematic to explain the earth's seasons by astronomical characteristics of the earth's orbit and the inclination of the earth's axis towards the plane of the orbit (ecliptic) (line of solstice: summer and winter point; line of equinox: day and night of same length; spring and autumn point) (after SCHÖNWIESE 1994)

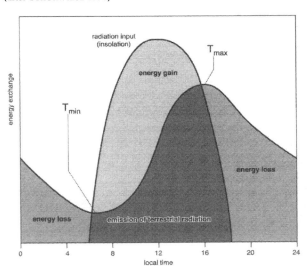

**Fig. 20.** Scheme of the daily course of solar radiation reaching the earth's surface and the emission of radiation from the surface. The sine curve also corresponds to the change of temperature on clear days (after SCHÖNWIESE 1994)

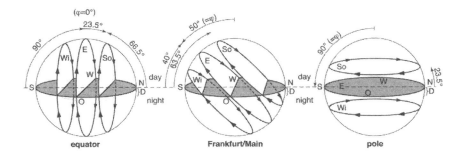

**Fig. 21.** Apparent orbit of the sun in the sky in different geographical latitudes (after Schönwiese 1994)

Seen from the earth, the **change of day and night** appears quite different depending on the geographical latitude, because of the astronomical characteristics (orbit of the earth around the sun with 365.25 days and inclination of the earth's axis with a rotation of 24 h). The apparent orbit of the sun in the sky (Fig. 21) is almost always 12 h on the equator; in the middle latitudes, for example Mainz in Germany, the difference in the length of the day between summer and winter is considerable, and at the pole, it is night for half of the year and day for the other half of the year (polar night, polar day).

In fact, only in spring and autumn, when the sun is exactly above the equator, are the days 12 h long (equinox).

These astronomical characteristics, discussed very briefly, are the preconditions for the climate which, through considerable modifications by the atmosphere, lead to the distribution of temperatures at the earth's surface, to fluctuations in air pressure, to air streams and thus to short- and long-term changes in the weather. This also determines the events in small areas, the micro-climate, such as in the area of the leaf. However, here the light factor is very rarely a minimum factor which determines the vegetation. Factors such as temperature and water have a much more direct influence.

## Temperature

Besides the water factor (see p. 42), temperature conditions play an important role in a habitat. Life can only happen in certain temperature ranges. Extreme temperatures are tolerated by different organisms to different degrees. The **heat resistance limit** of most plant species is between 50 and 60 °C. Up to a certain extent, plants are able to protect themselves against heat stress by reflection

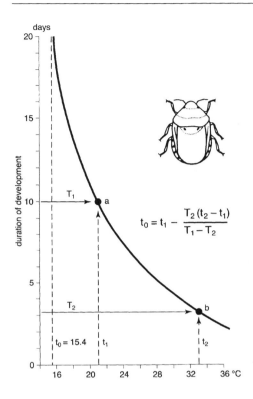

**Fig. 22.** Dependence of the duration of development of the bug *Eurygaster maura* (Pentatomidae) on temperature (after TISCHLER 1984)

$$t_0 = t_1 - \frac{T_2(t_2 - t_1)}{T_1 - T_2}$$

of radiation, by cooling through transpiration, and also physiologically (heat shock proteins).

The **cold resistance limit** is not as sharp as the heat resistance limit. There are certain sensitive plants (usually of tropical origin), frost-sensitive plants (which avoid formation of ice in tissues, for example by increasing the concentration of the cell sap) and frost-tolerant plants (they often possess many small vacuoles instead of one large vacuole. In these small vacuoles damage to the membrane by ice crystals can be kept to a minimum).

For animal organisms we distinguish, with respect to temperature, cold-blooded or poikilothermic species, whose body temperature, and thus the temperature of their plasma, depends on the external temperature and changes with it (Fig. 22), and warm-blooded or homoiothermic species, possessing their own body temperature which is, to a large extent, independent of the external temperature and rather constant. For these organisms it makes no sense to measure the external temperature in

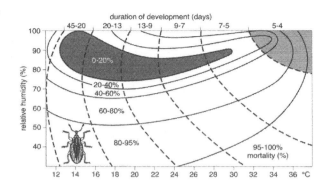

order to find a direct relationship with the metabolic processes in the protoplasm.

All plants are poikilothermic organisms (even if at
times they are able to generate heat, for example the spa-
dices of lords and ladies and other Araceae). The temperature of the surrounding air, therefore, gives an indication for the temperature required in the plasma. Certain
smaller deviations caused mainly by strong radiation can
occur at times. This must be considered in ecophysical research, because, for example chloroplasts or mitochondria
may have a temperature 10 K higher than the surrounding
air during the day. For ecological listings it usually suffices to give the air temperature.

For most poikilothermic animals the development depends very much on the temperature, usually modified by
the water factor, for example, the humidity of the air. In
the case of insects the development related to temperature
etc. can be given, often very precisely, by a mathematical
function (see Fig. 22), for example the hyperbolic function. The example in Fig. 23 shows the duration of development of bug embryos as well as the mortality of eggs in
relation to the air temperature and the relative humidity.
It can be seen that the optimum range is given at a certain
range of temperature and depends on relative humidity.
Correspondingly, it is easy to imagine that the rate of reproduction for some species of insects in certain biotopes
can differ greatly from year to year, even without any
other biotic interactions, depending on the outside temperature.

Freezing is tightly coupled with the behaviour inside
the tissue and the water in the cell. Freezing of the
vacuole usually means significant tearing of the membranes and thus considerable damage to the cell. Addi-

tional further supplies of water are blocked so that during longer frost periods plants usually dry out rather than are damaged by frost (**frost-induced drought**).

Various zonobiomes (see listing of zonobiomes on p. 79) are characterised by the temperature factor. However, mean values are less important than extreme values. Thus, it depends whether frosts in a region happen regularly in the course of the seasons, or episodically. One frost in 20 years in the coffee plantations of Brazil causes an increase in the price of coffee on the world market.

Plants prepare for the annual occurrence of the cold in winter. The important frost categories on the earth are shown in Fig. 24. Freezing of water below 0 °C and the simultaneous increase in volume are of special significance for organisms. The limit of 0 °C and with it the occurrence of frost, thus has a decisive effect on the characteristics of biomes.

The following applies to individual zonobiomes: Zonobiomes I to III are free of frost (except in higher mountain regions). In zonobiomes IV and V light frosts can occur occasionally (episodic, sometimes periodic). In zonobiome VI typical winters occur regularly, even if short and not very

**Fig. 24.** Occurrence of **frost** on the earth (after LARCHER 1994). **A** Frost free regions; **B** episodic frost up to −10 °C; **C** winter cold regions with mean annual temperatures between −10 and −40 °C; **D** mean annual minimum below −40 °C; **E** polar ice and permafrost. *dashed lines* 5 °C minimum isotherm; *dotted lines* −30 °C minimum isotherm

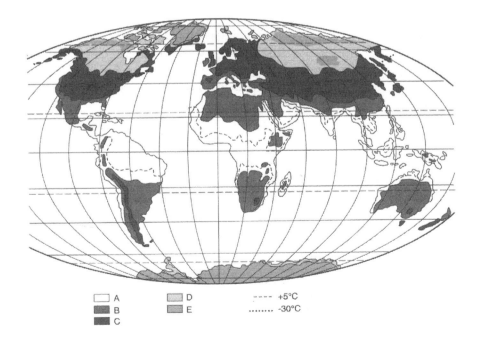

☐ A    ☐ D    ---- +5°C
■ B    ▨ E    ........ -30°C
■ C

hard. In zonobiome VII, however, winters are very marked and at times very hard (cold semi-deserts and deserts) because of the continental climate. In zonobiome VIII, in the taiga, winters can last many or several months and can be very hard; the tundra (ZB IX) is marked by the winter, it is by far the longest season of the year. The occurrence of frost determines the growth of plants with different resistances to frost. In the equatorial zone with minima not below +5 °C, cold-sensitive plants predominate. In zone D (Fig. 24), on the other hand, only completely frost-tolerant plants can survive, whilst in zones C and B plants and trees with a limited tolerance to frost can survive, or at least those protected by frost depression and good supercooling. Only 30% of the land area of the earth is free of frost, regular strong frosts occur in 42% with an annual mean minimum of below −20 °C.

# Water Relations of Plants and Vegetation – Poikilohydric and Homeohydric Plants and Adaptation to Water Deficit

### a  Water as a Factor

The most important habitat and environmental factors involved in a bio-classification of the geo-biosphere are temperature and water relations. Light is never a minimum limiting factor even at the poles, since the long polar nights coincide with the winter resting period of the plants.

As discussed earlier, temperature decreases fairly steadily from the equator towards the poles. In determining the boundary between the tropics and outer tropics the occurrence of frost is especially important. However, water plays an even greater part in the distribution of plants. Precipitation is very unevenly distributed (Fig. 25). Mean annual values vary from 10,000 mm (Assam) to almost zero (in the most extreme deserts).

In comparison, Fig. 26 illustrates the vegetational zones for which both precipitation and temperature are especially important; these are particularly obvious in relation to the latitude.

Temperature and water not only affect plant cover of a region, but local differences in water supply and temperature also have a strong differentiating effect. Water plays a very special role in the life of plants, much greater than in the life of animals, because plants are tied to their location. At the level of the cell as well as that of the whole plant and the eco-system, water relations can be described quantitatively.

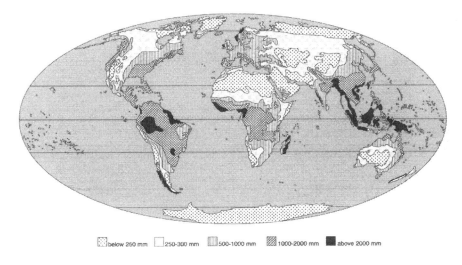

below 250 mm　　250-300 mm　　500-1000 mm　　1000-2000 mm　　above 2000 mm

## b Dependence of Plant Types on Water, and Their Resistance to Drought

According to the supply of water in a habitat, plants are differentiated into hygrophytes, mesophytes and xerophytes. **Hygrophytes** growing in equally moist or wet habitats (as many shade-loving plants in woods) hardly ever suffer deficiency of water. **Mesophytes** are somewhat better adapted to particular dry periods. Most species of the temperate latitudes belong to this category. **Xerophytes** developed many strategies of adaptation to the more or less strong and long lasting periods of water deficiency in their habitat. Levitt (1972) has shown various strategies: most plants avoid drought through spatial or temporal avoidance: real drought tolerance requires special adaptation, as will be shown for some examples of xerophytes (see p. 47).

**Fig. 25.** Annual rainfall map. Precipitation in mountains was not taken into account

## c Soil Water

The availability of water for plants not only depends on the water content of the soil. Distribution and size of particles and thus the volume of pores and the size of capillaries in the soil have a large effect on availability of water. The maximum volume of water taken up by the soil depends on the volume of pores. However, if all the space in the soil is taken up by water, there is no air in the soil and thus no oxygen. Because of gravity part of the water drains to deeper zones; the water remaining is called 'field

The amount of water retained in the soil against the force of gravity is called **field capacity**.

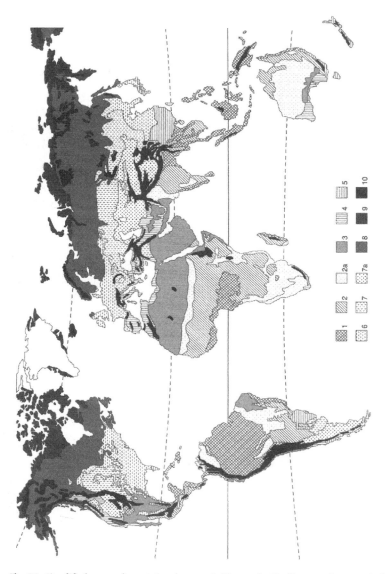

**Fig. 26.** Simplified map of vegetational zones (without edaphically or anthropogenically influenced vegetational regions). *1* Evergreen rain forests; *2* semi-evergreen and wet season green forests; *2a* savannas, grassland, dry woodlands; *3* hot deserts and semi-deserts (at 35°N, S transition to 7a); *4* sclerophyllic woodlands with winter rain; *5* moist warm temperate woodlands; *6* deciduous (nemoral) forests; *7* steppes of the temperate zone; *7a* semi-desert and deserts with cold winters; *8* boreal coniferous zone; *9* tundra; *10* mountains

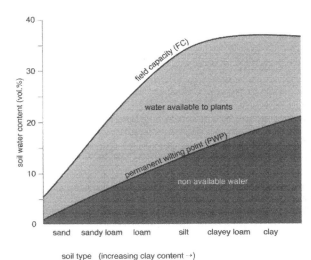

capacity'. Figure 27 shows the dependence of field capacity on particle size.

Part of the water in soil is retained by absorption to soil particles (for example electrostatically) and part by capillary action (absorption and cohesion) where it is retained in very small pores. Both fractions of water are not available to plant roots. If the soil only contains these retained parts of water, the so-called **permanent wilting point** is reached.

Although the water content in very fine-grained clayey soils can be almost 20% there is still no water available for the plant because of the small soil particles. However, the limit of water availability is not the same for all plants. Xerophytes and halophytes are able to take up water because of the very high 'suction power' (low water potential) of their roots. Thus, the permanent wilting point is different for different plant species. The response of plants to water is as complicated as the response of animals to temperature.

The amount of water in the soil available for plants is between the 'permanent wilting point' and the 'field capacity'. It is greatest in silty soils.

### d  Water Relations of the Cell – Hydrature

A distinction must be made between plants able to change their water content (poikilohydric) and those maintaining a constant water balance (homeohydric). The cell content, the plasma, is physiologically active only when its water content is high, that is, when it is in a hydrated or im-

Relative activity of water (a) expressed as relative vapour pressure compared to pure water expressed as percent (%) is called **hydrature**.

bibed state. When cells dry out the plasma goes into a latent condition and shows no detectable sign of life and eventually dies. According to the thermodynamics of imbibition, the degree of hydrature depends on the relative activity of water (a), where $a = p/p_o$, i.e. equivalent to the relative water vapour pressure. Pure water has a hydrature of 100%. In the air, hydrature corresponds to the relative humidity (also in percent). Above salt solutions a certain water vapour pressure occurs which is lower than that of pure water and thus leads to lower hydrature.

Since vital processes depend, to a large extent, upon the degree of hydrature of the protoplasm, it is essential to know the hydrature of the protoplasm (and the activity of water). In poikilohydric organisms such as single-cell bacteria or organisms with few cells, e.g. algae, fungi and lichens, hydrature depends entirely upon the humidity of the surrounding air, when these organisms are out of water. If these organisms are in contact with water or if the surrounding air is saturated with water vapour, then their protoplasm is almost fully hydrated and active. In dry air, however, dehydrature takes place and the protoplasm passes into a resting condition without dying. Cells of such organisms have no vacuoles, or only very small ones, so that the change in volume of the cell contents during desiccation is small and the protoplasmic structure remains undamaged. The **lower limit of hydrature** (humidity of the air) at which growth is still detectable is very high for most bacteria, usually 98–94%, for single-cell algae and moulds it varies and sinks to 70% for only a few organisms; 70% is a value which corresponds to the extreme lower level of hydrature compatible with any sign of life.

The productivity of these poikilohydric organisms is small and thus they contribute little to the entire terrestrial plant biomass. For this reason, they have been afforded little attention, although they are often much more widely distributed than is generally assumed, particularly in deserts. Poikilohydric organisms were probably widespread on wet areas before the conquest of land by higher plants, just as they are today on occasionally flooded clay soils in deserts (Takyre, see p. 218). These areas cannot be colonised by higher plants due to lack of root space. Since, however, fossilised remains of lower plants are exceptionally rare in older geological formations, we have no fossil evidence to support this assumption.

Terrestrial homeohydric plants are of much greater importance and include all cormophytes, which developed originally from green algae. Their cells have a large central vacuole with the cell sap in contact with the surrounding protoplasm; the hydrature of which is thus not so immediately dependent upon the state of water outside the cell. The vacuolar sap constitutes an internal watery milieu for higher plants, as previously mentioned. It is this 'internal milieu' and the cell wall of cellulose which provides an 'external watery milieu', which in the course of the phylogenetic development, made possible the transition from an aquatic to a terrestrial way of life as well as a steadily improving adaptation to arid conditions. As long as terrestrial plants are able to keep the cell sap concentration of the vacuole low (a high osmotic potential), the protoplasm remains imbibed, which means well hydrated, independent of the moisture of the surrounding air. The more water in the soil, and the better the supply to the plant via the roots and transport system, the greater the degree of independence from atmospheric moisture. Since the vascular system is only imperfectly developed in mosses they are, therefore, generally confined to very damp habitats. Ferns, too, have a relatively inefficient transport system and thus avoid dry habitats. Moisture is required particularly during the very moisture-dependent gametophyte stage. Those mosses and the few ferns (for example *Ceterach, Notholaena, Cheilanthes*) and *Selaginella* spp. that have penetrated into desert regions have had to change to a **secondary** poikilohydric way of life in order to survive dehydrature during times of drought. They achieved this ability to withstand desiccation, not normally possessed by plants with strongly vacuolated cells, by a diminution of cell size with a reduction of the vacuoles, which solidify as a result of even a small water loss, thereby avoiding deformation and damage to the protoplasm by desiccation.

### e Hydrature in Xerophytes

The most complete adaptation of water relations to the terrestrial way of life has been achieved by the angiosperms which have even penetrated into extreme desert regions. Measurement of their cell sap concentrations shows that they are able to keep the concentration of the cell sap low and thus, maintain the hydrature of the cytoplasm without greatly impeding the gaseous exchange required for photosynthesis. An increase in the cell sap concentration (= decrease in the osmotic potential) and a resultant dehydration of the cytoplasm and increased osmotic adaptation by the synthesis of particular substances (compatible solutes) is not an appropriate adaptation for desert plants, but a sign of damaged water relations and a danger to their existence. The measurement of external

**Fig. 28.** Flux of water through a plant from soil to atmosphere is represented by an electrical resistance analogue. The flux (I) is driven by the difference in potential (E), here given by the difference in water potential between soil and atmosphere. The flux occurs against the additive resistances (R) in the plant which are in some cases constant and in others variable (stomatal resistance as a means of regulating water flux). Ohm's law (E = R.I.) is applicable (after HILLER 1980)

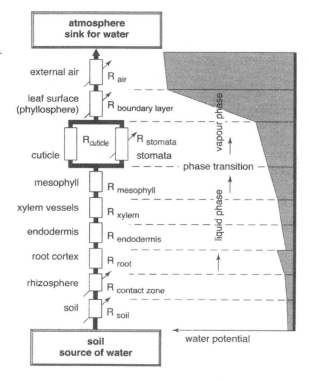

factors such as precipitation, humidity and water content of soil offers little information about the activity of water in plasma (hydrature) just as measurement of external temperature offers little information in studying warm-blooded animals.

Determination of the **cell sap concentration,** which is related to the osmotic pressure and the osmotic potential and thus related to the relative water vapour pressure, is the only means of determining whether the plant, and particularly the degree of hydration in the plasma, has been affected by the alteration of external conditions, especially by drought. Measurement of the water potential must be made if information on the movement in the plant from the roots to the transpiring organs is required. This can be demonstrated by the resistances in the plant in the hydraulic flow diagram (Fig. 28). Some of these resistances are constant, others more or less variable. Stomatal resistance should be mentioned, as this resistance is the most important factor in regulating water loss. Corresponding to Ohm's law, the flow of water (current) depends on the resistance and the potential (voltage). Total 'potential' corresponds to the difference in water potential between soil and atmosphere. This difference in water potential is almost always very big, even in moderate climates.

Water potential does not give any information about the hydrature of the plasma on which the course of all life depends. However, they are very tightly linked as described in the equations for osmotic characteristics.

To characterise a habitat, the usual information on external factors is required, but additional information on the cell sap concentration and its changes is also required to characterise the hydrature of the protoplasm, particularly when discussing arid regions where the water factor plays a dominant role. Therefore, adaptation to drought and osmotic conditions have to be discussed further.

The species under investigation in experiments are regarded as stable entities, although, over longer periods they appear to be very variable. Plants continuously adapt morphologically to the prevalent environmental conditions in order to survive. These adaptations are linked to growth and are noticeable only after weeks or months. Ecologically, they are particularly important and in arid regions very striking if a plant is investigated after a rainy period or during ensuing drought.

When considering the adaptation to water deficiency, various osmotic states of parts of plants should be considered:

- water potential ($\Psi$),
- osmotic potential ($\pi$)
- and the turgor pressure (P)

The following equation applies:

$$\Psi = \pi + P.$$

Measurements are expressed as pressures (MPa) and $\Psi$ and $\pi$ are always negative as they reduce the activity compared to pure water.

It is important to consider the significance of the various parameters for the water relations of plants. If only the more physical process of water flow through the plant, (from the soil to the atmosphere), is considered, then $\Psi$ must be measured. If, however, the study is concerned with the biological processes of adaptation linked to growth, then $\pi$ is the appropriate parameter since it is directly related to the hydrature of the plasma, and this is what controls growth processes in the plant. Even though

Values from older research expressed in atm and from newer research in bar is expressed as MPa (1 atm = 1.013 bar; 1 bar = 750 Torr = 750 mm Hg = $10^5$ Pa = 0.1 MPa; 1 MPa = 10 bar $\approx$ 10 atm; for Physical Units, see p. XVII)

**Fig. 29.** Various forms of *Encelia* leaves. *Above*: sparsely hairy, hygromorphic leaves; *below right*: mesomorphic; *below left*: xeromorphic; *middle*: twigs with terminal buds (all leaves have dropped)

these adaptations only become obvious after a considerable time lag, they could be regarded, from the cybernetic point of view, as the feedback regulation of the water balance under changing conditions required to maintain water relations of the plant. This is the controlling factor. The interfering factor is the increasing dryness during drought and the target is balanced water relations of the plant, i.e. water uptake should equal water loss. The living plasma should be regarded as the sensor during imbalance of the plant's water relations; $\pi$ decreases as a consequence of the rising cell sap concentration, resulting in a decrease of the hydrature of the plasma, which may be regarded as a set value. When $\pi$ decreases, changes occur in the meristem cells of the shoot and root. Newly formed organs are better adapted, e.g. internodes are shorter and the newly formed leaves are smaller and more xeromorphic, thus plants transpire less and the water balance is restored (WALTER & KREEB 1970)

An example from the Sonoran desert provides an illustration. The Compositae *Encelia farinosa* is a semi-shrub growing to a height of about 50 cm. During the rainy season, the plant has large, soft-green, hygromorphic leaves with a light covering of hairs. The leaf cells have $\pi$ of $-2.2$ to $-2.3$ MPa. In the dry season, when water supply becomes a problem, $\pi$ decreases to $-2.8$ MPa resulting in a slight decrease in hydrature of the protoplasm of the meristematic cells. Leaves which subsequently form from these meristems and replace the hygromorphic leaves are smaller, more mesomorphic and have a denser covering of

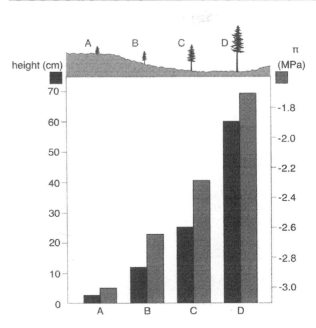

**Fig. 30.** Relationship between osmotic potential π (*right*) and vertical growth in *Solanum eleagnifolium* (*left*). **A–D** Samples from the plants are shown

hairs. With continuing drought π drops to –3.2 MPa and the next leaves are even smaller, thicker, and more densely covered with white hairs (Fig. 29), all of which tends to reduce transpiration even further. If the drought lasts for an extremely long period all the leaves are shed as soon as π drops to –4.0 MPa. Only the terminal buds containing rudimentary leaves are left. Water losses are now reduced to such an extent that the plants are in a state of water balance, even with the minimum of water available from the soil.

As soon as the next rainy season begins, the potential osmotic pressure (π) rises to the original value of about –2.0 MPa, the hydrature of the meristem cells rises, and the newly formed leaves are again large and hygromorphic. As a result of the ensuing intense photosynthesis, rapid growth takes place, transpiration values are high, and water balance is maintained. Thus the ever-recurring cycle of events is completed.

Figure 30 shows the close relationship between size and π for *Solanum elaeagnifolium*, an annual species that develops in the Sonoran Desert on clay soils after a heavy rainfall. In small hollows where water collects and the soil is wet down to a considerable depth, plants achieve a height of 60 cm. Toward the perimeter of the hollows, as

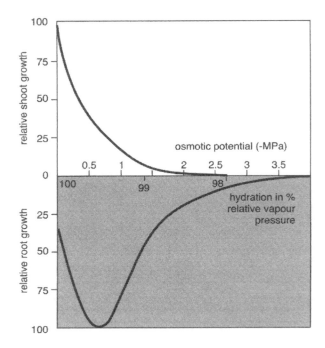

**Fig. 31.** Growth of pea seedlings at different, but constant hydration and osmotic potential

the soil eventually becomes increasingly dry, plants become smaller and smaller, down to a dwarf form only 1 cm high. The differences in water supply are reflected in the π values. The two curves run almost parallel (Fig. 30). Dwarf ephemerals of this type are seen everywhere after a poor rainy season.

It is interesting that roots react differently to a decrease in hydrature of the meristem cells than shoots. Roots become thinner, but longer and do not form side roots. Inhibition of growth only sets in after more severely decreased hydrature while the growth of shoots is immediately inhibited (Fig. 31).

*Brassica* seedlings grown in sands of different water content show this effect very clearly (Fig. 32). This kind of reaction is also a means by which water balance and survival are maintained in ephemerals. A smaller shoot means less water loss, and elongated roots enable plants to reach deeper soil that may still be wet even if the rains have been light.

A most remarkable phenomenon is seen in the barrel cacti, the various sides of which react differently according to exposure. The south-western side receives more

**Fig. 32.** *Brassica* seedlings grown in sand of various degrees of wetness (5 days after germination). Water content of sand: **a** 15.5%, **b** 6.7%, **c** 4.3%, **d** 2.5%, **e** 1.3% (three seedlings from each group are shown)

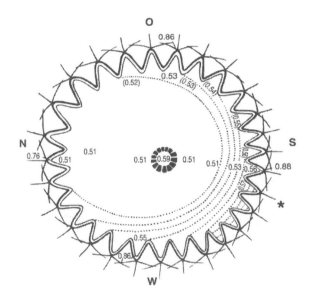

**Fig. 33.** Distribution of osmotic potential ($\pi$) in the cross section of *Ferocactus wislizenii*. Isomoses, lines of equal potential (cell sap osmotic potential in -MPa, lowest potential at * on the south-western side)

warmth than the rest of the plant and transpires the most, whereas the exact opposite holds for the north-eastern side. An example is provided by *Ferocactus wislizenii* which is widespread in the Sonoran Desert. A transverse section shows that the isosmoses (lines of equal $\pi$) clearly reflect the asymmetry of water demand; $\pi$ is lowest on the south-western side and highest on the north-eastern side. Correspondingly, the south-western side is the more xeromorphic (narrower, closer ribs, woody parts thicker; see Fig. 33). Vertical growth is also reduced on the south-western side so that the plant bends in this direction (Fig. 34) and older specimens often fall over.

**Fig. 34.** *Ferocactus wislizenii* in the Sonoran Desert, leaning towards the southwest (Photo: E. Walter)

**Fig. 35.** *Pachycereus pringlei* with flower buds on the southwestern side (one flower already open) (Photo E. Walter)

Another interesting point is that often the first flower buds form on the south-western side and not at all on the north-eastern side (Fig. 35). It must be concluded from this that the lower π favours reproductive growth at the expense of vegetative growth. This is seen in the ephemerals, since dwarf plants with their higher cell sap concentration always come into flower first and confirms the observation of gardeners that a poor water supply results in a larger number of blossoms whereas growth is mainly vegetative if water is plentiful.

## f  Water Relations of Ecosystems

The formation of a sufficiently large root system is particularly important for survival in dry regions. Perennial plants are able to survive several dry periods if the volume of soil for root growth is large enough and the water in lower layers can be utilised. Some plants with particularly deep roots appear to be able to lift the water ('hydraulic lift') so that even other plants can profit, as demonstrated by CALDWELL et al. (1991). The root volume of soil exploited by roots is, for example, in olive trees larger in dry than wet areas. In Tunisia, farmers have enlarged the distance between trees from north to south based on experience.

Generally, the water relations of a biogeocenosis, a certain area of a section in the landscape or a complete country, can be quantified with an equation. In places without access to the groundwater table, the available water is the precipitation plus storage in the rooting volume. Water losses include evaporation, transpiration and drainage. The full water balance equation for an ecosystem is:

$$P = \pm\delta W + (E + I + T + S + D)$$

where P is precipitation, I interception, S surface runoff, E evaporation, T transpiration, D drainage, and $\delta W$ stored water supply in the system.

Loss can occur in several ways, on the one hand, through evaporation **E** (from the soil) and through transpiration **T** (through the plant) as well as the interception **I** (surface wetting of leaves and vapour loss), and loss from runoff from the surface **S** and by drainage **D** into the ground (underground loss into the groundwater). The soil itself as well as the complete ecosystem, have a certain supply of water as storage, $\delta W$, which can increase (+) or decrease (−). E and T together (with I) are often called evapotranspiration (**ET**). The surplus of water which cannot be returned by **ET** to the atmosphere benefits the groundwater, and thus, the supply of neighbouring sources and finally, forms systems of brooks and rivers. At sites affected by groundwater, an allochthonous water supply to the system can occur so that besides precipitation, rising water is added and the loss of water through drainage is inverted. In dry areas most water is lost by **ET**, and a supply of groundwater no longer occurs (see p. 23 arid and humid areas).

**Halophytes and Saline Soils, Halobiomes**

**Halophytes** are plants which are able to grow on mildly to strongly saline soils (halobiomes)

Salt plants or halophytes are an important group occurring in deserts. They are dependent on the occurrence of saline soils. Many halophytes are succulent. However, they should not be grouped together with real succulents. The succulence of halophytes is a consequence of very high sodium chloride or chloride storage; therefore their cell sap concentration is often very high and the osmotic potential, $\pi$, lower than –5 MPa. The mesembryanthemums show a transition from real succulents with a low cell sap concentration to halophytes with a very high cell sap concentration. They can be extremely succulent, but often occur on non-saline soils and yet contain some or even large amounts of chlorides. Halophytes colonise saline soils on sea coasts and in deserts. Saline soils were probably invaded relatively late by terrestrial plants, for in these areas they not only have to cope with the water problem, but also have to adapt to the physiological effects of salt.

It is useful to start with a definition of these plants: true halophytes are plants that store large quantities of salt in their organs without being damaged and even benefit from the salt if its concentration is not too high. The salts involved are usually NaCl, but can occasionally also be $Na_2SO_4$ or organic Na salts. The concentration of vacuolar sap cannot be lower than that of the soil solution (and in saline soils, the latter is usually very high), otherwise a drastic drop in hydrature of the plasma would occur if other osmotically active substances such as sugar were formed in the cell sap. Thus, the problem is solved in a different way: salt is taken up from the soil into the cells until equilibrium with the soil solution is attained. Electrolytes ($Na^+$, $Cl^-$) are taken up in the vacuole and the plasma is not dehydrated; rather an additional hydration

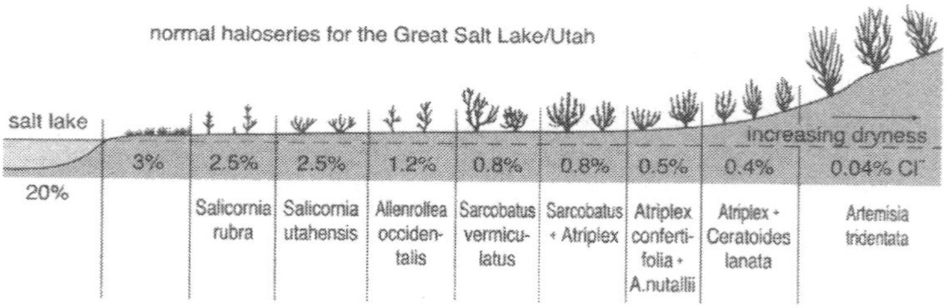

**Fig. 36.** Profile of vegetation for the Great Salt Lake (Utah, USA) with chloride content in the soil (% dry matter) in individual belts of vegetation (Kearney et al. 1914; Breckle 1976)

normal haloseries for the Great Salt Lake/Utah

| salt lake | | 3% | 2.5% | 2.5% | 1.2% | 0.8% | 0.8% | 0.5% | 0.4% | 0.04% Cl⁻ |
|---|---|---|---|---|---|---|---|---|---|---|
| 20% | | | Salicornia rubra | Salicornia utahensis | Allenrolfea occidentalis | Sarcobatus vermiculatus | Sarcobatus + Atriplex | Atriplex confertifolia + A.nutallii | Atriplex + Ceratoides lanata | Artemisia tridentata |

increasing dryness

occurs which causes the succulence of the organs. In contrast, in the cytoplasm additional osmotically active compounds are synthesised which balance the osmotic potential and do not affect the function of the cytoplasm, i.e. the materials are compatible solutes. These substances are frequently of quite a different chemical nature. They are often typical for certain plant families or species and thus taxon-specific (POPP 1995). Salts are very toxic in higher concentrations. Thus, halophytes have to be salt-resistant, which is only possible up to certain limits, and soils with very high salt contents remain bare of vegetation (Fig. 36, for example on the bank of a salt lake).

Because of the different behaviour of plants towards high salt stress from the soil, plants are classified according to different types of salt adaptation. The **non-halophytes (halophobes)** – the majority of plants – die of a loss of water because of their inability to adapt osmotically if subjected to salts. Salts are toxic for salt-sensitive species. They are, therefore, not able to grow on saline soils. The **facultative halophytes (pseudo-halophytes)** are, to a certain degree, able to adapt their uptake system in the root osmotically to take up salt and keep it in the area of the root and thus, keep their shoot relatively free of salt. Such salt-tolerant plants are able to withstand a moderately high salt concentration, but develop better on non-saline soils.

For all halophytes, as will be discussed for mangroves (p. 199), roots work as ultrafilters and take up only almost pure water from the saline soil solution and transport this through the vessels to the leaves. In the vessels of halophytes, a high cohesion tension has been observed. The root system also acts as an ultrafilter in **euhalophytes** and only allows a little salt into the transport system. However, these salts slowly deposit in the shoot system causing **halosucculence**. Leaf succulence (for example *Suaeda*) or shoot succulence (for example *Salicornia*) arises from an alteration to the formation and differentiation of tissues. The growth of euhalophytes is stimulated, to a certain degree, by salt. In normal soils with only traces of NaCl, they take this up so that even under these conditions, their salt content is relatively high. The stimulating effect is due to chloride ions, which cause swelling of the proteins and thus a hypertrophy of the cells because of the high water uptake, i.e. the succulence of the organs. The succulence is expressed more, the higher the chloride content of the cell sap. Only the chloride ion has this effect, not the sulphate ion which has the effect of decreasing the water content of proteins. Certain halophytes that store larger quantities of sulphates as well as chlorides in the cell sap are not succulent, or only barely so. A distinction must be made, therefore, between **chloride**

**Fig. 37.** Concentration of inorganic salts in the cell sap of transpiring organs of halophytes and non-halophytes. Figures on the *right*: concentration -MPa corresponding to the cation equivalent calculated as NaCl. *1-5* Chloride halophytes (all except the salt-excreting grass *Distichlis* are succulent); *6-8* alkali halophytes (high content of organic anions, often oxalic acid); *9* a slightly halosucculent pseudohalophyte, often with high K instead of Na content, similar to a non-halophyte *10* (with very little Na) (after ALBERT 1982)

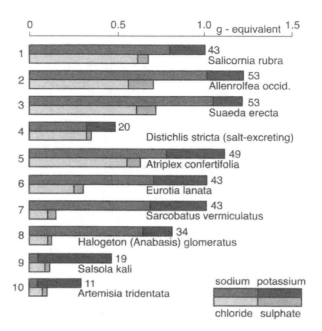

halophytes and sulphate halophytes, although they can exist side by side on the same type of soil. Salt uptake is usually species-specific (BRECKLE 1976). Investigations on halophytes are quite clearly incomplete if only the total salt content of the soil is determined, since, for plants, only those salts that come into contact with the plasma are important. Concentration and composition of salts in the cell sap must, therefore, always be measured. Figure 37 demonstrates how different the composition of salts in the cell sap of halophytes and non-halophytes can be.

Even the euhalophytes have an upper limit for the concentration of salt tolerated in the cell sap, varying from species to species. If this rises too much, the plants wilt; in the case of the Chenopodiaceae, this is usually accompanied by the plants turning red (N-containing pigments: betalaine) and finally dying. In a further group of halophytes, the Na in the cell is at a higher equivalent concentration than that of the Cl and $SO_4$ put together, so that the Na ions must be balanced by anions of organic acids. When such plants die, carbonic acid is produced as a result of the breakdown of organic acids, and the Na reaches the soil in the form of $Na_2CO_3$ (soda), which leads to increased alkalinity. These plants are termed alkali halophytes.

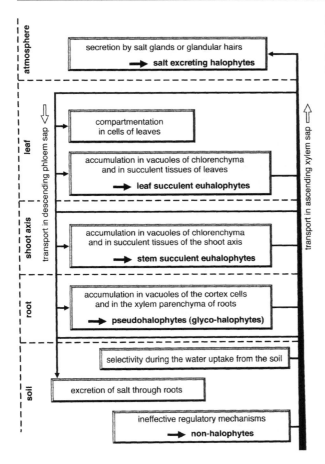

**Fig. 38.** Schematic of different types of halophytes according to the regulation of the internal salt content (after BRECKLE 1976)

There are also **salt-excreting halophytes,** for example *Limonium, Frankenia, Reaumuria, Glaux, Spartina* and other halophilic grasses which are usually non-succulent species with salt glands, continuously excreting the absorbed salt. The well-known tamarisk tree (*Tamarix*) of which there are many species in arid regions, also has salt glands. Salt dust rains down if the branches of these trees are shaken. Since *Tamarisk* primarily excretes NaCl, sulphates predominate in the cell sap, and the leaf organs are not succulent.

Salt excretion is also possible by deposits in isolated glandular hairs (*Atriplex* species etc.) which form a coating and can be shed. Excretion of salt is also possible by shedding older salty leaves. This is also known for facultative ha-

lophytes, for example *Juncus* where the leaves very early
turn yellow or in rosette plants (*Limonium* etc.) where
new leaves are formed continuously. Halophytes are described by an auto-ecological system (see Fig. 38) as well
as by a characterisation based on the ecological distribution
of different halophytic types: obligate halophytes – facultative halophytes – site indifferent halophytes – non-halophytes. The latter characterisation is, of course, mainly
the same as the characterisation of auto-ecological types.
Along the salt gradient in an area, for example around a salt
lake, halophytes usually occur in certain zones. At the highest salt concentration succulent euhalophytes predominate,
towards the outside leaf succulents follow, then there is often a zone with many special salt-excreting halophytes,
and further outside pseudo-halophytes and at the outside
(without salt stress) non-halophytes. Such a halo-catena is
formed best in areas where there are many different halophyte species, such as in central Asia (BRECKLE 1986).

For many halophytes of arid regions water is not the
problem, as explained previously, as they grow on wet salt
soils of the pans (hygrohalophytes), but salt balance is the
problem. However, there are also halophytes which grow
on dry salt soils and frequently suffer under water deficit
irrespective of the strong salt storage (xerohalophytes);
*Atriplex, Haloxylon, Zygophyllum* species are of this type.
Others often progressively reduce the transpiring surface
during the drought period in proportion to the available
water, for example *Zygophyllum dumosum* sheds the pin-
nules first, then the leaf bases. Others shed the young terminal shoots or even the green bark of leafless shoots
from the previous year.

In all dry regions, there is the continuous danger of
**salinisation of the soil.** The incoming rainwater only supplies a small amount of salt (rainwater contains on average 0.001% NaCl), but in the long term considerable
amounts accumulate if there is not a corresponding outflow as occurs more or less in all arid regions (according
to the definition: potential evaporation is larger than the
precipitation). Arid regions are correspondingly geomorphologically characterised (see Fig. 7). They form endorheic basins, the outflow usually does not reach the
oceans, but flows into local basins. There the salt of the
water from precipitation, from dust and from salt washing
out of the surrounding rocks is released and accumulated
(salt pans or salt lakes, for example the Dead Sea, the Aral
Sea, the Great Salt Lake in Utah etc.).

In all arid regions, irrigation even with water containing only 0.02% NaCl (=200 ppm) and thus of the best
quality, causes a slow salinisation, if no precautions are

taken to wash out the accumulated salt from the fields. This natural desalinisation occurred over thousands of years in Egypt as a consequence of the annual floods of the river Nile, before the construction of the Aswan dam.

In arid regions the vegetation mosaic is very much influenced by the salt content of the soil. Various biomes are characterised by the salt burden. Often a very marked gradient of increasing salt concentration (and decreasing particle size of the soil) occurs towards the bottom of the basin. An example is given in the areas of the Great Salt Lake in Fig. 36. Further examples are given in the discussion of the arid zonobiomes III and VII (see pp. 218, 399).

In arid regions, rivers carry water only periodically, or even sporadically, as a result of the long periods of drought. Since potential evaporation is many times larger than annual rainfall, all of the water running down into depressions in the arid regions evaporates. Soluble salts in the water accumulate, as mentioned above, during the course of time. A saturated brine may form from which salt then crystallises out. **Salt lakes** or salt pans are characteristic for arid climates. Ultimately the ocean is also a closed lake into which over billions of years everything soluble has been transported. The largest part of the soluble salt consists of NaCl, since hydrocarbonates deposit earlier as $CaCO_3$ after the loss of $CO_2$ and sulphates precipitate a little later as gypsum ($=CaSO_4$). Potassium salts crystallise last, if at all, thus the typical sequence of these evaporites is also the sequence of sediments.

Sodium ions are released through weathering from silicates. However, chloride-containing minerals are rare, although chloride ions occur in sea water at almost 20 g/l (sulphate only 2.7 g/l). Few chloride ions are released through weathering of minerals, despite this, NaCl is observed in river water. It could be that during the long history of the earth NaCl was deposited in river waters by the HCl-containing exhalations of volcanoes.

The NaCl of saline soils in arid regions can be of various origins:

1. Marine salts embedded in rocks formed as marine sediments and deposited as marine sediments (**evaporites**). During the weathering of these rocks the salt is washed out by rain water and carried into the undrained depressions. In deserts with marine sedimentary rocks of the Jurassic, Cretaceous, or Tertiary periods, the northern Sahara and the Egyptian desert for example, saline soils are common, whereas in arid regions with underlying magmatic rock or terrestrial sandstone, hardly any saline soils are found.

2. Brackishness in areas that in the most recent geological past were lakes or marine beds which slowly dried out, for example the area around the Great Salt Lake (Utah; Lake Bonneville is a glacial lake), around the Caspian Sea and Aral Sea in central Asia, around lake Tuz Gölü in central Anatolia, the Dead Sea in the Middle East, Lago Enriquillo (Hispaniola) and others.

3. Seawater turned into a fine spray by the force of breakers along arid coasts. The small droplets dry and form a salty dust which can be blown many kilometres inland. This salt is then either washed into the soil by rain or fog or simply deposited. This process can occur in humid regions, but here the salt is continuously washed out and returned to the sea via the rivers (cyclic salt). In arid regions with no outflow, however, the salt concentrates and in this way leads to such brackishness as is encountered in the outer Namib Desert and the arid parts of Western Australia. Once salt accumulates on the surface of depressions, it can be blown still further by the wind. However, away from the coast, rain (with up to 10–20 ppm NaCl) continuously brings traces of salt.

4. Brackishness can also occur if salt-containing spring water comes to the surface, for example in the northern Caspian Lowlands. In this case, the salt originates from marine beds which became desiccated in marine basins during earlier geological times and formed large salt deposits at a considerable depth (Permian, limestone). In arid regions these salts accumulate. In contrast to humid regions, salts from salt springs, for example, in Bad Salzuflen, Salzdetfurth, Salzgitter, Salzburg in Germany and Austria, are quickly returned again to the sea.

In deserts, salts are washed from higher elevations to lower-lying parts with each rainfall so that, for the most part, only the depression soils become brackish. If the sedimentary rocks contain much salt and rainfall is very scanty, as it is around Cairo-Heluan or in Central Iran, then the soil of the plateau habitats also contains salt. There is no rainfall in the central Sahara, so that lower areas do not receive any additional salt.

As far as plants are concerned, it is the salt concentration of the solution surrounding the roots that is important and not the salt content of the soil calculated on a dry-weight basis. In slightly saline, but dry soils, the concentration is often higher than in very saline but wet soils.

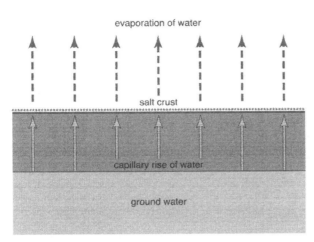

**Fig. 39.** Formation of a salt crust by capillary rise (*continuous arrows*) of groundwater and evaporation of the water (*broken arrows*); salt enrichment at the soil surface (after WALTER 1990)

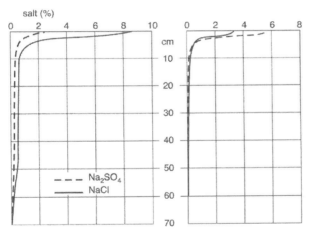

**Fig. 40.** Salt content at various soil depths in a watered plot (*left*) with rising groundwater and on an unwatered plot in the Swakop Valley (Namibia). *Solid lines* NaCl, *broken lines* $Na_2SO_4$. Salt deposits only at the surface (after WALTER 1990)

Evaporation from the soil surface in places where the groundwater is less than 1 m below the soil surface can also lead to salt accumulation. Water rises to the surface by means of capillary forces (Fig. 39), bringing salt to the surface, even if the groundwater contains only minute quantities of salt (Fig. 40). A salt crust forms wherever the capillary water column ends, which is at the highest point of the microrelief (Fig. 41).

The occurrence of a salt crust in dry periods does not necessarily interfere with plant growth as long as the roots

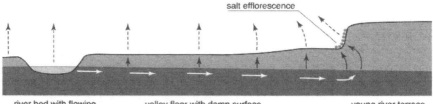

river bed with flowing                  valley floor with damp surface                    young river terrace
non-brackish water

**Fig. 41.** Salt accumulation in Swakop Valley (Namib Desert). *Solid arrows* indicate the direction and magnitude of water flow in the ground, *broken arrows* evaporation. The salt concentration increases towards the sides of the valley; salt efflorescence occurs at the foot of the terrace, where the water flow ends (after WALTER 1990)

have access to non-brackish groundwater. In the Pampa de Tamarugal in the Atacama Desert *Prosopis* trees grow through holes in a salt crust 0.5 m thick. This is possible only because the roots of the trees reach fresh underground water.

Each field irrigated in arid regions without at least a certain degree of drainage, is a basin without outflowing water and must necessarily turn brackish with time, even if the water used for irrigation purposes contains only a small quantity of salt. Extensive cultivated areas in Mesopotamia and the Indus region, have thus been transformed into salt deserts. So far, this has not happened in the undrained cotton fields of the Gezira in the Sudan, but only because the water of the Blue Nile, which is used for irrigated farming, contains hardly any salt. Small quantities of salt are removed in the crop itself with each harvest.

The following rule applies for sustainable irrigation in arid regions to remove salts accumulating over time: **no irrigation without drainage**.

If the rule "no irrigation without sufficient drainage" is not heeded, the cultivated land will become wasteland within a few decades, as has been and is still being demonstrated by many 'short lived' development projects.

## Mineral Supply and Soils

Besides the effects of salt (NaCl), the effects of other ions in salty habitats must also be considered, for example hydrogen carbonate (alkaline soils, sodic soils), sulphate, borate. The type of vegetation is not only affected by the more particular stress caused by salt (NaCl) in soils of arid regions, but also more generally by the soil factor (that is the edaphic base of the mineral supply for plants) which is an important precondition for the well-being of plants and thus for the normal development and formation of ecosystems and their characteristics. Not only the availability of essential nutrients for plants, but also the water holding capacity (dependent on pore volume and

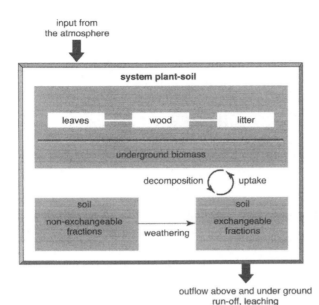

input from
the atmosphere

**Fig. 42.** System plant/soil
with the interlinkage of
compartments

outflow above and under ground
run-off, leaching

the distribution of particle sizes in the soil) greatly affects
their growth, particularly because the availability of water
varies at different sites (see p. 45).

The essential mineral substances for plants (and animals) are made available from the parent rock from which
they are released through weathering of individual minerals. By increasing surface area and by chemical reactions,
nutrients become available. In the process of soil formation (in interaction with plants) metabolic processes of
the ecosystem are supplied. Continuous losses because of
leaching into the groundwater (see Fig. 42) or by dust loss
through wind, require that new supplies must be pro-
cured, mainly through weathering. Only then will the ecosystem be sustained for long periods. The equilibrium of
materials in the system soil/plant is balanced in the long
term by input and weathering and by leaching (Fig. 42).

Input through dust has also been documented. Perhaps
a not insignificant part of the nutrients of the Amazon
rain forest could come from long distance transport of
fine dust particles (for example from the Sahara).

Around the globe there is a continuous and considerable transport of fine particles. Small particles released
during weathering, minerals etc. sediment for a time or
are transported further until they finally end up in the

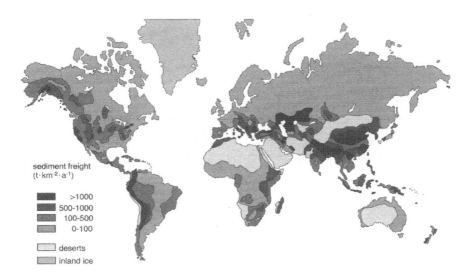

sediment freight
(t·km⁻²·a⁻¹)

■ >1000
■ 500-1000
■ 100-500
□ 0-100

□ deserts
▨ inland ice

**Fig. 43.** Global view of the
sediment transport from
medium-sized efflux basins
(after WHITE et al. 1992)

ocean or form large river deltas. The sediment transport
(Fig. 43) from various regions of the earth depends, on
the one hand, on the energy provided by the relief and
the differences of height, and on the other hand, on the
structure of the material. Thus, the easily erodable loess
from China is transported in large amounts across the Yellow River into the Yellow Sea.

Small particles created through weathering are either
transported by wind or in flowing water. Here, the speed
of the wind and the rate of flow as well as the size of the
particles to be transported are very important. With increasing particle size the process of sedimentation predominates more and more and can only be overcome by a
very high flow rate (Fig. 44).

Wind moves all particles of around 0.1 mm along the
ground and they sediment. They have a particularly low
critical displacement velocity (Fig. 45) so that saltation
(jumping of particles) becomes easier and leads to the formation of large sand dunes.

In the long term, processes of erosion and accumulation play an important role in the changing characteristics
of ecosystem habitats. The heavily increased erosion of
soil material on arable land leads to rapid changes and
can lead to a degradation of habitat which would not allow further cultivation. Table 3 lists the rates of erosion
for a humid region in the USA showing that a closed cover of vegetation is the best soil protection. The loss of valuable soil from 1 ha land in 10 years amounts to 1,500 t

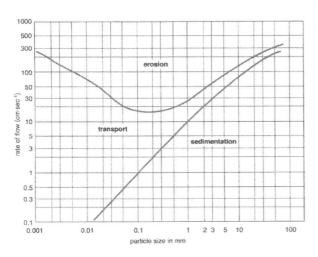

**Fig. 44.** Dependence of particle transport in flowing **water** on particle size (in mm) and the rate of flow of water (in cm s$^{-1}$) (after KUNTZE et al. 1994)

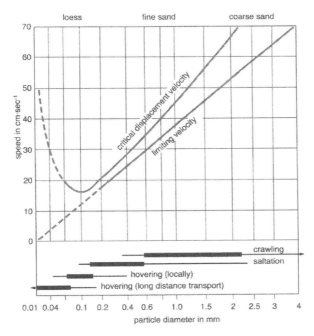

**Fig. 45.** Dependence of particle transport by **wind** on particle size (in mm) and the speed of soil particles (in cm s$^{-1}$). Particles with a size below 0.01 mm ("loess") remain longer suspended and can be transported over long distances. Fine sand with particle sizes between 0.1 and 0.5 mm are transported along the ground by saltation (dunes, ripple effect) (according to WHITE et al. 1992)

**Table 3.** Loss of soil cover through erosion for a layer of sandy loam. The period in years required to erode a 10-cm layer from a 10° slope in the south-east United States of America is given

| Cover of vegetation, use | Period in years |
| --- | --- |
| Cover of natural intact deciduous trees | 320,000 |
| Dense grass | 46,000 |
| Arable fields with changes in crop | 60 |
| Cotton | 25 |
| Maize | 20 |
| Without vegetation, bare soil | 10 |

in this example. However, it is slightly less in Germany, where precipitation is lower and slopes less steep.

The presence of minerals and the nutrients available in the soil because of the long-term formation of the soil essentially determine the productivity of ecosystems. However, parent rocks are not as important in older ecosystems with mature, deep soils (see black water regions, p. 120), but in younger habitats different vegetation units (and thus ecosystems) on chalk, crystalline, gypsum etc. can be observed.

Nutrients in the soil are obviously not responsible for the development of high biodiversity. On the contrary, very old, nutrient-poor regions (southwestern Australia, the Cape region of South Africa) with extremely poor quartz sands possess an unbelievable biodiversity. This contradicts the assumption of BARTHLOTT et al. (1996) of surprisingly poor biodiversity of the llanos in Venezuela and Columbia.

In humid climates leaching processes predominate in the soil (→ decreasing transport), in arid areas processes of accumulation predominate (→ increasing transport).

The availability of nutrients depends, on one hand, upon the internal surface area of soil minerals (and of humus) and on the other, of course, on the availability of ion exchange sites per unit internal surface area. This ion exchange is, in part, in equilibrium with the pH value, in acid soils an increasing part of the ion exchange sites is occupied by protons (Fig. 46). The proportion of exchangeable mineral nutrients (Ca, Mg, K) thus decreases. In more acid soils there is a further disadvantage, in that the tri-valent $Al^{3+}$ blocks further sites and thus, for example at pH 3, there are hardly any nutrient cations in such a soil. In cool, humid climate regions the formation of soil tends towards acid, nutrient-depleted soils (taiga, ZB VIII).

In hot, humid climates the weathering of the parent rock and the formation and transformation of clay minerals occurs much faster (Fig. 47). Whilst in moderate climates clay usually occurs as three layered materials in the soil (and

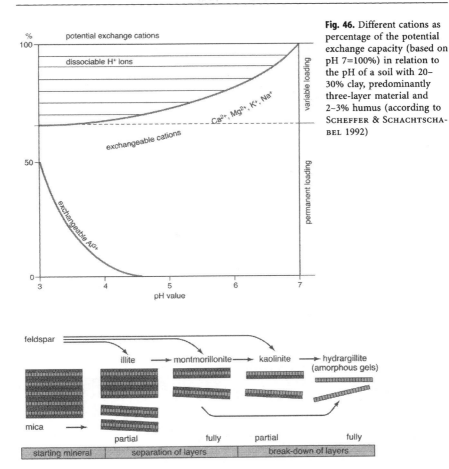

**Fig. 46.** Different cations as percentage of the potential exchange capacity (based on pH 7=100%) in relation to the pH of a soil with 20–30% clay, predominantly three-layer material and 2–3% humus (according to Scheffer & Schachtschabel 1992)

**Fig. 47.** Formation and breakdown of clay minerals. Illite and montmorillonite are three-layer clay minerals, kaolinite is a two-layer clay mineral (after Lerch 1991)

thus provides relatively large cation exchange capacity), tropical soils are usually characterised by the two-layer clay mineral kaolinite with only 5–10% of the ion exchange capacity compared to the three-layer clay minerals. This extreme cation depletion is one of the main reasons for the "ecological disadvantage of the tropics" (Weischet 1980).

In dry regions, accumulation at the surface of the soil or in the soil in certain depths leads to deposits which can be very hard and can also occur as chalk or gypsum crusts or also as laterites etc. In humid regions leaching processes lead in time to a depletion or acidification (podsolisation) of soils. Both processes characterise the formation of individual ecosystems decisively in the corresponding zono-

**Fig. 48.** Main groups of soils in an ecogram of humidity and temperature

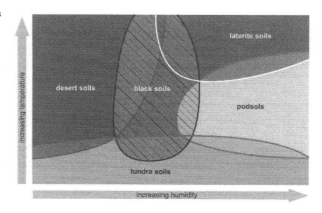

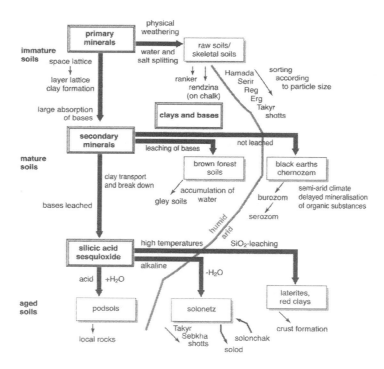

**Fig. 49.** Schematic of the genesis of soils from silicate and its dependence on different factors

biomes. The main types of soils are shown in the ecogram assigned to certain climatic factors (Fig. 48).

To distinguish between climate-typical ecosystems and those characterised more by pedological processes is only roughly possible because certain pedological processes themselves are zonobiome-specific. Many pedobiomes are thus zonobiome-specific biomes (for example in zonobiome II, where crust formation and laterites etc. occur). The formation of soil occurs from the parent rock in interaction with the developing vegetation. The processes of soil formation influenced by the climate thus lead to certain types of soil or groups of soil. However, the genesis of soil is a very long-lasting process. Some important processes are shown schematically in Fig. 49. It can be seen that for soils too, the historical aspect is important. Natural ecosystems very slowly create their specific type of soil which is in tune with the long-term conditions of the climate and the zonal vegetation of the habitat.

## 12  Ecotypes and the Law of Biotope Change and Relative Constancy of Habitat

Many plant species or phytocenoses (plant communities) are very widely distributed and, as can be seen if the areas inhabited by them are studied on the map, can apparently flourish under very different climatic conditions. There are two possible reasons for this:

1. The species as a taxonomic unit is often highly differentiated ecophysiologically, e.g. with respect to cold or drought resistance or the rhythms of the climate. Thus the pine species *Pinus sylvestris* is found from Lapland to Spain and as far east as Mongolia, with only slight taxonomic variations. The Spanish pine, however, cannot grow in Lapland because it is too sensitive to cold and the Lapland pine requires a winter resting period which is too long to be able to survive in Spain. It is therefore very important to know the provenance (origin) of seed used in reforestation programs. Most taxonomically uniform species are made up of many such ecotypes (races and varieties), or they may exhibit a quantitative gradation in ecophysiological properties, in which case they are termed ecoclines.
2. The second possible reason underlying a widespread distribution is the property, exhibited by a species or a phytocenosis, of changing its biotope if it extends into a different climatic region. If the climate of the northern limit of the plant's range becomes colder, for example, the species will no longer be found on the plains, but on locally warmer southern slopes. In other words, a change in biotope compensates for the change in cli-

mate, so that the habitat or environmental conditions remain relatively constant. Examples of this are seen everywhere. Toward the southern extremities of its range, a species increasingly occurs on northern slopes, or deep and moist canyons, or moves up into the mountains. If the climate becomes wetter, plants occur on drier chalky or sandy soils, rather than take to the heavy, wet soils or those with a high groundwater table if the climate becomes drier.

**Law of the relative consistency of habitat** and of the change in biotope. If within the range of distribution of a plant species or phytocenosis, the climate changes in any way, a change in biotope compensates as far as possible for the change in climate. In other words, the habitat or environmental conditions remain relatively constant.

Of course, it has to be remembered that in the southern hemisphere the warmer slopes face north and, on the equator, the eastern and western slopes are warmer. In arid regions, however, it is the sandy type of soil that provides the best water supply for the vegetation (see p. 214). This applies not only to the water relations in arid zones, but generally for all factors which are co-determined by climate.

The law of change of biotope also has to be considered when determining the altitudinal belts in mountainous regions. Even differences in the altitudinal limits at different exposure indicate the applicability of the law. Much more extreme special niches occur where intense irradiation and the runoff of cold air permit the growth of small stands of trees above the tree line and within the alpine belt. In West Pamir, isolated trees have been found even as high as 4000 m above sea level in gorges swept by wind which prevents the accumulation of cold air. Small shrubs have even been found in very protected niches on southerly slopes at 5100 m in the Hindu Kush. In contrast, in some parts of the eastern Alps, where the cold air is trapped (dolines) forest vegetation ceases at 1270 m above sea level. The lowest temperature ($-51\,°C$) in central Europe has been measured in one such region, Lunz am See, in Lower Austria.

Soil factors also play an important role. On the very weathered dolomites in the eastern Alps, fragments of alpine vegetation can be encountered in the middle of the beech belt. The paths of avalanches and landslides also provide special niches since competition from trees has been eliminated, and the dwarf *Pinus montana* of the subalpine belt is able to exist even within the forest belt. In special biotopes of this kind, it is not unusual to find relict species which had a greater range of distribution under previous climatic conditions. Nevertheless, before conclusively stating that a species is a relict it is essential to have historical evidence.

## Azonal and Extrazonal Vegetation

Zonal vegetation corresponding to the climate is found only in those areas where the typical regional climate is fully expressed. These biotopes are called **euclimatopes** (in Russian, Plakor areas). These are level, slightly raised areas with deep soils neither too permeable (as sand) nor tending to getting wet because of an accumulation of water. If the vegetation on the euclimatopes is termed zonal vegetation, then this is, following a biotope change, **extrazonal vegetation,** for which the local climate is decisive, not the climate of the region. If, for example forests along rivers extend as gallery forests far into an arid climatic region, then these gallery forests are extrazonal.

Extrazonal vegetation can provide information on the zonal vegetation of a more humid or colder zone or a more arid or warmer zone, if the zonal vegetation in that region was destroyed. It is a disadvantage of all area maps that the zonal and extrazonal distributions are not shown separately. This gives an incorrect impression of the climatic demands of plants, particularly as the limits of an area are often shown in relation to the climate lines.

The term zonal vegetation (p. 78 f.) should be used only for considering large spaces to subdivide the natural vegetation of whole continents. Only then does the influence of the climate become obvious and local differences caused by soil, relief and exposure are less important. On the other hand, even in natural conditions zonal vegetation can be lacking in large areas, if for example, the groundwater table is so high that swamps and bogs cover everything (western Siberia, Sudd swamps in the Sudan) or in the alluvial zones of large rivers. Also, on extensive lava sheets (Idaho) or on saline soils of basins without drainage (Aral Sea), a vegetation mosaic grows which is very dissimilar to the zonal vegetation.

Here, we deal with pedobiomes with an **azonal vegetation** which is much more strongly influenced by the special characteristics of the soil and where the climate has only little influence. This does not mean that azonal biotopes would look alike all over the world, they too are influenced by the climate, as different zones along the coastal areas show.

A vegetation mosaic which corresponds to a neighbouring vegetation zone and occurs as an island or tongue of land because of special favourable or unfavourable (local) climatic conditions is called **extrazonal vegetation**.

In azonal soils and **azonal vegetation** the consistency of the substrate has a stronger effect than the climate.

## Questions

1. What are the basic ecological factors to which all living organisms are exposed?
2. What is the meaning of the following statement: "Nitrogen is a minimum factor"?
3. Which ecological factors result in too much (maximum) as well as too little (minimum) stress?
4. What is a halophyte?
5. Why do endorheic salt lakes exist?
6. What does the law of biotope change and the relative habitat constancy mean?
7. What are zonal, azonal and extrazonal vegetation?
8. Where are the "hot spots" of biodiversity on earth and how can they be explained?
9. Competition or co-evolution – what is more important in ecosystems?

# Ecological Systems and Biology of Ecosystems

## Geobiosphere and Hydrobiosphere

The **biosphere** is a thin layer of land, water and atmosphere at the earth's surface to which all phenomena connected with living matter are confined. This comprises the lowest layer of the atmosphere permanently inhabited by living organisms into which plants extend, and on land the root-containing portion of the **lithosphere**, which we term the soil. Living organisms are also found in all bodies of water, to the very depths of the oceans. In the water medium, however, cycling of the material is achieved by means other than those on land, and the organisms (for example, plankton) are so different that aquatic ecosystems have to be dealt with separately. The biosphere is therefore subdivided into:

1. the geobiosphere comprising terrestrial ecosystems and
2. the hydrobiosphere comprising aquatic ecosystems which is the field of hydrobiologists (oceanographers and limnologists).

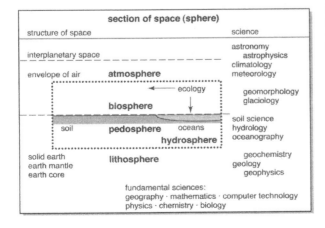

**Fig. 50.** Individual sections of the earth and the subject of ecology in the context of other sciences

Many results from other discipline are required in order to consider the most important processes in the biosphere. Research of ecological interrelations of large areas is only possible through interdisciplinary evaluation of results and ecology and has therefore become an interdisciplinary science extending far beyond its original area in biology, as shown schematically in Fig. 50.

 **Hydrobiosphere**

Within the context of this volume, we shall consider the hydrosphere only in outline even though 71% of the earth's surface is covered with water.

A quantitative distribution of the water on the earth's surface in the various compartments shows that only 3.5% of the total water is found on the terrestrial land mass (see Table 4). Most of the water is the salt water of the oceans. Even the amount of water in lakes and swamps is

**Table 4.** Quantitative values for the hydrosphere from a global point of view

| Region | Area (10⁶ km²) | Volume (10⁶ km³) | Percentage of total volume of water on earth | Percentage related to fresh water |
|---|---|---|---|---|
| Oceans | 361.3 | 338.0 | 96.5 | – |
| Land mass | 148.4 | 47.97 | 3.5 | |
| Groundwater | 134.8 | 23.4 | 1.7 | |
|   Of this fresh water | (134.8) | 10.53 | 0.76 | 30.1 |
| Moisture in soil | 82.0 | 0.015 | 0.001 | 0.05 |
| Polar ice, snow | 16.23 | 24.064 | 1.74 | 68.7 |
|   Antarctic | 13.98 | 21.6 | 1.56 | 61.7 |
|   Greenland | 1.80 | 2.34 | 0.17 | 6.68 |
|   Mountains | 0.224 | 0.041 | 0.003 | 0.12 |
|   Permafrost | 21.0 | 0.30 | 0.022 | 0.86 |
| Fresh water lakes | 1.236 | 0.091 | 0.007 | 0.26 |
| Salt water lakes | 0.822 | 0.085 | 0.006 | – |
| Swamps, moors | 2.683 | 0.0115 | 0.008 | 0.03 |
| Rivers | | 0.0021 | 0.0002 | 0.006 |
| Water in the atmosphere | (510) | 0.0129 | 0.001 | 0.004 |
| Biologically bound water | (510) | 0.0011 | 0.0001 | 0.003 |

(after Schönwiese 1994)

only a minute proportion compared to the total amount of water. However, the amount of water in the atmosphere is, despite its rapid turnover of only a few days, 13-fold greater than the total amount of water bound in living organisms in the biosphere.

If freshwater only is considered, two thirds is bound as ice. Frozen water, i.e. ice, occurs in considerable amounts in the colder regions of the earth. A distinction is made between compact ice which occurs in various forms in individual compartments (**cryosphere**) and the less permanent snow, which can be considered separately as **chionosphere**. The distribution in various parts of the earth is given in Table 5. The distribution in the two hemispheres is shown in the great imbalance between land and sea, which again is expressed in the asymmetrical distribution of the zonobiomes. The biomass of living organisms in the cryosphere is negligible.

**Table 5.** Quantitative values for the cryosphere and chionosphere

| Region | Area ($10^6$ km$^2$) | Volume ($10^6$ km$^3$) | Sea level equivalent (in m)[a] |
|---|---|---|---|
| Land ice | 14.44 | 32.44 | 81.2 |
| Antarctic | 12.2 | 29.32 | 73.3 |
| Greenland | 1.7 | 3.0 | 7.6 |
| Glaciers | 0.54 | 0.12 | 0.3 |
| Permafrost (without Antarctic) | | | |
| Consistent | 7.6 | 0.03 | 0.08 |
| Maximum | 17.3 | 0.07 | 0.18 |
| Sea ice | | | |
| Arctic | | | |
| (Winter) | 14.0 | 0.05 | – |
| (Summer) | 7.0 | 0.02 | – |
| Antarctic | | | |
| (Winter) | 18.4 | 0.06 | – |
| (Summer) | 3.6 | 0.02 | – |
| Snow | | | |
| Northern hemisphere | | | |
| (Winter) | 46.3 | 0.002 | [b] |
| (Summer) | 3.7 | <0.001 | [b] |
| Southern hemisphere | | | |
| (Winter) | 0.85 | [b] | [b] |
| (Summer) | 0.07 | [b] | [b] |

[a] Rise of sea level at complete melting
[b] Negligible (after Schönwiese 1994)

##  Classification of the Geobiosphere into Zonobiomes

Our studies are confined to the geobiosphere which constitutes the main habitat of humans and is, therefore, of special interest to us (WALTER 1976). The prevailing climate, being the primary independent factor in the environment, can be used as a basis for further subdivision of the geobiosphere since the formation of soil and type of vegetation are dependent upon it (see p. 19 f.) and it has not yet been substantially influenced by people. Furthermore, it may be measured exactly and in all locations with the increasingly dense network of meteorological stations (for the principles of classification, see WALTER 1976).

Climate is determined by planetary air currents in the atmosphere and meteorologists distinguish between seven genetic climate belts:

A. the equatorial rain zone (on both sides of the tropical convergence zone);
B. the summer rain zone on the margins of the tropics (trade wind zone);
C. the subtropical dry regions (high pressure areas of the horse latitudes);
D. the subtropical winter rain regions (partly only regarded as a transition region between C and E);
E. the temperate zone with year-round precipitation in the region of the predominant west winds;
F. the subpolar zone (with weak easterly winds);
G. the polar zone.

The effects of the planetary circulation system close to the ground can be seen in Fig. 51.

The climate system of the atmosphere is probably, basically, a very stable system.

Ecologists are mainly interested in the climate within the geobiosphere which can be characterised by the ecological climate diagram (see p. 20 f.). A further subdivision of the very large temperate zone of the meteorologists has proved to be useful, whereas the subpolar and polar zones have been combined. There are nine ecological climatic zones designated as **zonobiomes** (ZB), a biome being a large and climatically uniform environment within the geobiosphere. The term **humid** is used to describe a climate with much rainfall and **arid** to describe a dry climate with little rainfall (see p. 23). Where both terms are used, the first refers to summer and the second to winter conditions.

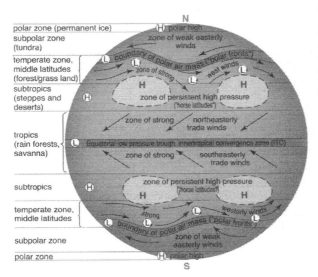

polar zone (permanent ice)
subpolar zone (tundra)
temperate zone, middle latitudes (forest/grass land)
subtropics (steppes and deserts)

tropics (rain forests, savanna)

subtropics

temperate zone, middle latitudes

subpolar zone

polar zone

**Fig. 51.** Schematic of the planetary atmospheric circulation near the ground and ideal sequence of climate zones (**H:** high pressure areas; **L:** low pressure areas, cyclones) (after SCHÖNWIESE 1994)

The nine zonobiomes (see also Fig. 26):

ZB I     equatorial ZB with diurnal climate, humid tropical ZB

ZB II    tropical ZB with summer rains, humido-arid tropical ZB

ZB III   subtropical ZB with desert climate, hot-arid ZB; very little rain

ZB IV   ZB with summer drought and winter rain, arido-humid (Mediterranean climate)

ZB V    warm temperate climate (oceanic), humid ZB; mild maritime

ZB VI   typical temperate ZB with short periods of frost, nemoral ZB

ZB VII  arid-temperate ZB with cold winters, continental ZB

ZB VIII cold temperate ZB with cool summers and long winters, boreal ZB

ZB IX   Arctic (including Antarctic), with very short summers, polar ZB

Each zonobiome is clearly defined by a particular type of climate diagram (p. 20f.), although, with a few exceptions, zonobiomes correspond largely to soil type and zonal vegetation. This is shown in Table 6. However, it must be stressed that there are many transition zones (ecotones; see p. 80).

**Table 6.** Types of soil and vegetation of individual zonobiomes

| Zonobiome (ZB) | Zonal soil type | Zonal vegetation type |
|---|---|---|
| I | Equatorial brown clays (ferralitic soils, latosols) | Evergreen tropical rain forest without seasonal change |
| II | Red clays or red earths (fersialitic soils) | Tropical deciduous forests or savannas |
| III | Serozemes, sierozemes (grey or red earths, raw soils, saline soils) | Subtropical desert vegetation (stony landscapes) |
| IV | Mediterranean brown earths (fossil terra rossa) | Sclerophyllic woody plants (sensitive to ground frosts) |
| V | Yellow or red forest soils, slightly podsolic soils | Temperate evergreen forests (lauriphyllic), frost sensitive |
| VI | Forest brown earths and grey forest soils | Nemoral broadleaf-deciduous forests (bare in winter, frost resistent) |
| VII | Chernozems to serozems (raw soils) | Steppe to desert with cold winters (frost resistant), short, hot summers |
| VIII | Podsols (raw humus-bleached earths) | Boreal coniferous forests (taiga), (very frost resistant) |
| IX | Tundra humus soils with solifluction (permafrost soils) | Tundra vegetation (treeless) |

# Zonoecotones

**Ecotones** are areas of eco-logical tension – transition regions, in which one type of vegetation is replaced by another.

Climate zones and hence the zonobiomes are not sharply defined (all attempts to draw borderlines are artificial), but are linked by broad transitional zones known as zono-ecotones (ZE).

Ecotones may be on a small scale, a forest edge chang-ing into meadows or a large scale, for example, in East Europe, deciduous forests gradually changing into steppes.

In zonoecotones both types of vegetation occur side by side under the same general climatic conditions and are in a state of strong competition. Which of the two types is successful depends upon the microclimatic conditions resulting from local relief or soil texture, so that there is either a diffuse mixture of the two kinds of vegetation or a mosaic-like pattern of the two. In crossing a zonoeco-tone, at first one kind of vegetation is better represented, then the two are more or less equally successful, and final-ly, the second type begins to take over, with the first becoming more sparse. When the latter eventually disappears, the next zonobiome has been reached.

Zonoecotones are designated according to the zonobiomes they link, i.e. ZE I/II, ZE II/III and so on.

Three-cornered zonoecotones are also found at the convergence of three zonobiomes (for example, the Pannonian lowlands (ZE VI/VII/IV). Important zonoecotones are discussed in short special sections at the end of the associated zonobiome.

The geographic distribution of the individual zonobiomes and zonoecotones is illustrated in the schematic world map (Fig. 26) and for individual continents in Figs. 65–70.

## Orobiomes

The geobiosphere can be divided horizontally and, because of the presence of mountains, also vertically and thus has to be considered three-dimensionally. Mountains differ climatically from the climate of the zone from which they rise and thus must be considered separately from the zonobiomes. Such mountainous environments are termed orobiomes (OB).

A characteristic of all orobiomes is that the mean annual temperature decreases with altitude. In the Euronorth Asiatic region, for each additional 100 m in altitude, the temperature drop is approximately the same as the mean annual decrease in temperature recorded over a distance of 100 km from south to north. Therefore, the width of the altitudinal belts in mountains is a 1000 times narrower than the vegetational zones in lowlands from south to north.

In Europe and North America certain similarities immediately catch the eye, but differences are invariably present. Apart from the drop in temperature, which means a reduction in the vegetational period with increasing altitude, the mountainous climate is different from that of the plains. For example, with increasing altitude, the length of the day and the position of the sun do not change, while the length of the day increases from north to south in summer, and the sun's position at noontime becomes lower. Direct radiation increases with altitude, and diffuse radiation decreases, whereas in moving north on the plains, day length changes and the change in radiation is exactly the opposite. Precipitation often increases in the mountains very rapidly with increasing altitude, whereas in the Arctic regions precipitation is lower.

**Orobiomes** are mountainous environments which can be vertically subdivided into altitudinal belts. Individual altitudinal steps are also called hypozonal or orozonal vegetation. It is the third dimension which is derived from the associated zonobiome.

**Altitudinal steps** in mountainous regions are, looked upon superficially, a small scale repetition of the planetary vegetation zones in the plain towards the poles.

**Fig. 52.** Schematic of the development of the föhn in mountains. On the windward side of the mountain, air mass rises (A) and leads to formation of clouds (B) possibly with precipitation. On the leeward side (C) the air becomes warmer, drier and more turbulent. The warming is due to the difference between the moist adiabatic temperature gradient (B; warming through condensation) caused by the lifting, and the dry adiabatic gradient (C, A) due to the sinking (with the addition of radiant energy input due to the clear air) (after SCHÖNWIESE 1994)

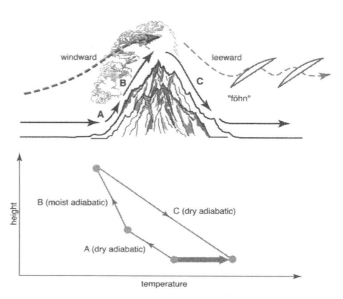

The flanks of a mountain are almost never symmetrical, and are also climatologically different, for example, because of the effects of föhns, warm dry winds on the lee side of mountains or hot southerly winds on the northern slopes of the Alps. Therefore, the temperature gradient is also no longer the same, due to the laws of physics (see Fig. 52); humid adiabatic and dry adiabatic warming (cooling) are energetically different. When air cools as it rises (and volume increases because of decreasing pressure), the dew point is eventually reached. Then condensation of water vapour ensues and clouds are formed ("cloud forest") and the humidity is "rained off". At the other side of the mountain, only a dry adiabatic warming occurs because the air sinks (with an increase in pressure) so the air becomes warmer and drier at the same position in the valley. This is **the föhn effect** (Fig. 52), caused by the release of latent heat of condensation (2.26 MJ kg$^{-1}$; compare the physical quantities p. XVIII) when the air rises.

Each mountain within a zonobiome is an ecological entity with a typical **sequence of altitudinal belts,** generally termed colline, montane, alpine, and nival, but exhibiting considerable differences according to the zone in which they occur. For example, the successions of altitudinal belts in the mountains of zonobiomes I, IV, and VI have little in common.

Therefore, a further subdivision of orobiomes is made according to the zonobiomes to which they belong, e.g.

orobiome I, orobiome II etc. In addition, a distinction is made between uni-, inter-, and multizonal orobiomes (mountains), depending on whether they fall within one zonobiome or between two, or range over many, as is the case with the Urals (from IX to VII) or the Andes (from I to IX). The Alps, the Caucasus and the Himalayas are examples of interzonal mountain ranges. These mountains usually form sharp climatic boundaries, and the altitudinal belts on the northern slopes differ from those on the southern side. In dealing with multizonal mountains, it is necessary to consider the constituent zones individually, each with its own particular sequence of altitudinal belts. The Andes are multizonal as well as interzonal (western and eastern slopes differ). The sequence of altitudinal belts may also be different in the innermost parts of the mountain valleys, with little precipitation and continental conditions (intramountane sequence of altitudinal belts).

## Pedobiomes

Not only orobiomes stand out from zonobiomes, but also certain areas with extreme types of soil and azonal vegetation. These are termed pedobiomes (PB), i.e. environments associated with certain types of soil. Soils have been altered significantly by human activity only in regions where soil erosion, i.e. loss of the upper levels or all of the soil has occurred, or where soil has been cultivated, or built upon. The unmitigated effect of the overall climate is seen only in euclimatopes (Russian "plakor"), which are flat areas where the soil is neither too light nor too heavy, so that precipitation does not run off but penetrates the soils and is retained as interstitial water. Since the water does not seep into the groundwater quickly, it is at the full disposal of the vegetation. This is not the case with extreme chalky soils, however, and they constitute biotopes that are warmer and drier than the general climate. On the other hand, the soil may contain harmful substances such as salts (NaCl, $Na_2SO_4$) or may be extremely poor in nutrients so that the vegetation differs from that typical of the zonobiome.

The vegetation of pedobiomes is influenced to a greater extent by the soil than by the climate, and thus, the same vegetational forms may occur on similar soils in a number of zones. This vegetation is termed azonal vegetation (see p. 73). Pedobiomes are designated according to soil type: lithobiomes (stony soil), psammobiomes (sandy soils), halobiomes (salty soils), helobiomes (moor or swamp soils), hydrobiomes (soils covered with water), peinobiomes (soils which are poor or deficient in nutrients; from peina, Greek=hunger, deficiency), and amphibiomes (temporarily wet soils) and others.

Pedobiomes often occupy vast areas, as for example, the lithobiome of the basalt-covered areas of Idaho (United States), the psammabiome of the southern Namib, the Rub-al-Khali in Saudi Arabia or the Karakum Desert in central Asia (35,000 km$^2$) the helobiome of the Sudd swamp region of the Nile (150,000 km$^2$), the bog region of western Siberia (over 1 million km$^2$, Fig. 279, p. 439). The ecology of these areas has to be given special consideration when dealing with the relevant zonobiome.

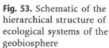

# 7 Hierarchy of Ecological Systems

Bearing in mind what has been said above, a scheme may now be constructed depicting the hierarchy of larger and smaller ecological units (see Fig. 53). In the central column, the main series of climatic biomes are shown, in the left column the biomes adapted to mountains (orobiomes, OB) are shown as orographic subseries and in the right-hand column the biomes adapted to specific soil conditions (pediobiomes, PB) are shown as pedological subseries. This hierarchical scheme of special units of ecosystems is the basis of the division of the geobiosphere.

**Fig. 53.** Schematic of the hierarchical structure of ecological systems of the geobiosphere

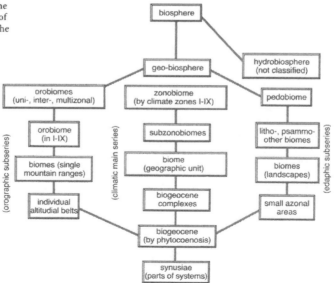

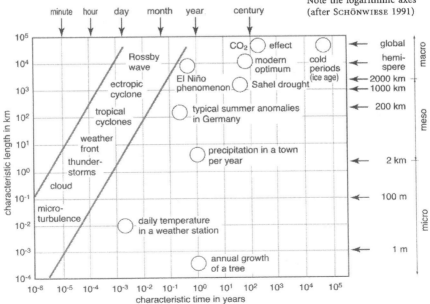

| geography | | diameter | | meteorology | |
| --- | --- | --- | --- | --- | --- |
| term | example | km | m | term | example |
| zone | climate zone | $10^4$ | | Rossby wave | macro α |
| region | Central Europe | | | | macro β |
| | | $10^3$ | | ectropic cyclone | meso α |
| large landscape | mountain range | $10^2$ | | tropical cyclone | |
| landscape | | | | föhn | meso β |
| | town | 10 | | weather front* | |
| | | | | thunderstorm | meso γ |
| terrain | valley | 1 | | | |
| | slope of mountain | 0,1 | 100 | cumulus clouds | micro α |
| | | | | tornado | |
| | gap in mountain | | 10 | thermal | micro β |
| | | | | small dust whirls | |
| site | tree | | 1 | air turbulence | micro γ |
| | | | 0,1 | | |
| | leaf | | 0,01 | heat haze | |
| interface | leaf surface | | 0,001 | | |

\* vertical to direction of flow ( in parallel considerably greater scale)

**Fig. 54.** Horizontal spatial scales in biology, geography and meteorology. Note the logarithmic scale (after SCHÖNWIESE 1994)

**Fig. 55.** Atmospheric and climatological phenomena in a distance-time diagram. Note the logarithmic axes (after SCHÖNWIESE 1991)

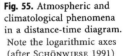

Note should be taken of the very different units used when characterising ecosystems. This not only applies to the spatial sizes of the structures but also to the time scale. In particular, atmospheric and meteorological phenomena determine the size of the scale. Figure 54 contrasts the scales used for ecological systems with the meteorological processes. Because of the differences in scale, such a comparison can only be made in the logarithmic scale. This also applies for the time scale in which certain phenomena occur (Fig. 55).

The layers of air near the soil and their dynamics and the atmospheric-biospheric interaction play a decisive role when considering individual biomes. This was already pointed out in 1927 by GEIGER.

##  Biome

**Biomes** are habitats which correspond to a concrete uniform landscape.

The word biome (without a prefix) is used for the fundamental unit of which larger ecological systems are made up. A biome as an environment is a uniform area belonging to a zonobiome, orobiome, or pedobiome. The central European deciduous forest, for example, is a biome of zonobiome VI, Kilimanjaro is a biome of orobiome I, the salt desert of Utah (United States) is a biome of the pedohalobiome VII and so on. In the following global survey, biomes are considered predominantly as the smallest unit of a region. The term "biome" used in the Anglo-American literature is much wider and defined less clearly.

## Smaller Units of Ecological Systems: Biogeocenes and Synusiae

After the classification of the land surface into nine zonobiomes, a further subdivision into even smaller units, biomes, can be undertaken according to the information available; if this information is lacking, such subdivisions cannot be carried out.

These smaller ecological units are best defined on the basis of vegetational units. In a geographical area with uniform features corresponding to a biome, even the slightest differences in water and soil conditions influence the vegetation and the ecosystems themselves. Thus, the typical ecosystem mosaic of a landscape is formed. Direct measurements of their combined effect cannot be made. Nevertheless, we can justifiably assume that the natural vegetation, which is in a state of equilibrium with its environment, provides us with an integrated expression of the effect of such environmental factors. Even the smallest change in one of them can bring about a qualitative, or at least quantitative, change in vegetation.

In view of the fact that human influence is now detectable almost everywhere, to a greater or lesser degree, the effects of natural and anthropogenic factors have to be carefully distinguished from one another, and as far as the latter are concerned, the role of man in the past has also to be taken into consideration. In the case of forest societies, this is still detectable centuries later (the effects of clear-cutting, type of rejuvenation, grazing, utilisation of litter etc.). It is often thought that the herbaceous layer of the forest gives a good idea of the natural situation, but the herbaceous layer itself depends to a large extent upon the composition and nature of the tree layer (amount of shade, degree of competition from roots, kind of leaf litter) and is less deeply rooted than the tree layer, so that only the upper horizons of the soil are of any significance to it. Any change wrought on the tree layer by man has its effect on the herbaceous layer too. Even the removal of hollow tree stumps decaying on the ground represents severe interference with the ecosystem. In densely populated regions, it can be assumed that there are no ecosystems that have not in some way been influenced by man.

The position of the **biogeocene** in the hierarchy of sizes is shown in Fig. 53. It corresponds to a plant community with the rank of an association, which is the basic unit of phytocenology and cannot always be clearly defined. Opinions differ as to the correct definition based mainly upon the association. Whereas some advocate a definition based mainly upon the dominant species, others place more weight upon the characteristic species. Some would like to set wider limits to an association, but others would draw the confines closer. These disputable, methodological questions will not be discussed here, nor the different syn-systematic or the so-called sigma-systematic systems (DIERSCHKE 1994).

Within an ecosystem, cycling of material, energy flow, phytomass and production are determined by the dominants, which in forests, are tree species. Characteristic species, even if represented by only a few specimens, are useful in identifying the community, but they may have not the slightest influence on the ecosystem. For this reason, it should be required that each type of ecosystem be defined with respect to the dominants.

Limits of a plant community can only be laid down in the field after thorough study of the history of the individual stands and careful examination of the entire area,

The basic unit of smaller ecosystems is the **biogeocene** corresponding to a concrete plant community comparable to an association; it may e.g. be an ecosystem of ca. 20×20 m.

The **plant community** (→ biogeocene as the real spatial unit) is somewhat different from the **plant society** (a theoretically defined construct). The association as the syn-taxonomic unit (basic unit) of the plant societies is a type with the higher categories as follows: alliance, order, class.

including habitat conditions and nature of the soil profile down to limits of rooting. Only actual stands (usually heterogeneous) can be investigated ecologically, not the abstract associations (plant societies) used by plant sociologists.

Although the biogeocene is the fundamental unit of the ecosystem, it is not the smallest. A number of **synusiae** can be differentiated within it. Synusiae are "functional communities" of species of similar developmental forms and ecological behaviour. However, synusiae should not be confused with ecosystems: they are merely a part of ecosystems, since they do not have a material cycle of their own, but contribute to that of the ecosystem of which they form a part. Although production of the individual synusiae is only a fraction of the total production of the ecosystem, it is important because turn-over in synusiae is often more rapid than that of the ecosystem as a whole.

Typical examples of synusiae are the various groups of species with a similar rhythm of development and similar ecological requirements, such as the spring geophytes (*Allium ursinum, Corydalis, Anemone, Ficaria* etc.) which exploit the light phase on the floor of broadleaf forests before the trees come into leaf. Other examples are herbaceous plants that survive the shady phase of summer and plants with evergreen leaves. Synusiae of lower plant forms are formed by lichens on tree trunks or mosses at the base of trees.

Between the biomes, on the one hand, and the biogeocenes, on the other, is a wide gap which has to be filled by units of intermediary rank, namely biogeocene complexes often corresponding to a particular kind of landscape. These have a common origin or are connected with one another by dynamic processes. As an example, a biogeocene sequence on a slope with lateral material transport (catena) or a natural succession of biogeocenes in a river valley or a basin with no outlet are cited here. A biogeocene complex can be a temporal succession of biogeocenes (such as secondary succession) or a spatial succession, in which the biogeocenes form an ecological series as a result of gradual alterations in habitat factor (sinking groundwater table, increasing depth of soil etc.). The extension in an area of such biogeocene complexes can be very different. Descriptions for individual types are very divergent. The neutral expression biogeocene complex suffices here.

Ecological units are real. Just as a doctor is only able to examine and treat living individuals and not types of human beings, the ecologist can only carry out investigations on real ecosystems and not on abstract units. Only a sufficiently extensive collection of data can serve to formulate **theoretical models.** Summaries composed and designed at the desk, on the basis of experience, must always start from certain presumptions and will thus, never completely describe the real ecosystem. However, because theoretical models provide an overview and are comparative, they will ease the understanding of ecosystems and the processes within them. If they are based on sufficient data they will even enable a prognosis for future developments (WISSEL personal communication). Models based on matrices have proved particularly suitable and adaptable to biological processes. Using these models, it is possible to formulate biological and ecological rules which can be adapted to the real facts.

# 10 Biology and Characteristics of Ecosystems

After discussion of the smaller ecological units, the principle, structures and processes in ecosystems will be explained in detail with the example of a deciduous forest of zonobiome VI of a size that is easy to comprehend and follow.

If the stand consists of a certain, limited and homogeneous society, for example, a forest, a moor etc. it is more useful to term this as a biocenosis or, in short, a **biocene.** A unit of plants and animals, including the soil with all roots and the layer of air near the soil into which plants have grown, is called a biogeocenosis or, for brevity, a **biogeocene.** The biogeocene "deciduous forest" describes the static appearance, spatial structure and organisms. In such a plant community, material cycles and flow of energy occur continuously. Plants form a dynamic structure with animal organisms and the inorganic environment, an **ecosystem,** which is not a closed system as the external energy supply in the form of radiation from the sun and material supply through precipitation, gas exchange and dust deposits take place. At the same time, output of energy occurs (in the form of heat) and of material (water runoff or seepage, gas exchange etc.). The dynamic picture of such a spatial section, that is the essential structures and processes, is investigated by the **biology of ecosystems.**

The role of groups of organisms in the ecosystem is differentiated as follows:

1. **Producers:** autotrophic plants which during photosynthesis store light energy as chemical energy by forming

The total of plant dry matter in a biogeocene is the **phytomass,** that of animals the **zoomass.** Together they form the **biomass.**

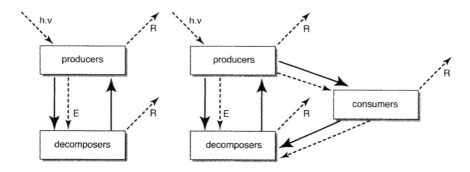

**Fig. 56.** Schematic of the simplest ecosystems (*left*) with material cycles and energy fluxes (*E*: energy fluxes; energy=ability to do work); *R*: respiration, respiratory energy; the same applies to the compartmentation of consumers (*right*)

organic compounds from $CO_2$ and $H_2O$ and taking up mineral nutrients and water from the soil.

2. **Consumers:** heterotrophic animal organisms (phytophages) which use plants as food and metabolise a small part of it into animal substances. Predators which eat phytophages (food chain, food net) are also part of this group.

3. **Decomposers** (mineralisers): They are usually in the soil (saprophages, bacteria, fungi), degrade all plant and animal residues into $CO_2$ and $H_2O$ and mineralise organic material, thus closing the cycle.

It is possible to imagine the simplest ecosystem as an interaction of producers and decomposers (without consumers; Fig. 56). In fact, there are many terrestrial ecosystems in which consumers only play a very subordinate role.

The total organic material produced annually by plants through photosynthesis is called **gross photosynthetic production** (GPP); the material remaining after subtracting the amount used by plants in respiration is called **net photosynthetic production** (NPP) or primary production; material formed by animal organisms is called **secondary production**. The latter is much smaller. Usually only a few percent of the primary production is eaten by consumers (long-period metabolic cycle, Fig. 56), the largest part enters the soil and is completely degraded by decomposers (short period metabolic cycle, Figs. 56, 57). During this process, $H_2O$ and $CO_2$ and mineral salts are released. Dead organic mass (litter) is first eaten by lower animals, saprophages and thus reduced in size. This material is then degraded and $CO_2$ is released from the soil in the process called soil respiration. The **short-period material cycle** is quantitatively the most important in terrestrial ecosystems (Fig. 57 A).

The **long-period material cycle** takes place via the consumers, i.e. herbivores or phytophages and also via zoo-

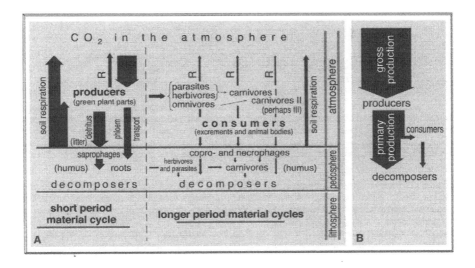

Fig. 57. **A** Schematic of short- and long-period material cycles in a deciduous forest biogeocene. *R*: Respiration. The thickness of the *arrows* indicates the rate of turn-over. **B** Scheme of the energy fluxes

phages, predators and omnivores, which consume both animal and plant matter. Also included are consumers of the second and third order, however, their turnover is relatively low (Fig. 57). Parasites of plants are also included in the group of consumers. Excretions and bodies of animals are also returned to the soil and are prepared by animal organisms (coprophages, necrophages) for degradation by micro-organisms. The long period metabolic cycle is quantitatively less important, however, it is much more important in the regulation of the equilibrium in the total ecosystem. Thus, it is possible to consider consumers as regulators. As soon as a certain plant species increases disproportionally, the number of consuming animals also increases. This reduces the population density of the plant species and, in consequence, leads to a decrease in phytophages. Achieving equilibrium is hardly ever constant. Cyclic oscillations of population densities are usually observed, although generally they are not in phase (i.e. phases are shifted). Such shifts also exist between the phytophages and their zoophage enemies. These regulatory processes, which may be regarded as cybernetic regulatory cycles with feed-back coupling, allow ecosystems to remain in a dynamic equilibrium (steady state). Population densities will always be subject to certain fluctuations, but only to a certain limited degree. Such fluctuations are usually caused by changing weather conditions in individual years, sometimes favouring the competitiveness of one plant species, at other times, that of another.

*Witheringia solanacea* is mentioned as a striking example from the tropical rain forest in Costa Rica. The berries contain a natural laxative so that birds empty their guts in less than 10 min and thus distribute the seeds on the soil of the forest. After this short passage through the gut, 70% of the seeds are still viable, after a longer stay in the gut, only 20%.

After an epidemic of gypsy moth caterpillars (*Lymantria dispar*) in oak woods the mass may increase considerably: For $10^6$–$10^7$ caterpillars per hectare their dry mass is 75–50 kg ha$^{-1}$, and 1–2 t ha$^{-1}$ dry leaf material is destroyed and 500–1000 kg ha$^{-1}$ excrement produced. This upsets the equilibrium of the complete ecosystem. This applies only to monocultures of the same age in forests. Deciduous forests usually recover, pine forests may die in certain conditions.

Interlinked regulatory cycles are based, on the one hand on direct control mechanisms through animals, such as pollination (zoogamy) or distribution of fruit and seeds (zoochory), and on the other hand on food chains starting with the herbivory. The long period cycle consists of several such food chains which should be called more precisely **food nets**. Despite all fluctuations, these food nets provide the ecosystem with a large degree of stability. Humans disturb these food chains particularly by the destruction of predators and by other interference and thus, upset the complete ecosystem and cause it to break down or be replaced by another (GIGON 1974).

In future, it will remain an important task of zoo-ecologists to focus less on the quantitative relations of secondary production and more on the different food chains in all their details. Phytophages and predators are often very strictly specialised, with certain species providing their food. Despite their lower density, they are very important for regulation, and a wealth of different adaptations play a role.

Another of the many surprising examples from the tropics is the very close dependency of ants and plants. Leaf cutting ants carry parts of leaves into their nests and breed fungi on them (thus acting as farmers); the fungi are their main source of food. The very close dependency of some ants of the species *Cecropia* (of the Neotropics) and the species *Macaranga* (of the Palaeotropics) should also be mentioned. The ant queen settles in the hollow stems of the fast growing pioneer trees through holes made by the plant. As the young trees grow, the ant community grows and settles in new internodes. Additionally, protein and fat-rich food particles are formed by the plant at the base of the leaves which also attract ants. This investment pays dividends for the plant because the ants keep plants free of other herbivores, by acting as protective policemen. As can be seen from these examples, it is not only the quantitative dimension of certain processes but also their qualitative importance which are essential for the stability of natural ecosystems by the inter-netting of processes.

Energy flow runs parallel to the material cycles. The sun's energy is metabolised in photosynthetic processes of producers into chemical energy, and used by the producers themselves as well as the decomposers to maintain life. In the respiration and fermentation of micro-organisms, chemical energy is lost continuously as heat until the energy is used up after the complete degradation of material. This energy flow is shown in Fig. 57 B.

The structure of the ecosystem can only be shown as a model. There are many ways how this can be achieved,

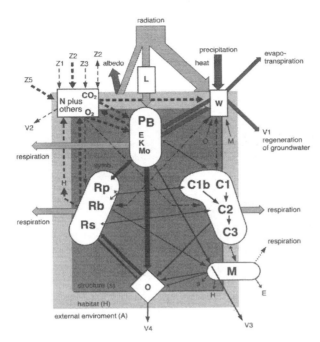

**Fig. 58.** Simplified schematic of the structure and processes in an ecosystem (after ELLENBERG et al. 1986)

one example is shown in Fig. 58 demonstrating the functional compartments and their interlinkage.

In comparison, quantitative values for different ecosystems are helpful. Table 7 shows the most important parameters of a beech forest ecosystem. The phytomass of this forest community is high, because of the matter stored as dead wood in the stems (hard wood 150 t ha$^{-1}$). However, even without this, the phytomass is more than 1000 times higher than the zoomass. The following numbers are given for European forests: reptiles 1.7 kg ha$^{-1}$, birds 1.3 kg ha$^{-1}$, mammals (predominantly smaller species, rodents) 7.4 kg ha$^{-1}$. The dry mass of invertebrates, mainly underground, is much larger (up to 14 kg ha$^{-1}$ dry mass, up to 90% dipterous larvae). Numbers are also available for an American deciduous forest with *Liriodendron* (REICHLE 1970), as dry mass (each in kg ha$^{-1}$); above ground: phytophagous arthropods 2.43, predatory arthropods 0.61; in litter: larger invertebrates 8.42, smaller invertebrates 3.42; in soil: earthworms (*Octalasium*) 140, smaller invertebrates 2.2.

For the oak mixed forests in eastern Europe it was found that after oak leaves were completely stripped by

**Table 7.** Important parameters of the ecosystem of an oak forest in the Belgian Ardennes with a hazel shrub layer (Querceto-Coryletum) and a sparse herbaceous layer

| | | |
|---|---|---|
| Leaf area index | | |
| Tree layer | | 3.87 |
| Shrub layer | | 1.83 |
| Total | | 5.7 |
| Phytomass (t ha$^{-1}$) | | |
| Above ground | | 260.8 |
| Of these | Leaves of trees | 3.5 |
| | Branches and twigs | 58.3 |
| | Stems | 180.2 |
| | Shrub layer | 18.1 |
| | Herbaceous layer | 0.7 |
| Below ground | | 55.4 |
| Total phytomass | | 316.2 |
| Primary production per year (t ha$^{-1}$ a$^{-1}$) | | |
| Above ground | | 15.3 |
| Of these | Total litter | 6.2 |
| | Lost through herbivory | 0.5 |
| | Increase in trees | 5.9 |
| | Increase in shrubs | 2.1 |
| | Increase in herbaceous layer | 0.6 |
| Below ground | | 2.3 |
| Total production | | 17.6 |
| Dead inorganic material in soil (t ha$^{-1}$) | | 122 |

(after DUVIGNEAUD 1974)

caterpillars, the growth of wood in ash and lime trees increased because of the better light conditions and an over-compensation ensued. In the 4 years after the caterpillar epidemic, the increase in wood production was 10%. Even after an epidemic with *Dendrolimus pini* in a spruce forest of different ages, compensation took place after a certain time by stimulation of the suppressed and less affected trees. The growth in wood in the second year was reduced to 76% and in the third year to 56%, but increased in the fourth and fifth year to 150 and 194% (see WALTER & BRECKLE 1991). Even moderate grazing of grassland stimulates the vegetative growth of grasses substantially so that the total annual production increases particularly if the amount eaten is considered (see WALTER & BRECKLE 1991, p. 43). This also applies to tropical forests when individual trees are almost completely defoliated by leaf-cutting ants.

**Primary production** is particularly important for ecosystems. As shown by analyses of production, the amount depends less on the rate of photosynthesis or the leaf area

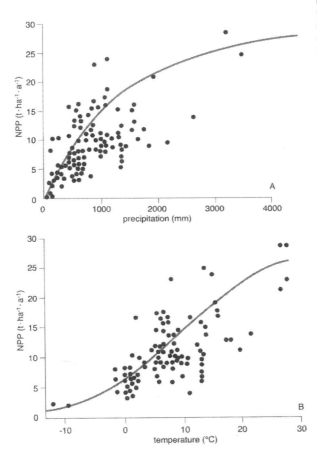

**Fig. 59.** Net primary production of forests depends on annual precipitation (**A**) and average annual temperature (**B**) (from EHLERS 1996)

index, than on the **assimilate balance** of the producers (WALTER 1960), that is, how assimilates are used during the vegetative period. If they are used productively by continuously forming new assimilating leaves, growth increases exponentially. If they are used unproductively to build up lignifying organs, the usefulness of which would only be noticed after years, this would correspond to a long-term strategy. However, this occurs very differently in individual biotopes and depends on the corresponding life forms.

Different investment strategies can be explained in the following example: If single seeds of beech (*Fagus sylvatica*) and of sunflower (*Helianthus annuus*) are sown in

The higher the temperature, the higher the productivity. The higher the precipitation, the higher the productivity.

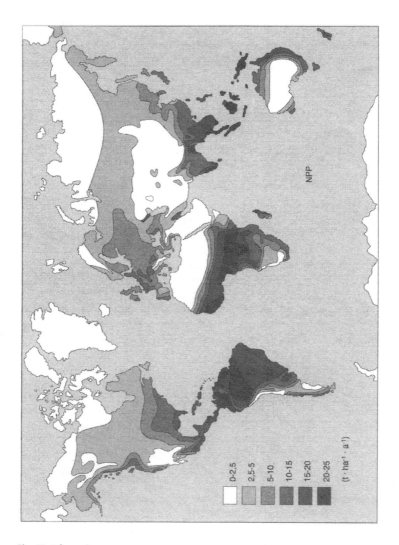

**Fig. 60.** Schematic representation of net primary production (t ha$^{-1}$ a$^{-1}$) on earth (from Schultz 1995, after Lieth 1964)

central Europe on good soil under similar conditions, the beech seedling produces only 1.9 g dry matter in the first year, whilst the sunflower seed produces about 800 g, even in a relatively unfavourable climate for this species. This is because sunflowers produce large continuously assimilating leaves, whilst the beech seedling suffices with two to three small leaves in order to use the assimilates for the production of a long primary root and a woody stem.

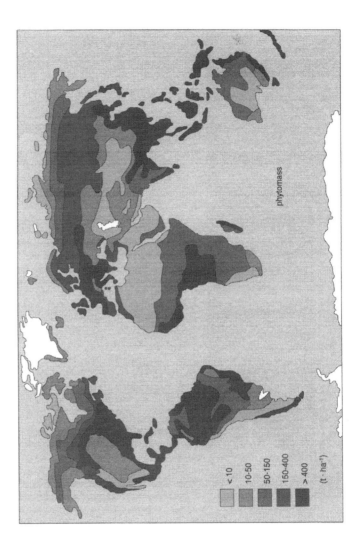

phytomass

< 10  10-50  50-150  150-400  > 400  (t · ha⁻¹)

**Fig. 61.** Rough schematic of the distribution of phytomass (t ha$^{-1}$) on earth (from SCHULTZ 1995, after BAZILEVICH & RODIN 1971)

The rate of photosynthesis is twice as high in sunflowers than beech, nevertheless this does not explain the 500 times higher production. The effect of "compound interest" of the investment in production organs ("carbon partitioning") plays the decisive role here. The most important life forms basically differ in this investment.

The interrelation between **net primary production** (NPP) and the most important ecological factors such as temperature and precipitation can be demonstrated for the different woods in Fig. 59 A, B. This, of course, only applies for a limited range; there are upper limits for both dependencies, as shown in Fig. 59 A, B although, in this case, the interrelations are not particularly strongly expressed.

On a global scale NPP shows the high productivity of warm-humid zonobiomes where the vegetative period may, in certain conditions, last the whole year. The shorter the growth period, and the colder and drier an area is, the smaller the NPP. The mean value of $10-15 \, t \, ha^{-1} \, a^{-1}$ for the central European beech forest is a relatively high value in a world-wide comparison, most dry areas are much smaller. In the tropics, however, values of up to $25 \, t \, ha^{-1} \, a^{-1}$ are achieved (Fig. 60).

The standing phytomass above ground (Fig. 61) in the tropics is often considerably above $500 \, t \, ha^{-1}$, the phytomass below ground adds a further 20–30%. In humid temperate latitudes, total phytomass is often just as large as in the tropics, whilst the phytomass below ground can be significantly greater. Forested areas of the earth are usually above $50 \, t \, ha^{-1}$, in deserts and semi-deserts however, the values are often under $10 \, t \, ha^{-1}$ (see Fig. 61).

## Highly Productive Ecosystems

In the hot humid tropics, NPP reaches average values of up to $25 \, t \, ha^{-1} \, a^{-1}$ as shown in the previous chapter. However, plant communities of tall perennial herbs achieve larger NPP values. Similarly to annual plants, such perennials also predominantly form assimilating leaves during the vegetative period and form flowers and seeds only at the end of the vegetative period. Since they possess relatively large reserves stored from the previous year, they are able to grow leafy shoots within a short time, whereas seedlings of annuals require a long lag period before the leaf surface is able to attain its maximum size. This is why summer grain requires 10 weeks for the production of the first quarter of the dry yield, 2 weeks for the second quarter and only 1 week for the last half (corresponding to the usual exponential growth curve).

Perennials, however, are able to take advantage of the entire vegetative period very productively. This explains the large amount of phytomass above ground and the substantial underground reserves stored in the autumn for the following year. A precise production analysis of a stand of tall perennials is given in Table 8. Primary pro-

**Table 8.** Production values of a monoculture of the adventitious shoots of golden rod (*Solidago altissima*) in a river meadow in Japan; vegetative period from April to October (in t ha$^{-1}$ a$^{-1}$)

| | |
|---|---|
| Increase in growth above ground | 12.01 |
| Increase in rhizomes and roots | 2.94 |
| Plants parts which had died during the vegetation period | 2.83 |
| Total production | 17.78 |

IWAKI et al. (1966); WALTER (1979)

**Fig. 62.** The upper Kamchatka river valley. A narrow bear path can be seen directly along the river bank, behind it is the perennial growth with *Filipendula camtschatica* and the gallery forest with *Salix sachalinensis* with dead branches. On the hill is a forest with *Betula ermanii* (Photo: E. HULTEN)

duction was determined through monthly measurements of the above- and underground phytomass. This amounts to an annual NPP of approximately 18 t ha$^{-1}$ and corresponds to that of a central European mixed oak forest (see p. 332). It is somewhat below the production of a 50 year old evergreen *Castanopsis cuspidata* forest in the warm temperate climate of Japan with an above ground annual primary production averaging 18.3 t ha$^{-1}$ a$^{-1}$. According to the investigations by MOROZOV & BELAYA (see WALTER 1981), the production of the natural giant perennials in the permanently wet and nutrient-rich soils of the river valleys of Kamchatka and Sakhalin is even higher.

Kamchatka lies within the subarctic zone with low growing *Betula ermanii* forests. The vegetative period lasts 90–110 days (the average frost-free period is only 64 days). The average temperatures are 3.5 °C in May, 10.6 °C in June, 14.3 °C in July, 13.3 °C in August and 7.2 °C in September. The perennials attain a height of 3.5 m, whereby *Filipendula camtschatica, Senecio cannabifolium* and *Heracleum dulce* are the dominant species (Fig. 62). According to HULTEN (1932), bears sleep in the vegetation during the day and emerge at night to feed on the salmon in the river. The standing phytomass attains a maximum of 31 t ha$^{-1}$ of which 10 t ha$^{-1}$ are underground. Since a portion of the shoot mass dies during the vegetative period, NPP is higher than the maximum herbaceous phytomass. In spite of the short vegetative period, NPP is estimated to be more than 16–20 t ha$^{-1}$ a$^{-1}$. This vegetation is used as silage for animal production. Even higher values were found on southern Sakhalin, which is much further south (approximately 45 °N), and has a warmer climate with mixed deciduous and coniferous species. The frost-free period is 145–155 days and the average temperature in the warmest months is 18 °C. There, the perennials grow to 4.5 m in height and their composition is similar to those in Kamchatka, although they are more heterogeneous, since certain species are locally dominant. In stands where *Filipendula* is the dominant species, a leaf area index of 13–14 is typical. Where *Polygonum sachalinense* is dominant, an index of 18–21 is possible if the perennial stand receives additional light, such as may be reflected from the river. This probably accounts for the enormous amount of above ground phytomass produced annually, which attains 30 t ha$^{-1}$ where *Polygonum* is the dominant species (total phytomass 70 t ha$^{-1}$). This NPP may, therefore, reach a record value of 38 t ha$^{-1}$ a$^{-1}$ for small areas.

It is not known whether the NPP of high *Papyrus* stands in the tropics exceeds these values. Because of the higher temperatures at night, it must be kept in mind that losses due to respiration in the tropics are very high, so that in spite of the high gross production, the net production is reduced substantially. Luxuriant stands of perennials are also found in the subalpine level of the western Caucasus Mountains (WALTER 1974). Stands of *Alnus viridis*, which are not quite so high, are also found in the subalpine regions of the Alps. They indirectly assimilate atmospheric nitrogen, thereby improving the soil. High perennial grasses in wet, nutrient-rich areas also produce a large phytomass annually. For 2.3 m high stands of reeds (*Phragmites*) on the lower Amudarya, for example, an above ground phytomass of 35 t ha$^{-1}$ a$^{-1}$ has been recorded.

## Special Features of Material Cycles in Different Ecosystems

Autotrophic algae suspended in water are the producers in aquatic ecosystems, if one disregards the littoral zone, the narrow zone between land and water. Algae are part of the plankton and are able to reproduce quickly through division. Since they need light to photosynthesise, they appear only in the upper layer of the water. They provide food for the animals of the micro- and macro-plankton, which in turn, are food for larger animals, such as fish and mammals living in water and also for predatory birds which catch their food in water. All dead organic waste material is mineralised by the decomposers in the water or the mud layer at the bottom of the water, the so-called benthic zone. The phytomass in water is relatively small, even though the primary production may be very high because of the very high reproductive rate of algae. However, the zoomass is relatively high in comparison to the phytomass, since this primary production serves as food for animals and is then incorporated to a considerable degree into their body substance (secondary production). The situation is quite different in terrestrial ecosystems. Average values are given as a comparison in Table 9.

In terrestrial ecosystems much unproductive biomass is accumulated in producers, whilst in aquatic ecosystems more biomass is accumulated in consumers. The schematic for ecosystems of deciduous woods (Fig. 58) does not apply everywhere. There are certain deviations and because of their importance, some examples are given. Most forest trees and most herbaceous plants (except Brassiceae), but in particular Ericaceae and Orchidaceae, form **mycorrhizae** with fungi. This may be interpreted as having the function of extending the root system and spreading it and thus facilitating the uptake of mineral nutrients from soils containing much humus. The mycorrhizal fungi can also supply their host plants with organic material, as is illustrated by the saprophytic orchids (e.g. *Neottia, Corallorhiza*), the Pyrolaceae (e.g. *Monotropa*)

**Table 9.** Proportion of phytomass and primary production of terrestrial and aquatic ecosystems

|                        | Phytomass : Primary production |           |
|------------------------|--------------------------------|-----------|
| Terrestrial ecosystems | 10–20                          | : 1       |
| Aquatic ecosystems     | 1                              | : 300–400 |

and other families. The effects of hormones probably also play a role. It has not yet been demonstrated that the mycorrhizal fungi of the forest trees and the Ericaceae provide their hosts with organic substances; however, this might well be the case in stands growing on extremely poor, sandy soils with a raw humus layer. In this case, the short cycle would be shortened still further since litter would not necessarily have to be mineralised.

A remarkable case of an ecosystem entirely lacking in producers has been discovered in the dune regions of the Namib fog desert (see p. 234): The organic matter, which is a prerequisite for the cycling of material, is blown by the wind from the neighbouring regions into this vegetation-free area and accumulates on the leeward slopes of the dunes or in sand hollows. It serves as food for saprophages (e.g. types of beetles), these in turn are eaten by smaller predators (e.g. reptiles) which again are the food of larger predators. Thus, a rich fauna has developed with very strange adaptations to life in moving sand even in the absence of vegetation, an open ecosystem without producers.

## Role of Fire for Ecosystems

Plant species requiring episodic fire to survive or to reproduce are often called **obligate pyrochores**. The reproduction of these plants is linked to fire events in ecosystems.

It is a well-established fact that fire can often replace decomposers and bring about very rapid mineralisation of the accumulated litter. In this respect, fire has a special effect on the material cycle of ecosystems. Natural fires caused by lightning have always occurred, even in the carboniferous forests and are characteristic for all areas with a period of drought, for example grasslands of the tropics and subtropics, for steppes of temperate climates and cold regions, for woodlands of the winter rain regions, and for all conifer regions, even without the "helping" hand of man. Fire is, in fact, a necessity for the vegetation if the decomposers are unable to break down all of the dead litter. After fires had successfully been prevented in the Grand Teton National Park, bark beetles were able to multiply to large numbers in the accumulated dead wood of the *Pinus* forests with catastrophic consequences. Now, since the natural fires have once more been allowed to take their course, the equilibrium of the ecosystem has been restored. Again and again, large fires over large areas occur at different intensities causing a fire mosaic in the landscape. Steppes and prairies (as well as grasslands and savannas) within national parks which are totally protected against fire, degenerate owing to the accumulation

of litter that would otherwise be periodically mineralised. In some Australian heath regions, cycling of material comes to a stop if the dead organic plant remains are not burned at least once every 50 years. Without the action of fire, all mineral nutrients are stored in the large woody fruits of *Banksia*, as well as in the hard, dead leaves of the grass trees (*Xanthorrhoea*), and cycling stops (see p. 280). Many *Eucalyptus, Banksia, Grevillea* and *Hakea* species in Australia only renew themselves after a fire. Many annuals use the open, newly fertilised soil after the next rain and germinate. Also, many geophytes suddenly form new shoots at the same time, and new shoots grow almost synchronously from the many burnt down stumps (Fig. 63).

After a fire the material cycle is stimulated by the components of the ash. It is similar for the large *Protea* stands around Cape Town, in the fynbos, where even shorter fire periods occur naturally than, for example, on the slopes around Jungershoek, where this phenomenon was researched. The natural occurrence of fires is relatively frequent there, on average every two to three decades. Not so much litter and dead material collects during that time, so that at many sites, the fire is not too hot and is thus, not as devastating. Thus, *Widdringtonia* (Cupressaceae) is able to regenerate time and time again, as this plant is only able to remain competitive with other shrubs and trees because of the effects of fire. The fynbos (Fig. 64) thus remains a species-rich mosaic of different age groups.

Thus, fire is often an important natural factor in maintaining the equilibrium of an ecosystem. For the years 1961–1970, exact statistics are available concerning the forest and grassland fires caused by lightning in the Unit-

**Fig. 64.** Proteoid fynbos near Junkershoek (South Africa) with sclerophyllic hard-leaved shrubs of different Proteaceae and *Widdringtonia* (Cupressaceae) which is only able to renew itself after fire (Photo: S.-W. BRECKLE)

ed States. Lightning was responsible for 34,976 fires (37% of all fires) in the Pacific states, for 51,703 (57%) in the Rocky Mountains, and for 13,733 (2%) in the south-eastern states, but only 1167 (1%) in the humid northwest (TAYLOR 1973). However, now fires which are laid by man for fire clearance, especially in the tropics, have increased enormously and it is possible to site thousands of fires every night in the satellite picture. Smoke particles are distributed in the whole atmosphere, where they have a large effect, albeit difficult to assess quantitatively, on the absorption of radiation and thus on the world's climate.

## Individual Zonobiomes and Their Distribution

Zonobiomes follow each other on both sides of the equator, however not symmetrically, because there is less land mass in the southern hemisphere and the climate is oceanic and cooler than in the northern hemisphere. It should also be noted that zonobiomes VI to IX in the southern hemisphere are much smaller. Zonobiomes VI and VII are only slightly represented whilst ZB VIII is completely missing. ZB IX is only represented in the sub-antarctic islands and the southern tip of South America, disregarding the Antarctic which is completely iced over and lacking in vegetation.

The sequence from the equator to the poles does not always correspond to the numeric sequence, thus ZB VII in Eurasia is partly inserted between ZB V and ZB VI and displays a very dry deviation (KRUTZSCH 1992 calls this climate facia regions), even having the precipitation of ZB

IV, but with cold winters and continental climates. The large zonobiomes are often divided into subzonobiomes (sZB) because of particular deviations. Before the discussion of individual zonobiomes, their distribution on the five continents is represented in Figs. 65–70 (pp. 105–110). Additional marks demonstrate smaller deviations within the zonobiomes.

**Fig. 65–70.** Large-scale ecological classification (after WALTER et al. 1975). *Roman numerals* I–IX indicate the zonobiome (ZB). *White spaces* between *shaded areas* are zonoecotones. Mountains are shown in *black*. Further distinctions within the individual zonobiomes are indicated as follows: a: relatively **arid** for that particular ZB; h: relatively **humid** for that particular ZB; oc: climate with **oceanic** or maritime tendency in extratropical regions; co: climate with **continental** tendency; fr: frequent **frost** in tropical regions, for example IIfr at higher altitudes; wr: prevailing **winter rain**, in ZB in which this is anomalous, for example Vwr, but also IIIwr; sr: prevailing **summer rain**, for example IIIsr; swr: two **rainy seasons**, for example IIIswr (or occasional rain at any season); ep: **episodic** rain, in extreme deserts; nm: **non-measurable** precipitation from dew or fog in the deserts; (rIII): **rain as sparse as in ZB III**, e.g. I(rIII) equatorial desert; (tl): **temperature curve as in I**, for example II(tI) = diurnal climate

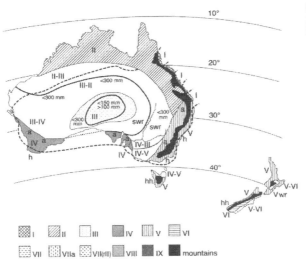

**Fig. 65.** Australia and New Zealand with zonobiomes I–VI

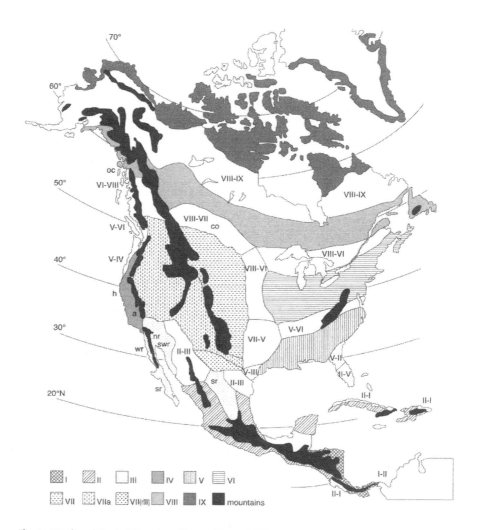

**Fig. 66.** North and Central America with zonobiomes I–IX

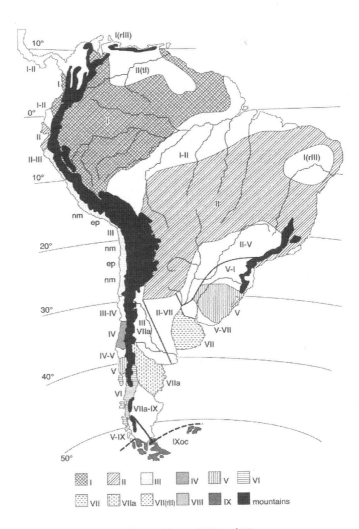

**Fig. 67.** South America with zonobiomes I–VII and IX

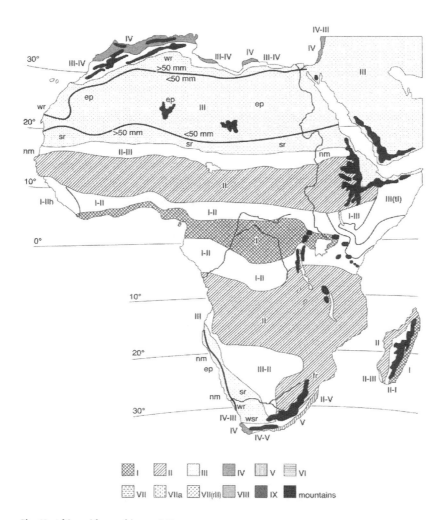

**Fig. 68.** Africa with zonobiomes I–V

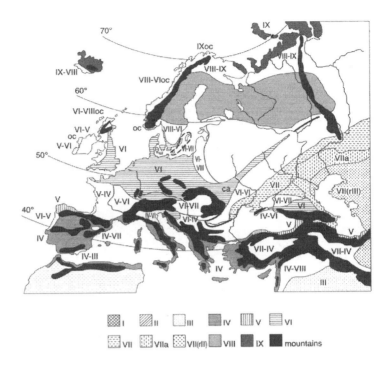

**Fig. 69.** Europe with zonobiomes IV–IX, plus western Asia. Because of the influence of the gulf stream zonobiomes run more from north to south in western Europe, whereas in eastern Europe they take the normal east-west course. From north to south: zonobiome IX (tundra zone) with zonoecotone VIII–IX (forest tundra); zonobiome VII (boreal coniferous zone); zonoecotone VI–VIII with zonobiome VI, both of which thin out toward the east (mixed forest and deciduous forest zone); zonobiome VII (steppe zone). Zonobiomes IX, VIII and VII continue eastward into Asia (Fig. 70). Southern Europe belongs to ZB IV (Mediterranean sclerophyllic region), offshoots of which are still detectable in Iran and Afghanistan. Zonobiome III is lacking altogether in Europe; only zonoecotone IV/III occupies a small desert like area in the southeast of Spain, which is the driest part of Europe. In central Europe, zonation is greatly disrupted by the Alps and other mountains. The situation in the mountainous Balkan peninsula is also complicated.

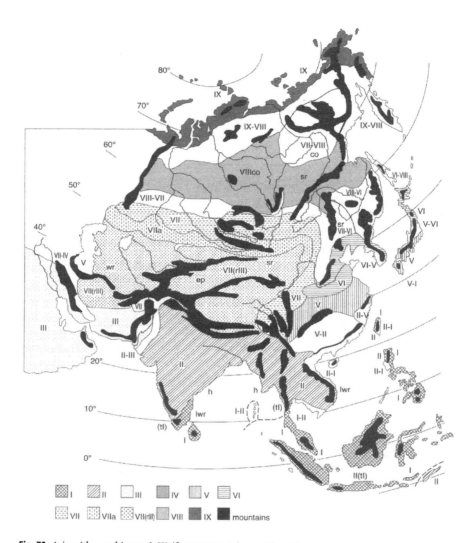

**Fig. 70.** Asia with zonobiomes I–IX (for western Asia, see Fig. 69)

## Questions

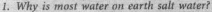

1. Why is most water on earth salt water?
2. How many days/weeks/years does it take to turn over the water in the atmosphere once?
3. Why are zonobiomes defined according to types of climate and not to floral or soil regions?
4. What is a triangular zonoecotone?
5. What are the most important differences between the sequence of altitude and the belts of vegetation towards the poles?
6. What is the difference between a plant community and a plant society?
7. Which minimum compartments must be included in a functioning ecosystem?
8. Why are the quantitative importance of the material turnover as well as the qualitative relations between organisms in an ecosystem important for its maintenance?
9. In which ecosystems does fire play a natural role as an ecological factor?
10. What are the basic differences between terrestrial and aquatic ecosystems?
11. Is it possible that consumers are able to increase production in an ecosystem?
12. How do the rate of photosynthesis, leaf area and relocation of assimilates affect growth and competitiveness of a plant?

# Special Section

# I Zonobiome of the Evergreen Tropical Rain Forest (Zonobiome of the Equatorial Humid Diurnal Climate)

## Typical Climate in Zonobiome I

In the wettest of all vegetational zones, a month with less than 100 mm rain is considered to be relatively dry. Only on the Malay Peninsular and in Indonesia are there large areas which are wet throughout the year; in the Amazon basin there is merely a small area on the Rio Negro, in central America there are a few areas on the rain-side of mountains. Two seasons with slightly less rain can usually be observed in the Congo (Fig. 71).

One or two short dry seasons invariably occur annually in South India. Bogor (Buitenzorg) in Java has an extreme rain forest climate (see Fig. 5, p. 20 f.). Mean monthly temperatures only vary between 24.3 °C in February and 25.3 °C in October, the mean annual rainfall is 4370 mm, with 450 mm falling in the wettest month and 230 mm in the driest. Daily temperatures can vary on sunny days by more than 9 °C (diurnal climate); on overcast days the variation is an insignificant 2 °C. Correspondingly, the humidity also hardly changes (Fig. 72).

Frost never occurs, only in high mountains, but even there the tropical diurnal climate dominates (see p. 145). Rain usually falls in the afternoon as short, heavy downpours, in the evening the sun shines again. When the sun is at its zenith, radiation is extremely strong, which means that the leaves directly exposed to its rays heat up several degrees (up to 10 °C) above the already high air tempera-

**Zonobiome I** (tropical rain forest) exbibits a typical diurnal climate. Daily changes in temperature are much larger than the annual changes of average monthly temperatures.

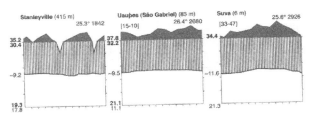

**Fig. 71.** Climate diagrams of stations in tropical rain forest regions: Congo, Amazon basin, New Guinea

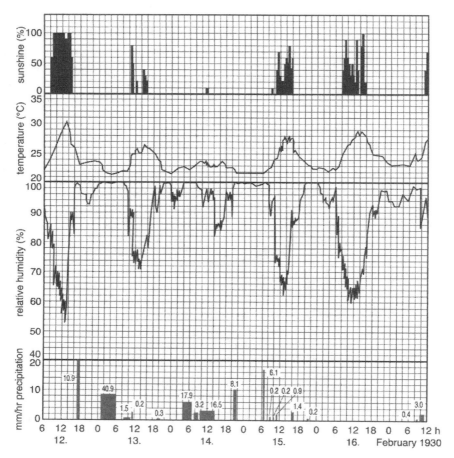

**Fig. 72.** Daily course of weather factors in Bogor (Java) during the rainy season (compare the sunny February 12, when the air humidity sank to almost 50%, with the overcast February 14). The figures for rain indicate absolute quantities in millimetres (after STOCKER, from WALTER 1990)

ture. Consequently, even in water vapour-saturated air, large vapour saturation deficits occur at the surface of the leaves (Fig. 73). Overheating of as much as 10–15 °C has been recorded for unshaded *Coffea* leaves on clear days in Kenya. Even in the permanently wet Bogor (Buitenzorg), clear days when humidity of the air sinks to almost 50% and the temperature rises to 30 °C are not that rare (Fig. 72). As a result, the saturation deficit over the over-heated leaves rises to almost 6 kPa. Even in the wettest

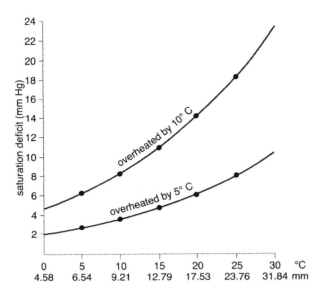

**Fig. 73.** Curves showing the water vapour saturation deficit at leaf surfaces if overheating reaches 5 or 10 °C relative to the air temperature in water vapour-saturated air (lower line of figures = saturation pressure in mm Hg)

tropics, therefore, the leaves are at times exposed to extreme dryness for hours on end. To human beings, with a body temperature independent of the temperature of the environment, the air feels continuously oppressive and damp.

Investigators who have spent years in the tropical forests, point out that even in the perhumid region of Borneo weeks without rainfall are observed time and again. The trees are thus exposed to recurrent dry periods, although the monthly means taken over many years give no indication of this. The same applies to the Amazon. It is not surprising, therefore, that the leaves are extremely well adapted to resist transpiration losses. They are equipped with a thick cuticle, are leathery but not completely xeromorphic (for example the rubber tree *Ficus elastica*, *Philodendron*, *Anthurium*), and can radically reduce transpiration by closing their stomata, thus preserving the hydration of the protoplasm. They are often lauriphyllic, but not sclerophyllic. Cell sap osmotic potential is usually only about −1.0 to −1.5 MPa. Some of these species can tolerate the dry air of heated rooms or buildings and are commonly found in Europe as indoor plants.

The situation of species growing in forest shade, however, is vastly different. The micro-climate prevailing in the interior of a rain forest is much more constant, especially on the ground itself where little direct sunlight falls. Variations in temperature are almost non-existent at this point, and the air is constantly water vapour-saturated. Because of the high air humidity, even the slightest atmospheric cooling at night regularly leads to the formation of dew on the tree tops, which serves to moisten the leaves of the lower layers as it drips down.

Light conditions are also of great importance to forest plants. The irregular contours of the tree canopy and the strongly reflecting, leathery leaves ensure that light penetrates deep into the interior of the forest, although its average intensity at the forest floor is very small. However, short periods of sun flecks on the floor play an important role in the availability and exploitation of light. Depending on the structure of the forest, on average 0.5–2% of the daily light reaches the herbaceous layer and the forest floor, at times even as little as 1% (similar to deciduous forests of temperate climates). If the numerous spaces in the forest stand, the reason for the very heterogeneous structure, are added and the light exploitation is integrated over a larger area, on average considerably more than 2% of the light penetrates to the ground. However, individual plants receive at times less than 1% of the light. Some of the very tender herbaceous plants have leaves which reflect blue light from their lower surfaces and thus appear blue (see p. 129).

## Soils and Pedobiomes

Ignoring recent volcanic soils and alluvia, the soils in rain forests are usually very old, often reaching back as far as the Tertiary period. Weathering effects penetrate many meters down in silicious rock; the basic ions and silicic acid are washed out, leaving the sesquioxides ($Al_2O_3$, $Fe_2O_3$), so that a so-called laterisation takes place and reddish-brown loam (ferrallitic soils or latosols) is formed, with no visible stratification into horizons. A comparison of the great variety of soil types establishes that about two thirds of the soils in the tropics are oxisols or ultisols, that is, soils with moderate to very low fertility. About 7% of the tropical soils are quartz sand-rich alluvial terraces or other strongly weathered leached areas (psammente or spodosol) and thus very nutrient-deficient. Present day agricultural practices are only possible on about 20% of the tropical soils, namely the younger volcanic soils (alfisols) and the rich alluvial soils in large river valleys (fluvente, aquepte). Litter decays very rapidly and wood is destroyed by termites, which are not obvious in the rain forest because they live in subterranean colonies. This is

their having been accumulated at a time when the rock was not so deeply weathered and the roots of the plants were still in contact with the parent rock and, on the other, on the episodic input of nutrients (for example long distance transport of dust from the Sahara to South America) which can be stored completely by the biomass. In totally depleted areas virgin forest can develop once more, if, as a result of soil erosion, the entire soil down to the underlying rock is removed and a new primary succession is initiated. This, of course, takes considerable time and presupposes the availability of the required propagules from the environment. If, however, the parent rock itself is poor in nutrients, as is the case with weathered sandstone or alluvial sand, then the nutrients only suffice to support a rather weak tree or heath population or a sparse savanna. These are a special type of pedobiome, termed peinobiomes, and may cover very large areas. They have true podsol soils with a raw humus layer 20 cm thick (pH 2.8) and bleached horizons, or even peat soils. The latter have been found in Thailand and Indomalaya, as well as in Guiana (*Humiria* bush, *Eperua* forest) and in the regions of the Amazon basin drained by the Rio Negro, the waters of which appear black owing to the presence of humic acid colloids from the raw humus soil. Peat soils have also been reported in Africa (in the Congo basin and in the heather moors of the island of Mafia), and have been studied in great detail in north western Borneo, where extensive $(14,600 \text{ km}^2)$ raised forest bogs (helobiomes) with *Shorea alba* etc. occur. These bogs commence immediately beyond the mangrove zone, and their peat deposits are up to 15 m thick (pH approx. 4.0). Heath forests (*Agathis, Dacrydium* etc.) with *Vaccinium* and *Rhododendron* are also found on raw humus soils. The total area occupied by tropical podsol soils is estimated at 7 million ha.

At the other extreme are the tropical limestone soils, i.e. lithobiomes, associated with peculiar topographical conformations such as are found in Jamaica and Cuba. Limestone dissolves easily in damp tropical climates, the softer parts disappearing completely, with the harder parts left in the form of sharp ribs or ridges. The entire area turns into karst, and a honeycomb formation develops out of the original plateau. Circular depressions, or dolines, up to 150 m in depth, develop from the sinkholes caused by subterranean streams. If erosion continues still further, as it has in Cuba, all that finally remains of the network of ridges are solitary towers with almost vertical faces such as the "mogotes," or organ-pipe hills or the "moros" in northern Venezuela. The floor of the dolines is covered with bauxitic "terra rossa" soil upon which a wet ever-green forest develops. The bare limestone rock of the honeycomb ridges presents a heterogeneous habitat, depending on whether an alkaline soil (pH = 7.7) can accumulate in the few depressions or not. This explains the extremely interesting flora, ranging from rain-forest species to cactus deserts. In the above-mentioned areas, the annual rainfall is less than 1000 mm. A "limestone" vegetation has not been reported from true rain-forest regions.

The subject of halobiomes (mangroves) will be discussed on p. 195.

# 3 Vegetation

## a Structure of the Tree Stratum, Periodicity of Flowering

The most conspicuous feature of a tropical rain forest is the large number of species constituting the tree stratum. As many as 40, or even over 100, species can be counted on 1 ha. On the other hand, there are also forests containing species of only a few families, as in Indomalaya, where Dipterocarpaceae frequently dominate, and Trinidad, where the upper tree storey consists of *Mora excelsa* (Fabaceae). Large floristic differences exist between the forests of South America, Africa, and Asia (Fig. 74). Forest types are also very dissimilar, but we shall only be able to discuss the features which are more or less common to them all. Palms, for instance, are almost completely absent from African rain forests, although they are abundant in Central and South America (particularly in wet habitats). The tree stratum reaches a height of 50–55 m, occasionally even 60 m, and three storeys, an upper, middle, and lower storey, are sometimes recognisable, although these are by no means always distinguishable. The upper tree storey is not compact, but consists of solitary giants which reach far above the other trees. It is the middle or lower storeys which form a dense leaf canopy, and in such cases the trunk region is relatively free owing to the lack of light and thus, of undergrowth. This makes walking in the forest quite easy. However, the detailed structure of such forests varies greatly, and generalisations should be treated with caution. Examples of profiles from forest stands showing this are given in Figs. 75–77.

The trunks of the trees are usually slender with a thin bark; the crowns begin high up and are relatively small with irregular forms as a result of crowding. It is difficult

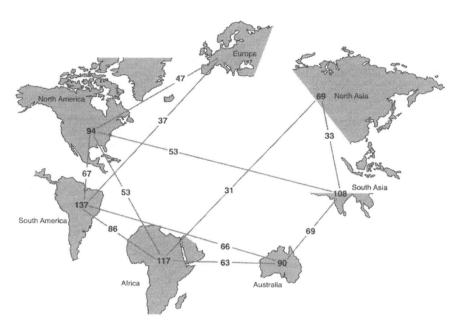

**Fig. 74.** Number of families of flowering plants on the continents (number in each region) and percentage of similarities (number between the larger regions) without cosmopolitan plant families (after TERBORGH 1991)

**Fig. 75.** A rain forest profile taken through the Shasha forest in Nigeria. The strip involved is 61 m long and 7.6 m wide. All trees taller than 4.6 m are shown (after RICHARDS, from WALTER 1973)

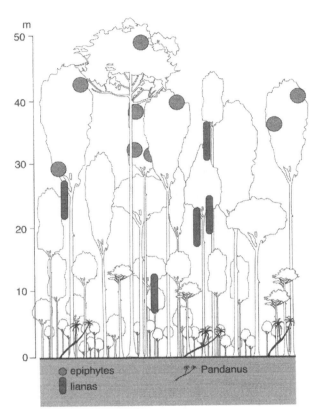

**Fig. 76.** Schematic profile through the Dipterocarpaceae rain forest in Borneo, length 33 m, width 10 m (after WALTER 1973)

epiphytes     Pandanus

lianas

to judge the age of the trees since annual rings are not present, but estimates based on rate of growth put them from 200 to 250 years for the older, thicker trees. The roots reach further down than was hitherto assumed: 21–47% are in the upper 10 cm, and the rest are mostly below this and down to 30 cm, but 5–6% reach down as far as 1.3–2.5 m (HÜTTEL 1975). A root mass of 23–25 t ha$^{-1}$ was calculated (49 t ha$^{-1}$ according to other methods). The giant trees achieve stability by means of enormous *plank-buttress roots* which reach, pillar like, as far as 9 m up the trunk. From the base of the trunk, they radiate outward, perhaps another 9 m, but are much reduced in thickness (VARESCHI 1980). They can become stilt roots when the soil is washed away.

The wetter and warmer the climate, the larger the leaves on the trees, although in a given species, the leaves

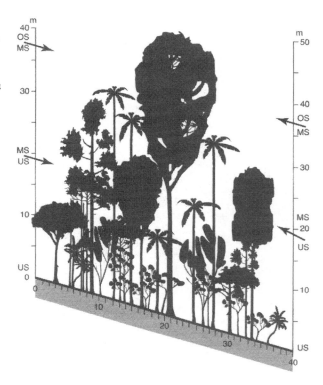

**Fig. 77.** Schematic profile through the tropical mountain rain forest in the Sierra de Tilaran (Costa Rica). *MS* Middle storey, *OS* upper storey, *US* lower storey (from SPRENGER & BRECKLE 1997)

that are exposed to light are always much smaller. For example, in an East African rain forest, *Myrianthus arboreus* showed a ratio of 8:1 (largest leaf 48×19 cm, smallest leaf 16×7 cm), and for *Anthocleista orientalis* the ratio was 28:1 (largest leaf 162×38 cm, smallest 22×10 cm). Both belong to the lower tree storey. Differences are much smaller for *Elaeagia auriculata*, from the mountain forest of Costa Rica, a species which always has very large leaves.

Any form of bud protection seems to be unnecessary for trees of the rain forest, although the young leaves are sometimes protected by hairs, mucilage, or succulent scales, or even by specially modified accessory leaves. Even though conditions are always favourable, the growth of shoots is periodic. The growing ends of the twigs often give the appearance of "**nodding foliage**". This is because at such a rapid rate of growth no supporting tissue is formed at first and the young leafy shoots droop downwards. They are white or bright red initially and turn green only later as they become stronger. The rapid differentiation of the leaf tips results in the formation of "**drip tips**," which are found, for example, in 90% of the undergrowth species in Ghana. Experiments carried out in the forest have shown that leaves with drip tips are dry within 20 min after a rain shower, whereas those without are still wet after 90 min (LONGMAN and JENIK 1974).

A special problem is presented by the periodicity of development and growth of plants in the continuously wet tropics, where there is no annual sequence of temperature. As has been mentioned, there is a periodicity in growth. Many trees show growth rings in their wood, often several per year. These rings are thus not true annual growth rings. Similar phenomena have also been observed for the rhythm of flowering. However, these phenomena are not confined to a particular season since external conditions are always constant.

On the Malay Peninsula, it is reported that in wet weather, the old leaves drop after the appearance of the young ones, but in dry weather, they drop even before new leaves have sprouted so the tree may be bare of leaves for a short time. Trees which shed their leaves developed this trait in a climate zone with a dry period. It can happen that, of two adjacent trees of the same species, one is bare while the other is in leaf or that the branches of one tree behave differently and put out leaves at various times. The same is true of the flowering period. Individuals of the same species may bloom at different times or the branches of one tree may be in bloom at different times. These are all manifestations of an autonomous periodicity which is not bound to a 12 month cycle. Periods of 2–4 months, 9 months, and even 32 months occur. This means that a rain forest has no definite flowering season, but there are always some trees in bloom. Although large and beautiful, the flowers are relatively inconspicuous against the predominately green background.

European tree species (beech, oak, poplar, apple, pear, almonds) were planted in tropical mountains without seasons. In general, they initially kept their annual periodicity of leaf fall, short growth and flowering, but with time the development of flowers became irregular, individual branches reacted differently and finally all seasons were observed on one tree, i.e. leafless branches, growing shoots, flowers and fruits. Central European tree species are usually long-day plants and therefore only flower in the long days (>14 h) of high latitude summers and thus they do not usually flower in the tropics. However, lower temperatures may replace long days, for example: *Primula*

The tropics differ from the temperate latitudes because of the continuously short daylength of around 12 h.

*veris* grows in Indomalaya at an altitude of 1400 m only vegetatively, but at 2400 m it blooms and bears fruit abundantly. *Fragaria* species do not bloom at low altitudes in the tropics, but produce many runners. In upland regions, the plants bloom and bear fruit if the runner formation is suppressed (for example in Sri Lanka). *Pyrethrum* plantations are cultivated at 1500–2500 m in Kenya because their flowers, which are harvested, do not develop at lower altitudes.

The endogenous rhythm of these plants adapts itself immediately to the climatic rhythm wherever there is one, for example in the wet tropics, which have a short slightly drier season. This is the case in many areas and thus a **timer** is present. On the mango tree, which is cultivated throughout the tropics, the few paler sprouting twigs on the otherwise dark crown are very obvious, but as soon as a dry season occurs, the growth and flowering of all the twigs and trees adapt to it. The teak, or djatti tree (*Tectona grandis*), is never bare in western Java which is always wet, but in the dry season of eastern Java it loses all its leaves.

However, even in the wet tropics, there are species, such as the orchid *Dendrobium crumenatum*, which bloom on the same day over a large area. In order to open, the buds need the sudden cooling after a widespread storm which is the timer. The buds of the coffee tree, too, open only after a short dry spell, and the reproductive organs of the bamboo develop only after a dry year, all blooming synchronously and then dying. In such a uniform climate, certain species are extremely sensitive to even small changes in the weather.

Tropical tree species often exhibit the phenomenon of **cauliflory**, where flowers develop on older branches or on the trunk. This occurs in about 1000 tropical species. Most of these species belong to the lower tree storey and are either chiropterogamous or chiropterochorous; i.e. fruit-eating or insect-eating bats pollinate their flowers or distribute their seeds as they can easily reach the cauliflorous flowers and fruits. Cauliflory also occurs on the widely distributed Mediterranean carob tree (*Ceratonia siliqua*) and the Judas tree (*Cercis siliquastrum*).

### b Mosaic Structure of Habitats

In tropical forests a rotation or cyclic rejuvenation of tree species takes place.

Rejuvenation of virgin forest in the tropics is difficult to research. When a giant tree falls, a large gap is left in the forest. When a large branch falls, a smaller gap occurs. In these gaps rapidly growing species of the secondary forest (balsa = *Ochroma lagopus* and *Cecropia*, in Central and

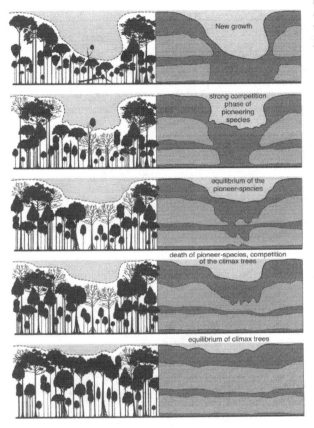

**Fig. 78.** Formation of a gap
and the ensuing strong
competitive growth resulting
in closure of the gap (after
TOMLINSON & ZIMMER-
MANN 1976)

South America, *Musanga* and *Schizolobium* in Africa, *Macaranga* on the Malay Peninsular immediately develop. *Ochroma* puts out annual shoots 5.5 m in length, with light wood; those of *Musanga* are 3.8 m, and of *Cedrela*, 6.7 m. These trees are later ousted by the upper tree-storey species (Fig. 78).

It has been observed that in the rain forest, a tree will often have no saplings of its own species growing beneath it. This led to the conclusion that such a forest is mosaic-like in composition, each tree species being replaced by a different one. Only after a lapse of several generations can the species return to its original site so that a sort of rotation or cyclic replenishment occurs. A similar process can be observed in the undisturbed primary woods of the tai-

The cyclic change of species in a particular site and heterogeneous mosaic formation is a generally valid principle for all species-rich, virgin plant societies in a dynamic equilibrium. It provides an explanation for the observation that none of the species competing is able to attain absolute dominance and a species-rich, mixed population is the rule.

**Fig. 79.** Profile diagram of three transects of the tropical sub-alpine oak woods from the Cordillera de Talamanca (Costa Rica), comparing differently aged secondary phases and primary structure (after KAPPELLE 1995)

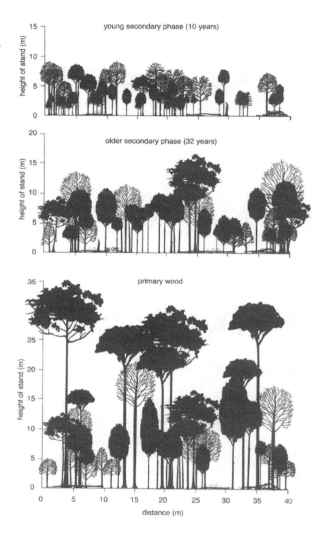

ga and virgin forests in eastern Poland, and also in the meadows and beech forests of central Europe.

An attempt has been made by JANZEN (1978) to explain the rotation of species in tropical rain forests by the herbivore hypothesis. Only around old trees are sufficient seeds, fruits and young plants available at particular times so that the reproduction of herbivores, appearance of parasites over an area, inhibition by mycorrhizal fungi or other factors, strongly limit the density of young growth. In palms it has been observed that when leaves up to 10 m in length and weighing several kilograms are shed, many young plants are cut down and damaged. This process leads to a reduction in the density of individual seedlings and young plants which is particularly strong in the neighbourhood of the parent tree. Only at a particular distance from the old tree is it then sometimes possible for a maximum density to develop. This has often been observed.

For the rather species-poor montane tropical oak forests in Costa Rica, KAPPELLE (1995) made a detailed examination of succession. From the transects in Fig. 79 it is possible to see, on the one hand, the considerable heterogeneity of the stand with its rather unclear storeys, the uneven upper crown, and also the "clumping" of particular species (in Fig. 79, middle). These species then grow up with time and many smaller or larger gaps occur. Initially, the number of stems per hectare (from 3 cm diameter at breast height) only decreases slowly. The thinning process during later succession is a sign of particularly strong competition during which some stems become dominant and out-compete the others.

## c Herbaceous Layer

About 70% of all species growing in a rain forest are phanerophytes (trees); in addition, phanerophytes are quantitatively absolutely dominant. It is difficult to distinguish the shrub and herbaceous layers from each other since herbaceous plants, for example, bananas, Heliconiae, Scitamineae can reach a height of several meters. Even in the areas of good light conditions on the ground, undergrowth is often lacking, perhaps because of competition with the tree roots for nitrogen and other nutrients. The lower herbaceous plants have to make do with little light, and as indoor plants in Europe, they also manage with very weak illumination (e.g. *Aspidistra*, *Chlorophytum* and the African violet *Saintpaulia*). The frequent occurrence of velvety or variegated leaves with white or red patches or a metallic shimmer deserves to be mentioned.

At high humidities, guttation plays an important role in the herbaceous layer, and the hydration of the protoplasm is correspondingly very high (cell sap osmotic potential only $-0.4$ to $-1.2$ MPa). The cell sap osmotic potential in ferns, which have a less efficient conducting system, is $-0.8$ to $-1.2$ MPa. Heterotrophic flowering plants, saprophytes or parasites occur, but play only a minor role. A

large variety of synusiae is undoubtedly present in connection with the different light and water conditions, but
so far they have not been investigated. The groups discussed below form a number of typical synusiae.

### d Lianas

In dense tropical rain forests, autotrophic plants have to
struggle primarily for light. The higher a tree, the more
light its leaves receive and the greater its production of organic matter. However, in order to reach the light, a trunk
has to be formed over many years, a process requiring the
investment of large quantities of organic substance. Lianas
and epiphytes attain favourable light conditions in a much

**Fig. 80.** Liana (probably *Ser-
janae polyphylla* and other
species) in seasonal rain
forest (in the background
*Myrcia* trees, Myrtaceae,
*Capparis* and *Clusia*) near
Arroyo Blanco, Dominican
Republic (Photo: S.-W.
BRECKLE)

**Fig. 81.** Epiphytes on the branch of a tree in the rain forest in Brazil. Rosettes of Bromeliaceae (*on the extreme right*), *Rhipsalis* species hanging down, and lanceolate leaves of *Philodendron cannaefolium* (Photo: H. SCHENK)

simpler manner. The former exploit the trees as a support for their rapidly upward-growing, flexible shoots instead of developing a rigid stem (Fig. 80), whereas epiphytes germinate in the topmost branches of the trees, which thus serve as a base (Fig. 81). Lianas attach themselves to the supporting tree by various means. Scrambling lianas climb among the branches of the trees using divaricating branches armed with spines or thorns which prevent them from slipping, as for example the climbing palm *Calamus* (rattan), *Smilax* or the *Rubus* lianas. The root climbers put out adventitious roots to fasten themselves to cracks in the bark or encircle the trunk (many Araceae). Winding or twining plants have long, rapidly growing, twining tips to their branches and very long internodes upon which the leaves are at first underdeveloped. The tendril-climbing lianas possess modified leaves or side shoots which are sensitive to touch stimuli and serve as grasping organs. Light is essential for the growth of lianas, and for this reason they are common in forest clearings and grow simultaneously as the trees re-grow until they finally reach the upper forest canopy. Tropical **lianas** are long-lived compared with those of temperate latitudes. Their axial organs are equipped with secondary thickening, but their stems remain pliable, enabling them to follow the growth of the supporting trees. The woody structure is not compact: the ligneous portions are permeated by strands of parenchymatous tissue or broad medullary rays (anomalous thickening). In cross section, the vessels are large and have no dividing walls so that, despite the small diameter of the pliable stem, the crown of the liana receives a sufficient supply of water. Should the supporting tree die and decay, the lianas still remain fastened to the tops of the other trees and their stems hang down like ropes. They often slip down to the ground and loop around their own base, but the shoot tips work their way up again. If this happens several times, the stem can attain a great length: a total of 240 m has been measured in *Calamus*. Complete deforestation is especially favourable for the development of lianas, so they are more numerous in secondary than in virgin forests, where they prefer the margins. Ninety percent of all liana species are confined to the tropics; in Central America 8% of all species are lianas. The difficulty of water transport is probably what confines lianas mainly to the wet tropics. In a dry climate, the low negative water potential built up in the leaves causes disruption of the water column due to a lack of cohesion in the wide vessels. In temperate climates, lianas are at their most abundant in wet floodplain forests. In Central Europe, few species of woody lianas are found: the root-climbing ivy (*Hedera helix*), the scrambling traveller's joy (*Clematis vitalba*), the twining wild grape (*Vitis sylvestris*), and the winding honeysuckle (*Lonicera* species). Although European blackberries (*Rubus* spp.) do not grow high above the ground, in New Zealand they grow as thick as a man's arm and reach up to the treetops.

### e Epiphytes and Hemi-Epiphytes and Stranglers

Particularly characteristic of the tropical rain forest are epiphytic ferns and flowering plants. These plants only occur, however, in forests which are frequently wetted (fog); high air humidity is not enough. A large number of interesting adaptations can be found (Fig. 81). In Liberia, 153 species have been investigated ecologically (JOHANSSON 1974).

Germination of epiphytes high up in the branches of trees guarantees favourable light conditions, but brings with it the problem of water supply, since the constant water reservoir otherwise provided by the soil is lacking. Epiphytic habitats can be compared with rocky habitats, and epiphytes do, in fact, grow well on rocks if light conditions are satisfactory. Since they can only take up water while it is actually raining, the frequency with which they obtain water is of greater importance to them than the absolute quantity of rain. On windward, mountainous slopes, the frequency of rainfall is greater than on flat land on account of the ascending air masses. This is why montane forests, especially the cloud forests (see p. 146) where leaves are constantly dripping, are rich in epiphytes. To be able to withstand the large intervals between rain showers, epiphytes must either be capable of resisting desiccation without undergoing damage, as is the case with many epiphytic poikilohydric ferns, or must store water in their organs, as the succulents of dry regions do. A whole series of cacti has changed over to an epiphytic way of life (*Rhipsalis, Phyllocactus, Cereus* species). Epiphytes conserve their water economically just as the succulents do; many orchids possess leaf tubers as water reservoirs, Ericaceae have woody nodes and the majority of orchids, bromeliads, peperomiads and other epiphytes have succulent leaves. The velamen of the aerial roots of orchids ensures a rapid uptake of water during showers; bromeliads are equipped with water-absorbing scales which take up the water collecting in the funnels formed by the leaf bases or serve to retain the water by capillary forces and then suck it up. The roots in epiphytic bromeliads serve only as adhesive organs and are completely absent in *Tillandsia usneoides*, which appears superficially similar to the lichen *Usnea*. *Myrmecodia, Hydnophytum* and *Dischidia* species develop special cavities which are sometimes inhabited by ants. Ferns, which cannot tolerate dehydration, can produce their own soil by collecting litter and detritus between their funnel-like, erect leaves (*Asplenium nidus*) or with the aid of special overlapping "niche" leaves (*Platycerium*). In this way, a soil is formed which is rich in humus and retains water so that the roots growing into it are well provided for. This can also be observed in many other species.

In a forest densely populated by epiphytes, epiphytic humus can amount to several tons per hectare. In this way, a new biotope is created far above ground level which can even be considered an almost closed ecosystem. Only recently have new techniques (areal ropeways, cranes) been used to supplement the rather destructive (helicopters, balloon nets etc.) or insufficient methods (canopy walkways) previously used to research these ecosystems.

In the canopy, dripping water and dust bring in nitrogen and other nutrients, ants colonise the area, build their nests, and also drag in seeds which then germinate and grow into flowering plants. Such "flower gardens" occur in South America and harbour a special fauna and microflora. Mosquito larvae, water insects and protista inhabit the funnels of the bromeliads, which often achieve considerable dimensions (phytotelmae). There are also many different types of insects.

It should be mentioned that the insectivorous *Nepenthes* (pitcher plants) as well as certain *Utricularia* species can also grow epiphytically. Epiphytes may be distributed by wind, e.g. the minute spores of ferns and dust-like seeds of orchids, or have membrane appendages. They may also be distributed by birds which eat berries (cacti, bromeliads); the seeds are then widely distributed and reach the branches of trees in the droppings. Many epiphytes are able to endure a long period of drought. Among them are orchids, which lose all their leaves, densely scaled *Tillandsia*, and poikilohydric ferns. Epiphytes are also found in the dry type of tropical forest. COUTINHO demonstrated the occurrence, in some epiphytes in Brazil, of diurnal acid metabolism (Crassulacean acid metabolism or CAM), i.e. the *nocturnal absorption* of $CO_2$ through open stomata which is bound as organic acids (usually malate). The organic acids are decarboxylated during the day, and the carbon dioxide thus set free is immediately assimilated while the stomata are still closed. By means of this process, the loss of water due to daytime transpiration can be avoided. Crassulacean acid metabolism has been frequently observed in the succulents of dry regions. MEDINA studied this phenomenon in *Bromeliaceae* as long ago as 1974.

Mosses and Hymenophyllaceae (filmy ferns) require constant humidity and are therefore the typical epiphytes of mist forests as well as the epiphyllic species (see below). Hemi-epiphytes occupy an intermediate position between lianas and epiphytes. Many Araceae germinate on the ground and then grow upward as lianas, usually as root climbers. In time, the lower part of the stem dies off so that the hemi-epiphytes turn into epiphytes although they still remain in contact with the ground by means of their aerial roots. Even more striking are the "strangling" trees, of which the numerous "strangling" figs are the best known (*Ficus* species). Such "stranglers" occur in many different families, as for example the *Clusia* species (Guttiferae) of South America, the New Zealand *Metrosideros* (Myrtaceae), and many others. They germinate as epiphytes in the fork of a branch and, at first, put out only a small shoot and a long root. The latter rapidly grows down the trunk of the supporting tree and enmeshes it. Root interconnections are formed (anastomoses, see Fig. 82). Only when the root reaches the ground does the shoot begin to grow; at the same time, the roots thicken and prevent secondary thickening in the supporting tree. The tree is thereby "strangled" and dies and the wood rots. The root mesh of the "strangler" unites to form a trunk bearing a broad crown (Fig. 82). Such trees can attain enormous dimensions, and it is no longer apparent that they began life as epiphytes. Palms, which do not have secondary growth, are not strangled and may survive

**Fig. 82.** Air roots and starting shoots of a strangler *Ficus* in the humid monsoon rain forest in northeast India, north of Siliguri (Photo: S.-W. BRECKLE)

until their leaves are too greatly overshadowed by the crown of the "strangler".

## f Epiphyllic Plants

Epiphyllic plants grow on the surface of leaves of other plants, they include microscopic algae, *Cyanophyceae* and other bacteria, *Azotobacter* able to bind nitrogen, green algae, yeasts, fungi, lichens and mosses (but most of all liverworts), *Selaginella*, and even small seed plants (Fig. 83).

Epiphyllic plants occur particularly in very humid tropical rain forests. Illumination, venation of leaves and their longevity are the basis for the occupancy by epiphyllic

**Fig. 83.** Coating of epiphyllic plants on a large leaf of *Cyclanthus* consisting of various Cyanobacteriae, green algae, mosses and lichens, Hymenophyllaceae, *Selaginella* and even with angiosperms; a *Begonia*, and a crawling *Peperomia* from the primary forest (Reserva Biol. San Ramón, Costa Rica) (Photo: S.-W. BRECKLE)

plants. Leaves thus suffer additional loss of light depending on the covering. Some epiphyllic plants even grow into the leaf tissue.

### g Biodiversity

Up to 300 different species of trees occur in the tropics per hectare; in Europe, north of the Alps up to the Ural Mountains, less than 50 native species occur.

In Central European forests only five to ten species of trees can be found, one or two of which are dominant and thus comprise about 90% of tree trunks. Temperate forests in North America or East Asia are not quite so poor in species, but even here only 15–40 species per hectare are found. In the tropics the number of species is very much greater. On the island of Barro Colorado in Panama (a research centre of 15 km$^2$), about 1400 higher

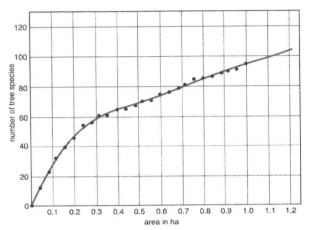

**Fig. 84.** Number of tree species (10 cm DBH) identified in an area increases with the size of the research area in the montane rain forest in the Sierra de Tilaran (Costa Rica). There is no minimum area. If the area is increased from 1 to 2 ha, 30 new tree species might be found (from WATTENBERG & BRECKLE 1995)

plants have been counted, of these 365 are species of trees. In the montane rain forest in the biological reserve north of San Ramón in Costa Rica 94 tree species per hectare were counted (DBH 10 cm and larger), all belonging to very different families (WATTENBERG & BRECKLE 1995). More than a third of these species is only represented by one trunk, which means that the minimum area to count the number of species is probably much greater than 1 ha (see Fig. 84) and cannot be determined. From other sites, for example in Peru (Yanamono area) about 300 species of trees were described in 1 ha, this is currently the "record of diversity". There, 63% of species are represented by only one trunk per hectare. However, the biodiversity of larger areas is so far not known, as it would take years to identify all species.

The tendency for the number of species per area to increase towards the equator not only applies to higher plants or trees, but also to reptiles and amphibians, birds, insects etc. (salamander and aphids are exceptions). MAC-ARTHUR has shown the differences in the number of bird species in a map of North and Central America (Fig. 85). The small country of Costa Rica has more species of land-breeding birds than the USA and Canada together, even though its landmass is only a small fraction (Fig. 86).

The greater structural diversity of tropical rain forests and the much tighter network with many more different sources of nutrition, the activities of organisms throughout the year, their narrow niches and specialisation and thus, the enormous variety of mutual interdependencies (symbioses) is one explanation for the greater diversity.

In tropical areas, the maintenance of a functional network of ecosystems with the immense diversity of tropical life forms requires disproportionally larger protected areas than in temperate latitudes.

**Fig. 85.** Number of breeding birds in the north of America in raster areas of $0.31 \times 10^6$ km$^2$. Despite the small area, more species of land birds breed in Costa Rica than in USA and Canada together. See also Fig. 86 (after TERBORGH 1991, from MACARTHUR 1972)

**Fig. 86.** The north-south gradient of species of land birds in northern America (*dots*) and in comparison, the area of land along the latitudes up to the equator (after REICHHOLF 1990)

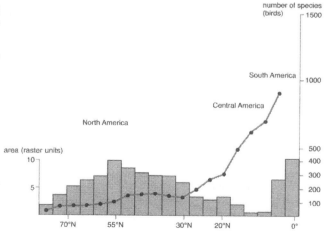

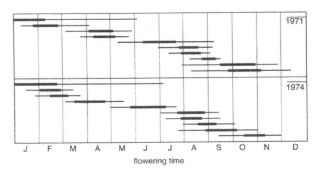

**Fig. 87.** Flowering of Helico-
neae in the rain forest of
Costa Rica occurs throughout the year and is similar
in different years. They provide continuous nectar for
hummingbirds (after TER-
BORGH 1991)

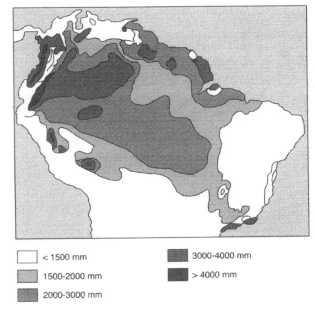

< 1500 mm

1500-2000 mm

2000-3000 mm

3000-4000 mm

> 4000 mm

**Fig. 88.** Present-day distribution of annual precipitation in tropical South America. Areas with more than
3000 mm/year probably
received enough rain
15,000 years ago (in excess
of 2000 mm) so that the
closed rain forests were able
to survive (after SIMPSON &
HAFFER 1978)

The tight functional network of many organisms is very
important. A relatively simple example is the spreading of
flowering periods of the *Heliconia* species throughout the
year (Fig. 87). This provides a continuous source of nutrition for different species of hummingbirds. However, for
such a network of relations between several Heliconiae
and several hummingbirds, a sufficiently large area is required. If they are too isolated, the network collapses at

**Fig. 89.** Correlation between the distribution of former rain forest refuges (ca. 10,000 years ago) and the present-day endemism of butterflies and birds (after Brown & As'Saber 1978)

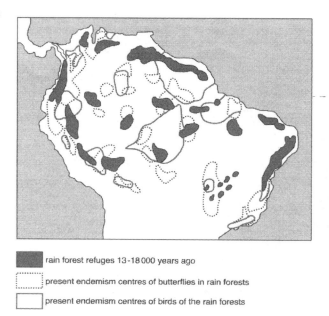

rain forest refuges 13-18 000 years ago

present endemism centres of butterflies in rain forests

present endemism centres of birds of the rain forests

some point with consequences for other hummingbirds and also for other Heliconiae.

During the glacial period, the area of the rain forests was drier than today, but deserts were wetter: pluvial periods. In earlier epochs, the extension of the Amazonian rain forests was probably restricted and broken up into smaller areas. The distribution of precipitation is shown on the map of Fig. 88. It is assumed that the areas with more than 3000 mm annual precipitation today would have had sufficient rain 15,000 years ago (more than 2000 mm), in order to maintain closed tropical forests.

These areas correspond well with certain refuge areas which are particularly rich in endemic species of birds, butterflies and reptiles, but also in flowering plants (Fig. 89). Thus, it may be concluded that rain forests alternated between shrinking and extending and the focal points of the biodiversity and the endemism correspond to those sites which have been continuously covered with rain forest. In between, there were probably larger covered areas with drier, seasonal rain forest. These changes, however, occurred very slowly, whilst the present day anthropogenic destruction happens at such rapid speed that organisms are unable to adapt. In African forests, for example in Upper New Guinea, Cameroon/Gabon and eastern Zaire, such endemic-rich refuge areas have been found. Only those areas were rich in tree species, with up to 140 species of trees per hectare, whilst in other regions of Africa the number of tree species is usually below 100, in Nigeria, for example, only 23.

The history of vegetation is different for the Malay Peninsular. During the Pleistocene, a large part of the continental shelf now under the sea was covered with rain forest. Probably the perhumid rain forest lying above sea level today survived. This would explain the extremely high biodiversity (with 180 tree species per hectare) and the lack of geographic-geological indications of earlier seasonal climates. On Mt. Kinabalu in northern Borneo there are as many species of ferns as in the whole African continent.

## Different Types of Vegetation in Zonobiome I Around the Equator

Typical climate diagrams for zonobiome I show a perhumid diurnal climate with two maxima in rainfall at the equinoxes, the rains occurring when the sun reaches its zenith, exactly at midday. This type of climate, however, is only found in some parts of the equatorial zone. Regions with wet monsoon winds (Guinea, India, South East Asia) exhibit a particularly well-developed rain maximum only in summer and, in addition, have a brief period of dryness or even drought (trend toward ZB II). The vegetation still consists of rain forest, although leaf fall and flowering are clearly connected with a particular season, so we can speak of this vegetation type as **seasonal rain forest**. Ghana, which is not affected by the monsoon, still has two rain maxima with an intervening period of drought, a situation similar to that seen in East Africa, where the monsoon winds are dry and rain falls when the wind changes. A distinction is drawn between a longer (and heavier) and a shorter (and lighter) rainy season. In Somalia, the rainfall decreases to such an extent that, in places, no wet season can be detected on the climate diagram and the vegetation is desert-like. This is a zonoecotone I/rIII.

Trade winds also affect the character of the climate. The southeast trade wind is wet and is responsible for the rain-forest climate with a maximum period of rain in south eastern Brazil, eastern Madagascar and in north eastern Australia from the equator to 20 °S and beyond. On the other hand, the northeast trade wind in the southern Caribbean only brings rainfall where the wind is con-

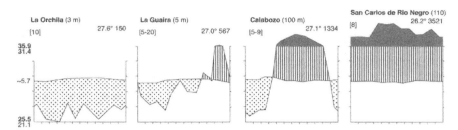

**Fig. 90.** Climate diagrams of a north-south profile through Venezuela: (after WALTER & MEDINA 1971). *Left* to *right*: Offshore islands; coastal stations; typical trade-wind climate (rainy season, 7 months); perpetually wet climate of the Amazon basin

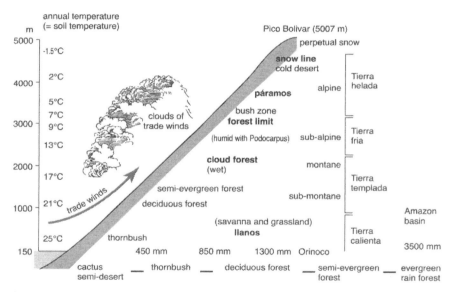

**Fig. 91.** Schematic representation of vegetational zones in Venezuela from north to south with mean annual rainfalls (in mm) and the altitude (in m) with mean annual temperature (in °C; *left*)

fronted with an obstacle such as mountains. Thus, mountainous Venezuela exhibits a considerable variety in climate and vegetation (Fig. 90). Similar conditions apply to Costa Rica. Venezuela lies between the equator and 12 °N and provides examples of every altitudinal belt from sea level up to the glaciated Pico Bolivar (5007 m). The northern part of the country is exposed to strong trade winds from November until March and it rains in the lowlands only during the seven calm summer months with ascend-

**Fig. 92.** Cactus-thornbush semi-desert with *Cereus jamaparu* between Barquisimeto and Copeyal, Venezuela, in February (the dry season) (Photo: E. WALTER)

ing air masses and numerous thunder storms. Only in the south, in the Amazon basin, are there no months with less than 200 mm rainfall. In Venezuela, the annual rainfall fluctuates between 150 mm (on the island of La Orchila) to more than 3500 mm in the south. In mountainous regions, precipitation increases rapidly on the windward side up to cloud level, but decreases again above this. At the same time, the mean temperature falls by 0.57 K per 100 m increase in altitude. The inner valleys of the Andes, situated in a rain shadow, are extremely dry. Figure 91 shows, schematically, the changes in vegetation with increasing altitude.

In the driest semi-desert areas cacti dominate; these succulents store so much water that they can easily survive a dry period of 6 months or more (Fig. 92). With a slight increase in rainfall, thornbushes and ground bromeliads appear, and impenetrable thickets similar to the north-eastern Brazilian caatinga are found. At an annual rainfall of 500 mm, umbrella-like thornbushes predominate (*Prosopis, Acacia*) and are accompanied by *Bursera, Guaiacum, Capparis* and *Croton* species, as well as by *Agave, Fourcroya* and others.

The tree cactus *Peireskia guamacho*, which has normal leaves and is considered to resemble the ancestral form of all cacti, occurs in the caatinga and during the dry season is leafless. These cactus semi-deserts and thornbush areas are used only as grazing land for goats.

As the rainfall increases, so too does the number of arboreal species, until the start of true deciduous forests, which is extremely rich in species. The tree layer is 10–20 m high, and only the Bombacaceae (*Ceiba* etc.), with their thick, water-storing trunks, and the lovely flowering *Erythrina* species extend above this level. During the dry season, such a forest has the appearance of a European deciduous forest in winter, although a few of the trees already start to bloom in this season. A distinction must be made between dry and wet tropical-deciduous forests; with a rainfall of up to 2000 mm, the latter attains a height of over 25 m and contains valuable timber species, for example, *Swietenia* (mahogany) and *Cedrela*. Deciduous forests are sometimes cleared for coffee plantations under shade trees and for the cultivation of maize, sugar cane and pineapple etc. Pasture land can also be produced by sowing *Panicum maximum*. Although the forests are poor in lianas, they are rich in epiphytes (drought-resistant ferns, cacti, bromeliads, and orchids). In the regions with more rain and a still shorter dry season, semi-evergreen forests occur, in which only the lower bush- and tree-layers consist of evergreen species. With even more rain, the evergreen-tropical rain forest commences (VARESCHI 1980).

A peculiar situation exists in Venezuela in the llanos region of the Orinoco basin which extends far into Colombia. Instead of the deciduous forests expected in this climatic zone, grasslands suddenly occur, dotted with small groves or solitary trees. These are true grasslands or savannas. Although these grasslands are used for grazing nowadays and are regularly burned, fire cannot be regarded as the primary reason for the absence of forests. The peculiar soil conditions prevailing in this area will be discussed later (pp. 181). The following vegetational formations of Venezuela are determined edaphically (pedobiomes), or by the relief of the terrain, rather than climatically: the mangroves on the sea coast and in the mouths of rivers, the beach and dune vegetation, the freshwater swamps and the aquatic plant communities, the floodplain forests and the vegetation of shallow rocky soils. Deciduous forests in Venezuela are extrazonal and caused by the dry trade wind which will be discussed together with the vegetation of ZB II. Orobiomes must also be considered separately, as the different altitudes in the mountains have different features. A wide variety of conditions can also be found in equatorial East Africa, with two definite, but short, rainy periods.

# 5 Orobiome I – Tropical Mountains with a Diurnal Climate

## a Forest Belt

In many tropical regions, mountains and volcanoes rise from the lowlands of the rain forest. HOLDRIDGE et al. (1971) developed a very artificial "hexagonal" classification of vegetational types (formations) for Central America related to mean annual temperature, potential annual evaporation and total annual precipitation. Altitudinal levels are shown parallel with climatic belts. The periodicity of the climate is not considered. This classification cannot be transferred to other regions. Mountains often have very different altitudinal belts. If the trade wind meets a mountain range across its path, the wind is forced to ascend, condensation takes place and clouds and rain result. Because the force of the trade winds drops in the evening, nights and the early hours of the morning are clear, but for the rest of the time a layer of cloud is present at a certain altitude, so that this altitudinal belt is shrouded in mist during the day. In addition to the rain from ascending air masses, there is also the condensation of fog droplets on the tree branches. Because the atmosphere is saturated with water, transpiration is entirely lacking.

The extremely humid climate, and as a consequence of the altitude the cooler climate, favours the development of hygrophilic tropical rain forests characteristic of all wind-exposed tropical mountain regions. The succession of altitudinal belts is determined by the increasing precipitation, whilst the decrease in temperature is only really noticeable at 2000 m. In northern Venezuela the following succession of altitudinal belts occurs:

- areas of glacier ice
  → climatic snow line
- cold desert
- alpine level (Páramos)
  → forest limit
- high montane woods with many *Podocarpus*
- cloud forests
- deciduous forests
- thornbush
- cactus semi-desert

This continuously dripping, wet-cool cloud forest differs from the hot tropical rain forest because of its vegetation. It favours a large number of tree ferns and epiphytic mosses which hang down from all branches as well as Hymenophyllaceae (filmy ferns) covering all branches and tree trunks. In the less humid high montane forest, which is often above the cloud cover, epiphytic lichens predominate.

Because of the ascending air masses, precipitation increases with altitude on the slopes of mountains, as long as they are not on the leeward side. If dry seasons occur in the plains, dry periods in the mountains become shorter and shorter with altitude and eventually disappear completely. Montane forests therefore are particularly luxuriant and rich with epiphytes because they are continuously sprayed with water.

Tropical mountain slopes are generally very steep, so the soil is well drained and swamps are absent, although they may be present in the plains. The change in vegetation with decreasing temperature is initially scarcely noticeable. Finally, in the **cloud layer,** where maximum humidity prevails, the cloud forests commence. They are not connected with a definite altitude, but rather with the cloud layer itself which, in turn, is dependent upon the humidity at the foot of the mountain. The greater the humidity at the base of the mountain, the lower the cloud level. In a climate with both a wet and dry season, the clouds are higher during the dry season. Cloud forests can be found between 1000 and 2500 m above sea level, or even higher, and the variety of temperature conditions which can prevail accounts for the floristic differences exhibited by such forests. Even the height of the tree stratum decreases with increasing altitude, and the trees of very high forests are gnarled and stunted by the wind (elfin forest). However, the common feature of all cloud forests is their profusion of epiphytes. Whereas the number of warmth-loving, epiphytic flowering plants decreases with increasing altitude, the ferns, lycopods and, above all, the filmy ferns (Hymenophyllaceae) and mosses become more abundant. The branches of the trees are typically draped with curtain-like mosses, themselves covered with water droplets, and filmy ferns. As soon as the humidity drops slightly below 100% the leaves of the filmy ferns roll up and form a green covering on the trunks and branches between the rising lianas. The forest floor is often carpeted with bright-green *Selaginella* species and tree ferns are numerous, as they require a damp, cool climate. In tropical mountain regions, the wettest altitudinal belt is often characterised by palms (South America) or dense bamboo groves (East Africa). Increasing altitude also brings changes in the soil with it. The reddish-brown loams of the lower belts are gradually replaced by more yellow or brown types; at the same time, a humus horizon appears, and the clay content decreases. Further up, a

slight podsolisation is detectable and eventually true podsols, with a leached horizon and raw humus, occur. In the
perhumid cloud belt, gley soils are found.

## b  Forest Limit

Precipitation decreases rapidly above the cloud belt
(usually above 2500–3000 m), and if the forest extends
further up the slopes, the leaves of its trees become smaller and more xeromorphic. In Venezuela conifers (*Podocarpus* species) occur, with narrow, leaf-shaped structures
in place of needles. Mosses are replaced by beard-like lichens and finally, at the upper forest limit, a shrub zone
commences. This occurs at a much lower level in the tropics than in the subtropics: an altitude of 3100–3250 m
above sea level is given for the Venezuelan Andes, for
Costa Rica 3200–3300 m. The shrub zone is narrow and
the shrubs themselves get smaller, the higher up they
grow. In Costa Rica, for example, *Escallonia, Weinmannia,
Myrrhodendron* (bushy Apiaceae) are replaced by bamboo
(*Chusquea*) species. In Venezuela, the last shrubs are
found at 3600 m in the shelter of rocks,.

Very detailed research on montane forests from the
central Cordillera in Costa Rica (Sra. de Talamanca, rising

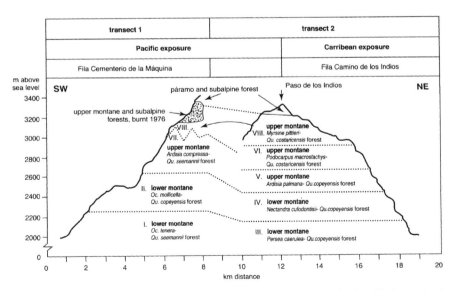

**Fig. 93.** Schematic profile of both slopes of the higher mountain regions in the Chirripo national
park (Costa Rica) with montane subalpine tropical oak forests (after KAPPELLE 1995)

**Table 10.** Plant families with woody species according to the number of species (given in brackets), from five different altitudes of the montane oak forests of the Sra. de Talamanca in Costa Rica

| 2000 m | 2300 m | 2600 m | 2900 m | 3200 m |
|---|---|---|---|---|
| Rubiaceae (31) | Lauraceae (20) | Ericaceae (14) | Ericaceae (9) | Asteraceae (11) |
| Lauraceae (27) | Melastomataceae (14) | Melastomataceae (114) | Rosaceae (9) | Ericaceae (9) |
| Melastomataceae (27) | Asteraceae (12) | Myrsinaceae (11) | Poaceae (8) | Rosaceae (6) |
| Asteraceae (15) | Myrsinaceae (12) | Loranthaceae (9) | Asteraceae (6) | Clusiaceae (5) |
| Myrsinaceae (14) | Araliaceae (11) | Poaceae (9) | Clusiaceae (6) | Poaceae (5) |
| Araliaceae (11) | Ericaceae (11) | Araliaceae (8) | Cunoniaceae (6) | Cunoniaceae (3) |
| Solanaceae (11) | Rubiaceae (10) | Asteraceae (8) | Loranthaceae (6) | Scrophulariaceae (3) |
| Ericaceae (10) | Solanaceae (10) | Lauraceae (8) | Araliaceae (5) | Clethraceae (2) |
| Euphorbiaceae (9) | Rosaceae (9) | Rosaceae (8) | Lauraceae (5) | Lauraceae (2) |
| Piperaceae (9) | Fagaceae (6) | Solanaceae (8) | Myrsinaceae (5) | Loranthaceae (2) |
| Rosaceae (9) | Poaceae (5) | Cunoniaceae (7) | Solanaceae (5) | Melastomataceae (2) |
| Loranthaceae (7) | Celastraceae (5) | Rubiaceae (7) | Caprifoliaceae (4) | Rubiaceae (2) |
| Myrtaceae (7) | Cunoniaceae (5) | Aquifoliaceae (4) | Aquifoliaceae (3) | |
| Poaceae (7) | Loranthaceae (5) | Caprifoliaceae (4) | Fagaceae (3) | |
| Clusiaceae (6) | Aquifoliaceae (4) | Chloranthaceae(4) | Melastomataceae (3) | |
| Moraceae (6) | Acanthaceae (3) | Fagaceae (4) | Rubiaceae (3) | |
| Celastraceae (5) | Caprifoliaceae (3) | Myrtaceae (4) | Clethraceae (2) | |
| Cyatheaceae (5) | Chloranthaceae (3) | Celastraceae (3) | Myrtaceae (2) | |
| Fagaceae (5) | Clusiaceae (3) | Clethraceae (3) | Polygalaceae (2) | |
| Smilacaceae (5) | Cyathaeceae (3) | Clusiaceae (3) | Rhamnaceae (2) | |
| Urticaceae (5) | Myrtaceae (3) | Loganiaceae (3) | Rutaceae (2) | |
| Cunoniaceae (4) | Onagraceae (3) | Rutaceae (3) | Scrophulariaceae (2) | |
| Flacourtiaceae (4) | Rhamnaceae (3) | Symplocaceae(3) | Symplocaceae (2) | |
| Mimosaceae (4) | Rutaceae (3) | | | |
| Theaceae (4) | Theaceae (3) | | | |
| 14 Families (3) | 11 Families (2) | 14 Families (2) | | |

**Table 10** (continued)

| 2000 m | 2300 m | 2600 m | 2900 m | 3200 m |
|---|---|---|---|---|
| 17 Families (2) | 35 Families (1) | 23 Families (1) | 25 Families (1) | 22 Families (1) |
| 26 Families (1) | | | | |
| 82 Families | 71 Families | 60 Families | 48 Families | 34 Families |
| 349 Species | 226 Species | 197 Species | 125 Species | 74 Species |

(after KAPPELLE 1995)

up to 3800 m (on Chirripó) was made by KAPPELLE (1995). In most cases oak species (see Fig. 93) and bamboo (*Chusquea*) dominate, thus enabling a characterisation according to the dominant species. Biodiversity decreases with increasing altitude, for woody species there is also a considerable change of importance, for example Rubiaceae decrease from 31 species at 2000 m to two species at 3200 m (Table 10).

The question of which factors are responsible for determining the tree line in the tropics is still unresolved. The fact that precipitation decreases with increasing altitude would suggest that aridity sets the limit. On the other hand, it might well be that frost is the limiting factor since at this elevation temperatures can drop below freezing point. Research in Venezuela, however, suggests that it is the soil temperature that is of ultimate importance, although, as always with such phenomena, a variety of other factors is involved. In the equatorial zone, where the climatic fluctuations are diurnal rather than annual, temperature variations do not penetrate very deeply into the ground. On shady ground the temperature at a depth of 30 cm is constant throughout the year and is the same as the mean annual air temperature measured by meteorologists. With only a spade and a thermometer, it is possible to determine the mean annual temperature at any spot in the tropics in a few minutes. In dense forests, the temperature is constant immediately below the ground surface and this temperature is decisive for the germination of tree seeds and the root system of the plants. Although the minimum temperature requirements for root growth of tropical trees are unknown, it is known that the enzymes responsible for protein synthesis in the roots have a temperature minimum well above 0 °C. This means that at temperatures considerably above freezing point, tropical species can be "chilled" and die slowly; *Ceiba* seedlings, for instance, only grow at temperatures above 15 °C. Assuming the minimum temperature for germination or root growth of trees situated at the tree line to be around 7–8 °C, this would coincide exactly with the temperature of the soil at the tree line in Venezuela, where the vegetation is made up of typical tropical species and is completely lacking in Holarctic species. If this assumption holds true, it would explain the higher tree line in the subtropics. In these regions, there is already an annual cycle of temperature, and in the summer the temperature of the soil rises considerably above the mean annual temperature. Arboreal species are able to exploit this favourable season, which is lacking in climatic zones where the fluctuations in temperature are of a diurnal rather than an annual nature.

### c Alpine Belt

In the wet tropics, the alpine belt is termed the **páramo**. It is described as being perpetually wet, misty, desolate, and cold. In the páramos of Venezuela there is little rain during the trade-wind period, between November and March. In January, there can be a whole week under a cloudless sky when the cloud forms at a lower altitude. The hourly temperatures reflect the lack of incoming radiation (Feb-

**Fig. 94.** Daily temperature curves, recorded in the meteorological screen on the páramo, 3600 m above sea level on 26 June and 27 July during the rainy season (diurnal fluctuations only 1.6 and 2.0 °C respectively) and on 10 February (hottest day) and 12 February (coldest day) during the dry season (fluctuations 17.0 and 17.5 °C respectively). Temperature maximum 14.5 °C; temperature minimum, – 7.5 °C

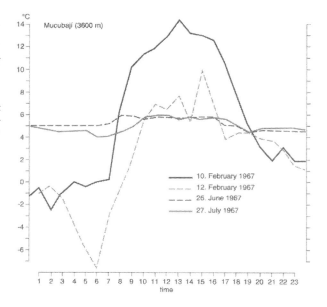

ruary 12; Fig. 94) or the intense incoming radiation (February 10) and high net outgoing nocturnal radiation (February 12). The coldest day of 1967 followed the warmest almost immediately. The air at 3600 m usually reaches a daytime temperature of 10 °C during the dry season, although the temperature goes down to freezing at night. Furthermore, plants themselves are exposed to much greater extremes than the thermometers in the meteorological screen. Clearly, this continuous freezing and thawing does the plants no harm, because it coincides with the main flowering season. At this season, the upper soil layers, harbouring the roots of the páramo plants, warm up during the day to temperatures above the annual mean. A rocky habitat is apparently more favourable than a wet soil. On the basis of soil measurements, the mean annual temperatures can be given as follows: at 3600 m, 5.0 °C (agrees with meteorological data); at 3950 m, 3.9 °C; and in the firn, or permanent snow, at 4765 m, –1.5 to –3.5 °C.

A decrease in temperature forces the plants to spread their roots nearer the surface which inevitably leads to sparser vegetation, until vegetation finally ceases altogether below the permanent snow line. This cold desert belt, with frost-rubble soils due to freezing and thawing, is typical of tropical mountain regions. In higher latitudes (in the Alps), the plants take advantage of a favourable season to grow in snow-free places even above the snow line (see p. 357 f.). The soil in the páramos itself remains moist during the dry season, so the vegetation does not suffer from drought and gives a hygromorphic impression. In Columbia, besides páramos with dry periods, constantly wet páramos with cushion plants, dwarf bamboos, grasses and mosses also occur.

The floristic composition of the páramos in South America, Africa, and Indonesia varies greatly, with each area having its own peculiarities. It is striking that apart from the low plants which give the impression of hugging the ground, there are also tall species, mostly Compositae, with a proper stem and large, bushy, upright leaves covered with thick, white hairs. In the Andes these plants are *Espeletias* (27 species), in equatorial African regions, *Senecio* tree species (Fig. 95), and in Indonesia they are species of *Anaphalis*. In addition to this form, the woolly, candle-like form of *Lupinus* and *Lobelia* is also conspicuous. On Kilimanjaro, extremely hairy species of *Helichrysum* grow as far up as 4400 m and it is assumed that their hairiness serves as heat insulation providing protection against sudden variations in leaf temperature. At such altitudes, the passage of a cloud on a sunny day invariably leads to a sharp drop in temperature. The upper sharp vegetation

limit lies at about 4400 to 4600 m and probably coincides with a mean annual temperature of about 1 °C. At such altitudes frosts occur daily.

It is especially remarkable that in the Venezuelan Andes, in the middle of the alpine belt, at an altitude of approximately 4200 m with a mean annual temperature of 2 °C, small stands of *Polylepis sericea* (Rosaceae) trees grow. They are invariably found on steep east- or west-facing talus slopes consisting of large boulders, and are exposed to the sun in the morning or afternoon. The roots of *Polylepis* sometimes go down as far as 1.5 m and the only possible explanation for this isolated occurrence of trees 1000 m above the tree line is that the talus slopes provide particularly favourable temperature conditions.

Incoming radiation warms up the air layer closest to the ground very strongly; the heavier cold air between the boulders flows out at the lower end of the talus, so that warm air is drawn in between the boulders in the upper part. This explanation finds support in the observation that the lower part of the talus slope is devoid of trees and is often even completely bare. Temperature measurements on talus slopes in Mexico have proven this to be the case. Small *Polylepis* forests often occur in Ecuador above the treeline on mountain slopes, as remnants of fire clearings.

Altitudinal belts on the African volcanoes (Mt. Elgon, Mt. Kenya, Kilimanjaro) are slightly less humid since they rise from a humid savanna zone. The soil temperature at the upper limit of the forest (*Hagenia, Podocarpus* species) is similar to that in Venezuela. Similarly to the *Polylepis* groves in South America, there are high growing *Hagenia* (Rosaceae) groves. *Erica arborea* plays a considerable role in the lower alpine belt, above the bushy Senecioneae (Fig. 95) and candle-like *Lobelia*. Interspersed are wet bogs with Cyperaceae, *Alchemilla* species, Gentianaceae and the massive *Huperzia saururus* (Lycopodiaceae) which often occurs in the southern hemisphere.

## Biogeocenes of Zonobiome I as Ecosystems

The tropical evergreen forest is one of the most complicated of all plant communities. Individual biogeocenes are still totally uninvestigated, perhaps it is impossible to find plausible borders for many of them. In one of the most thoroughly studied forests, in north eastern Australia, 18 lists of forest stands revealed 818 species taller than 45 cm, 269 of which were trees. A computer-based floristic analysis showed 6 floristic groups in the 18 lists, with 3 stands in each group. These floristic groups are the result of climatic differences, although nutrients in the soil, water supply and altitude also play a role. From a further 70 forest-stand lists, spread across $20°$ latitude, 24 life-forms and structural characteristics were considered. They corresponded to geographical regions (which were already known) and the combinations in which the plants occurred changed continually.

A classification into ecosystems is obviously an extremely complicated matter. The luxuriance of the vegetation and its large biodiversity would suggest a very high primary production. The preliminary estimates of about $100 \, t \, ha^{-1} a^{-1}$ dry weight (DW) proved to be too high. It must be borne in mind that the phytomass of tropical rain forests has a very high water content (75–90% for the herbaceous plants) and although the green leaves are able to assimilate $CO_2$ throughout the year, nocturnal respiration losses are especially large owing to the high temperatures. The phytomass of wood and leaves in tropical rain forests is two to three times higher than in temperate vegetation; the costs of maintaining this mass and the respiration losses in wood are four times higher and in leaves, six times higher. Tropical forests are forced to higher metabolic turn-over rates because of the high temperatures, therefore relatively little can be invested into the production of wood.

Tropical rain forests have a high turn-over rate because of the high temperatures. The very high productivity is balanced by particularly high respiration losses.

The leaf area index (LAI), i.e. the ratio of total leaf area to the total ground area covered, in a particular ecosystem is of great significance for the productivity of the biogeocene. For a tropical rain forest on the Ivory Coast LAI had a very low value of 3.16. However, this experimental plot cannot be considered representative. Gross production proved to be very large, but 75% of the organic substance produced was lost by respiration. Losses due to respiration in a Central European beech forest, on the other hand, amount to only 43% of the gross production. The primary production of this tropical rain forest is therefore no higher than that of a well-managed beech forest in central Europe:

- Tropical rain forest    $13.4 \text{ t ha}^{-1}$
- Beech forest    $13.5 \text{ t ha}^{-1}$

Wood production in tropical forest plantations can attain values of $13 \text{ t ha}^{-1}$, which is only about twice that of a good European beech forest and occurs only because the vegetational period is twice as long.

A forest in Thailand, with 2700 mm annual rainfall and an annual temperature of $27.2 \,^\circ C$ had an above ground phytomass of $325 \text{ t ha}^{-1}$, which probably indicates a total $360 \text{ t ha}^{-1}$. This value increased still further by $5.3 \text{ t ha}^{-1} \text{ a}^{-1}$, over the 3 years of the study. The LAI was 12.3; gross production was $124 \text{ t ha}^{-1}$, and respiratory loss was $95 \text{ t ha}^{-1}$ ($=76\%$), which gives an annual primary production of $30 \text{ t ha}^{-1}$.

In tropical forests, small mosaics consisting of three phases can be distinguished. The first is a young phase with rejuvenating stock and an increasing phytomass. The next is the optimum phase with a maximum, constant phytomass. Finally, there is the ageing phase with decreasing phytomass. The stand on the Ivory Coast was clearly a thin, juvenile phase. However, all three phases occur together as a mosaic. From the available data for the optimum phase of a luxuriant tropical rain forest the following can be deduced: total phytomass amounts to 350–450 t ha$^{-1}$, with a leaf area index of 12–15, and an annual gross production of 120–150 t ha$^{-1}$. This corresponds to a primary production of 30–35 t ha$^{-1}$, of which 10–12 t ha$^{-1}$ are accounted for by litter. Soil respiration corresponds roughly to the quantity of litter, but the larger part of the primary production is probably mineralised above ground (dead, upright trees and epiphytes). About 106 kg ha$^{-1}$ nitrogen is returned to the soil in the Amazon region annually, compared with only 2.2 kg ha$^{-1}$ phosphorus. Impoverishment of the secondary forest is probably mainly a question of phosphorus deficiency as phosphorus in the soil is quickly bound to Fe and Al and thus, no longer available for plants. Nitrogen is also added from the atmosphere in the course of the frequent thunderstorms.

## Fauna and Nutritional Chains in Zonobiome I

Only a few short notes will be made here; the diversity and gaps in knowledge are too great because organisms in the tropical rain forest are very tightly linked to each other. As mentioned earlier, this causes a great sensitivity to any intervention (see p. 15). For the animal world as well as for many plants, the top storey is a very important space of action. More than half of the mammals live in the crowns of trees and possess a prehensile tail; the number of birds is also very large and again their main site of activity is in the crowns of trees. So far, it has only been possible to give a rough estimate of the number of species. In particular, the number of invertebrates above and below ground is still almost unknown, and consequently so are their functional relations. Termites and ants are particularly important for processes within the ecosystem as they degrade a considerable amount of biomass; their zoomass is, however, despite their diversity, relatively small.

Sloths (*Cholopus, Bradypus*) are typical animals of tree crowns in the Neotropics. Their way of living has been studied intensively (MONTGOMERY et al. 1975). The total zoomass of these animals was 23 kg ha$^{-1}$, and they consume 53 kg ha$^{-1}$ of leaf mass annually, corresponding to 0.63% of leaf production. The excrement of these animals decays slowly and thus provides a nutrient reserve in the soil.

Leaf cutting ants (*Atta*) are particularly important because of their selective attacks. HAINES (1975) showed that they remove material from tree species of secondary forests and carry it as far as 180 m to their subterranean nests which have a diameter of up to 10 m. Here, they cultivate the cut leaves in fungus gardens, the fungus supplying the ants with nutrition, thus providing an example of perfect "microbiological agriculture". The cut leaf area can be as large as 4000 m². This increases the light in a stand by up to 7%. The number of ant species living on one tree is very often greater than the total number of ant species in a country with a temperate climate in Europe.

##  Human Beings in Zonobiome I

The tropical rain forest growing in poor soils is inhospitable to human settlements and is usually avoided by people. It is often a refuge of aboriginal tribes. In Africa these are the Pygmies, and in Latin America, original Indian tribes. Even in South-East Asia some of the original tribes have survived. In contrast to these habitats, the remnants of former rain forests on nutrient-rich, young volcanic soils are densely populated cultivated lands (for example Java, Central America). Only in these areas is it possible to maintain long-term agriculture. Tree felling in all other areas leads to a catastrophic loss of nutrients, expressing very clearly the "ecological" disadvantage of the tropics (WEISCHET 1980). Deforested areas are worthless after a few years, as they are subject to erosion and soon covered by *Gleichenia* or *Imparata* thickets of no value.

**Table 11.** Value of wood of marketable tree trunks ha⁻¹ on an experimental station in Amazonia (rain forest of Misana on the Rio Nanay, Peru) by an irreversible, single logging, in contrast to the yearly harvest and marketable value of fruits, raw rubber, resins and other sustainable useful products

|  | Single value of wood | Sustainable use (per year!) |
|---|---|---|
| Number of species | 27 | 12 |
| Volume of wood (m³) | 94 | – |
| Value of wood ($) | 1001 | – |
| kg raw products | – | 160 |
| Number of fruits | – | 5500 |
| Market value ($) | – | 698 |

(PETERS et al. 1989; REICHHOLF 1990)

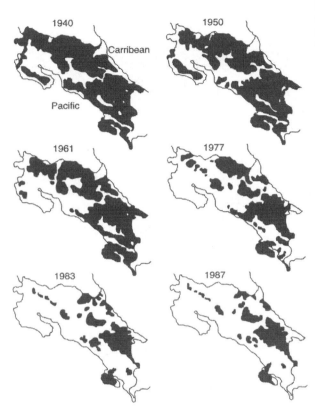

**Fig. 96.** Deforestation of Costa Rica during the last decades (from ELLENBERG & BERGEMANN 1990)

We have many good reasons to realistically judge the economic value of tropical rain forests. Tropical rain forests are worth much more than the wood growing in them, even without including the almost inestimable genetic resources of an unbelievable biodiversity which has so far not been clearly recognised. This is demonstrated in Table 11.

The devastating over-exploitation is shown in these figures: the loss of biodiversity is incalculable, the enormous wealth of all sorts of possible, even useable secondary material is not included in such calculations.

Deforestation has taken place at a rate which defies the imagination. An example is given from Costa Rica. At present, about 21% of the total area of this country is protected (national parks, reserves) and yet, the pressure on these protected areas is enormous, as hardly any forest

**Fig. 97.** Tropical rain forests decline continuously. The *straight line* shows the prognosis at constant deforestation, at the steady annual deforestation rate since 1990. The two other curves are based on the calculations of different organisations and consider the increasing demand. *LIN* Linear decrease, *FOE* Friends of the Earth, *WRI* World Resource Institute (after TERBORGH 1991)

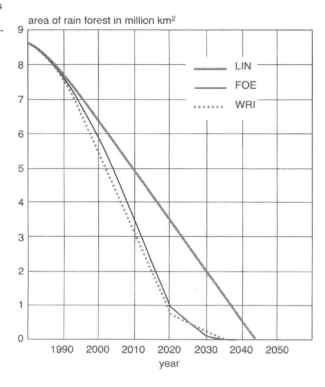

with large supplies of wood is still available. The forest area has declined at a frightening speed, particularly during the last decades (see Fig. 96). In the Dominican Republic on Hispaniola, the primary forest area has declined in half a century from more than 70% to less than 6%, in Haiti everything has been deforested.

The worst fears are expressed in any prognoses for the maintenance of tropical rain forests. In the year 2040, all of the approximately six million square kilometres of humid tropical forests still standing will be cut down if deforestation continues at the present rate. According to other calculations, this catastrophic situation will be a fact in the year 2025, as overpopulation and continuing poverty will increase the rate of deforestation (see Fig. 97).

This is not only a regional or national problem, but a global problem. In the United States and in central Europe, almost all forests and prairies have been destroyed and replaced by monotonous woods or maize fields etc. However, the climate in these areas is so favourable that it

An immediate major global effort to save rain forests is required because of their tight interlinkage with the global cycle, their unique and unbelievably high biodiversity with their irreplaceable genetic resources, sensitive soils and the irreversible damage to landscapes caused by logging.

is possible to build up efficient agriculture and forestry management and long-term use can be expected. The situation is, however, very different in the ecologically disadvantaged tropics. Every night thousands of fires are burning. The smoke still reduces the effects of increasing greenhouse gases by reflecting solar energy back into space and reduces global warming. Deforestation by fire must not go on like this. Natural fires almost never occur in the rain forests of ZB I.

## Zonoecotone I/II – Semi-Evergreen Forest

The semi-evergreen tropical rain forest is the zonoecotone between zonobiome I, evergreen rain forest, and zonobiome II, the tropical summer-rain region with deciduous forests. It is thus, a transitional zone with a diffuse mixture of both types of vegetation. In small areas this presents itself locally as vegetational mosaic, sometimes as a spotted carpet of different types of vegetation, albeit modified according to groundwater, structure of soil and availability of water and nutrients. On the very poor sandy soils of Venezuela and Guyana the periodically flooded Igapo forest grows near the river as a pedobiome. Higher up on the sand ridges, the caatinga develops with large sand deposits; it becomes the lower caatinga or even the sparse bana, even though precipitation is high (3300 mm). The sandy soils do not provide any nutrients and the capacity for water storage during the dry periods is very low. Considered on a larger scale with increasing annual precipitation and with increasing periods of drought, the following sequence applies to Venezuela: evergreen rain forest – semi-evergreen rain forest – deciduous forest.

This zonation can only rarely be observed within the equatorial climate zone, since a stepwise increase in rainfall, such as is found in Venezuela, is exceptional. As one leaves the tropics, however, such zones can be distinguished. Moving away from the equator towards the tropic of Cancer, the tropical climate zone with zenith summer rains occurs and absolute rainfall decreases continuously, so the rainy season becomes shorter and the dry season longer. In contrast to Venezuela, the annual course of temperature becomes more marked, and the dry season occurs at the cool time of the year. However, since the latter means a rest period for the vegetation, the temperature variations are of little significance from this point of view.

It has already been mentioned that when a short dry period occurs in the very wet tropical region, the endoge-

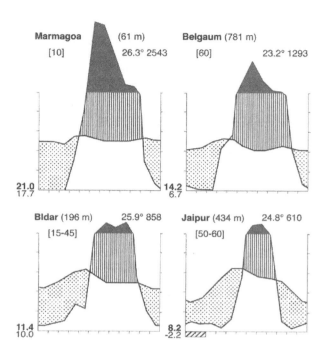

**Fig. 98.** Climate diagrams of Indian meteorological stations in the evergreen, semi-evergreen, moist- and dry-monsoon-forest

nous rhythm of tree species adapts itself to the climatic rhythm. The general character of the forest remains unchanged, but many trees lose their leaves at the same time, or sprout or flower simultaneously, so the vegetation does in fact exhibit definite seasonal changes in appearance (seasonal rain forest).

If the dry season becomes even longer, the type of forest changes. The upper tree storey is made up of deciduous species (in South America, these are the large, thick-trunked Bombacaceae and the beautiful flowering *Erythrina* species), whereas the lower storeys are still evergreen, therefore, it can be termed tropical semi-evergreen forest. With a further decrease in rainfall and a lengthening of the dry season, all of the arboreal species are deciduous and the forest is bare for shorter or longer periods of time. These are the moist or dry deciduous summer green tropical forests. This sharp transition can be observed from central Costa Rica towards the northwest (Guanacaste) over a very short distance. The transition can be particularly well observed in climate diagrams for forests of this type in India, in the summer monsoon-rain region as shown in Fig. 98.

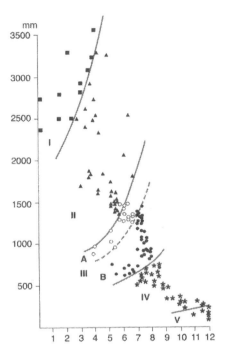

**Fig. 99.** The relationship between annual rainfall (*y-axis*) and length of drought in months (*x-axis*) for different types of forest vegetation in India: *I* evergreen tropical rain forest; *II* semi-evergreen tropical rainforest; *III* monsoon forest (*A* moist, *B* dry); *IV* savanna, thornbush forest; *V* desert (after WALTER, from a study carried out for UNESCO)

The question now arises as to whether the amount of rain or the duration of the dry season determines the structure of the forest. The diagram in Fig. 99 shows that both factors are ecologically important and neither should be considered alone. The course taken by the limiting lines indicates that for the wet type of forests the duration of the dry season is more important, whereas the absolute annual rainfall is of greater significance for the dry types.

In Africa, the above-mentioned succession of forests is difficult to recognise. Dense population and shifting cultivation have led to the deforestation of semi-evergreen and wet deciduous forest. These forests are easier to clear than the rain forests because, even in earlier times, they could be burned during the dry season. Besides this, the rainfall is sufficient to ensure an annual crop yield.

**Questions**

1. Why are tropical ecosystems more sensitive to human intervention than those in temperature latitudes?
2. How is diversity defined?
3. What is the difference between tropical and temperate ecosystems regarding their structure and processes?
4. Do deserts exist at the equator?
5. What are the conditions for forest and snow lines in tropical mountains?
6. Why are soil temperatures (at 50 cm depth) constant in tropical regions?
7. Why does the most competitive species not become dominant under natural conditions?
8. How do plants transport nutrients into their shoots at high humidity?
9. What functional interlinkages stabilise the ecological equilibrium in tropical rain forests?
10. What is the role of soil in tropical rain forests?
11. A tree is pollinated by hummingbirds, fruits are distributed by bats and roots take up nutrients via mycorrhizae. Which of these biological interactions is the most important for the tree?
12. How does the transition in the climate and the types of forests from the equator to the tropic of Cancer occur?

# II Zonobiome of Savannas and Deciduous Forests and Grasslands (Zonobiome of the Humido-arid Tropical Summer Rain Region)

## General

The tropical zonobiome II shows a distinct annual temperature cycle, with heavy zenithal rains in the warm, usually perhumid season whilst the cooler season is arid. ZB II is frost-free in low-lying areas which is similar to ZBI.

In America, this zonobiome covers a large area south of the Amazon basin and smaller areas extending beyond 20°N in Central America. It is also found extra-zonally in Venezuela. In Africa, zonobiome II covers enormous areas on both sides of the equator. In the southern hemisphere, in the uplands of the Zambezi, severe frost damage has been observed in the cold years, and this limits further extension of the zonobiome to the south. The cool tablelands near Johannesburg are largely grassland. In Asia, the main areas occupied by zonobiome II are India and South-East Asia. In Australia, this zonobiome is limited to the northernmost part of the continent (see Figs. 65–70). The humido-arid climate of zonobiome II corresponds to the zonal soils of the euclimatotopes. These store such a large amount of water during the rainy season that they do not completely dry out during the dry season, a prerequisite for the growth of the zonal deciduous forests which, although they significantly reduce transpiration losses in the dry season by shedding their leaves, must still take up a certain amount of water from the soil. Even the leafless twigs and branches lose so much water that the water stored in the trunk does not suffice for the entire dry season.

A distinguishing feature of zonobiome II, however, is the absence of the zonal forest vegetation in large areas where it is replaced by the vegetation type of the **savannas** (or savannahs). There are several different reasons for this (see p. 170f.), one of the most important being the presence of water-impermeable barriers (laterite layers for example) at various depths in the ground. Although their

The humido-arid zonobiome II is characterised by the sharp change from rainy to drought season. Where the drought season is short, deciduous forests often cover a whole area; with longer periods of drought grasslands and thorny savannas predominate, often modified by special soil conditions.

**Fig. 100 A–D.** Vegetation depending on the position of impermeable layers (*black lines*) in the soil; *dark grey* water-saturated soil above the impermeable layer. *T* Termite mound, *R* reeds. Detailed explanation in text (see TINLEY from WALTER & BRECKLE 1991)

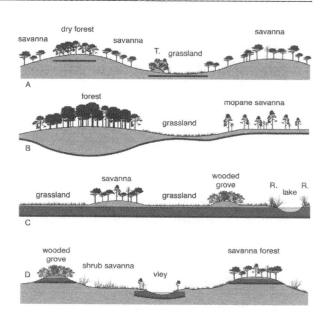

presence has been known for some time, their extensive distribution was first demonstrated by exact soil profile measurements by TINLEY (1982) on a 200-km-long profile in East Africa. He was able to determine the position of the impermeable layers by digging 7-m-deep trenches. These impermeable layers alter the water balance of the soil to such a degree that the development of the zonal forest vegetation becomes inhibited (Fig. 100). Savannas and grasslands are determined by the soil (edaphically) rather than by the climate and are, therefore, to be regarded as pedobiomes.

Laterisation occurs, on the one hand, because of slow dissolution of silicic acid and the accumulation and solidification of round pisolite nodules often consisting of aluminium, iron and manganese oxides which are slowly baked in the sun into a hard layer. On the other hand, leaching processes and soil erosion play a large role (see individual stages in Fig. 101). A wave-like, cement hard surface remains on which plants can hardly grow (see also Fig. 107).

In zonobiome II **peinobiomes** often developed on the old Gondwana shield areas: they are biomes which are influenced by the extreme nutrient deficiency of the old soils.

Leaching over long periods explains another edaphic factor, the low concentration of nutrients often found in the soils of zonobiome II. The land surfaces in Africa, as well as in Australia, western India, and especially in the Brazilian plate of South America are all parts of the Gond-

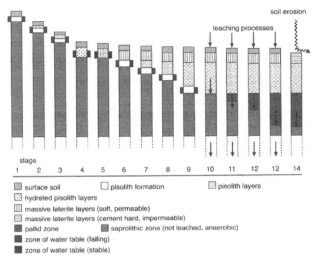

**Fig. 101.** Schematic of the individual stages and processes during laterite formation in variably moist savanna climates, caused by leaching and formation of a cement-hard laterite layer with soil erosion

soil erosion

leaching processes

stage
1  2  3  4  5  6  7  8  9  10  11  12  12  14

☐ surface soil    ☐ pisolith formation    ☐ pisolith layers
☐ hydrated pisolith layers
☐ massive laterite layers (soft, permeable)
☐ massive laterite layers (cement hard, impermeable)
■ pallid zone    ■ saprolithic zone (not leached, anaerobic)
■ zone of water table (falling)
■ zone of water table (stable)

wana shield, the ancient continent which separated into the respective continents millions of years ago (in the Mesozoic). The land has never been covered by oceans and the soils are ancient, never having been rejuvenated by ocean sediments. The exposed rock surfaces have been constantly subject to leaching and erosion. Wherever recent volcanic rock is absent, the weathered products of which the soil is composed have been leached out substantially so that the soils are too deficient in the essential minerals required by plants (phosphorus, trace elements) for a forest to be able to develop (see p. 186, campos cerrados).

On the large plateaux the scarcely noticeable deeper areas are flooded during the rainy season and their soils are waterlogged. Wooded groves only grow on the somewhat higher surfaces which are not flooded, while tropical grasslands occupy the wet areas. The result is a mosaic-like parkland with wooded groves and grasslands which are not savannas in an ecological sense. Savannas are ecologically homogeneous plant communities of scattered woody plants within a relatively dry grassland (see pp. 170). Most geographers employ the term savanna in a much wider sense.

Three types of vegetation, therefore, are characteristic for zonobiome II: (1) zonal deciduous forests, (2) relatively dry savannas and (3) parklands with a wet rainy season. Many laterite crusts and layers are of a fossil nature, having been formed in the Pleistocene, the geological

**Table 12.** Floristic diversity of some Neotropical savanna regions

| Region | Area (km²) | Number of species trees and shrubs | Number of species semi-shrubs and herbs | Number of species grasses | Total number of species |
|---|---|---|---|---|---|
| Llanos in Columbia | 150,000 | 44 | 174 | 88 | 306 |
| Llanos in Venezuela | 250,000 | 43 | 312 | 200 | 555 |
| Cerrados in Brazil | 2,000,000 | 429 | 181 | 108 | 718 |

(after SARMIENTO 1996)

period before the present, which was subject to several glaciations. In the desert zone of the Sahara, these ice ages resulted in pluvial periods with heavy rainfall. Recent pollen analyses have revealed that during that time the tropical zones experienced dry periods extending into zonobiome I. This resulted in the formation of laterite crusts and layers and relict savannas which can still be found surrounded by evergreen forests. The biodiversity in savannas is considerably lower than in tropical rain forests. Some examples of neotropical areas are given in Table 12.

## Zonal Vegetation

Zonobiome II is divided into dry and wet subzonobiomes, depending on the length of the dry season and the amount of annual precipitation. The respective climate diagrams for India are presented in Fig. 98. It is not useful to specify certain limiting climatic values for each of the continents, since the respective conditions are quite variable.

Corresponding to the climate, zonal tropical deciduous forests are distinguished as either wet or dry types. The zonal soils are not yet well enough understood to specify general characteristics to differentiate between wet and dry forests. Like the soils of zonobiome I, they belong to the red-ferrallitic group. Leaching of $SiO_2$ from the soil, which is wet only during the warm rainy season, however, is not as acute. In zonobiome I, the ratio of $SiO_2$ to $Al_2O_3$ is below 1.3, while in zonobiome II it is between 1.7 and 2.0. The absorbability of the zonal soils is also somewhat greater, meaning that they are more capable of retaining ions (by adsorption), and are, therefore, not quite so poor in nutrients necessary for plants.

The most conspicuous difference between the zonal vegetation of zonobiomes II and I is the shedding of foliage in an annual cycle. In every climatic zone, a type of foliage has developed which guarantees the greatest production under the respective climatic conditions. Leaves are always short-lived structures, since they age rapidly, and soon lose the ability to assimilate $CO_2$, which is their main purpose. The rapid ageing process is probably caused by the accumulation of 'ballast', which reaches the leaves dissolved in the transpiration stream, and by metabolic side products, such as tannins, alkaloids, terpenes etc. These may protect the leaves from herbivores.

The evergreen trees of zonobiome I also shed their old leaves soon after the younger leaves have become functional. Some species of zonobiome I have been found to be evergreen in years of plentiful rain, while during unusually dry periods they shed their leaves before the new leaf buds have sprouted and are, therefore, leafless for a short time. In zonobiome II, a long dry period is normal and the rainy season, in contrast, is extremely humid. Tree species consequently develop very large thin drought-sensitive leaves at the beginning of the rainy season, for which less material is required per unit leaf surface than for the thick leathery leaves of the species of zonobiome I. Although the thin leaves assimilate $CO_2$ only during the rainy season, the saving of organic material is advantageous considering the annual production balance. Besides the leaf surface area, the rate of assimilation (which is higher in thin leaves) is also decisive for the assimilation of $CO_2$ (i.e. the production of organic material).

Uptake and loss of water by trees in zonobiome II is well balanced during the rainy season, since the diurnal transpiration curve and the evaporation curve run parallel and have almost no midday depressions, which are always an indication of incipient water deficiency. The osmotic potentials of the leaves of different species are high, ranging from -0.7 to -1.9 MPa. At the onset of the dry season, the sugar concentration in the cells of the leaves increases six-fold (a decrease of 0.2 MPa in osmotic potential). Soon afterwards, the leaves turn yellow and become dry.

Frost damage has been observed in especially cold years in the southern limits of zonobiome II in southern Africa (ERNST & WALKER 1973). The growth of annual shoots and the development of leaves begin after the arrival of the first rains (Fig. 102). A striking factor, however, is that the buds of the blossoms of many tree species open before the first rain. Since the petals possess only a very low cuticular transpiration, they are subject to only an insignificant water loss. On the other hand, insects are

**Fig. 102.** *Colophospermum mopane* forest coming into leaf at the beginning of the rainy season near Victoria Falls (Photo: E. WALTER)

able to pollinate the blossoms of the leafless forest more easily. The factor responsible for the induction of flowering is most probably the maximum of the temperature curve, which occurs at the end of the dry season and before the onset of the rainy season.

The most expansive forests of zonobiome II are situated south of the equator in the less populated regions of Africa. These are the "Miombo" forests, located on the watershed between the Indian and Atlantic Oceans, and on the Lunda threshold south of the Congo basin, where there is no drinking water for settlements during the dry period. A conspicuous sight in the dry limit of zonobiome II is the baobab (*Adansonia digitata*) with its bizarre trunk, which may attain a circumference of more than 20 m (Fig. 103) and is able to store up to 120,000 l of water. It may, therefore, be assumed that such plants are able to survive the dry season in a leafless condition without the uptake of water from the soil. In South America, as well as in Australia, there are other baobab trees from the same family (Bombacaceae).

For information on the production see CANNEL (1982) and also Table 13.

MEDINA (1968) determined the soil respiration in a deciduous forest 100 m above sea level (mean annual temperature: 27.1 °C, annual precipitation: 1334 mm) in Venezuela. Respiration was three times as intensive during the rainy season as in the dry season. This corresponds to an average amount of catabolised organic substance of 11.2 t ha$^{-1}$ a$^{-1}$. The litter amounted to 8.2 t ha$^{-1}$ a$^{-1}$. The difference may be attributed to root respiration.

**Fig. 103.** A very large baobab (*Adansonia digitata*) east of Tsumeb (Namibia) (Photo: S.-W. BRECKLE)

Data from Thailand are given by OGAWA et al. (1971) for the following forest types:

1. A dry forest of Dipterocarpaceae at an altitude of 300 m with an open growth of trees about 20 m tall and a covering of grass 20–30 cm in height and
2. a mixed wet-deciduous forest with trees 20–25 m tall and a sparse covering of grass.

The following values were obtained for the phytomass and the annual primary production in t ha$^{-1}$ (LAI, leaf area index):

| Type of forest | Phytomass | Production | LAI |
|---|---|---|---|
| 1 | 65.9 | 7.8 | 4.3 |
| 2 | 77.0 | 8.0 | 4.2 |

**Table 13.** Quantitative comparison of two dry forests

1. Open miombo forest in Zaire (11°37'S, 27°29'E, 1244 m above sea level)
   Tree species: *Brachystegia, Pterocarpus, Marquesia* etc.
   Soil: latosols
   LAI: 3.5
   Phytomass (aboveground): 144.8 t ha$^{-1}$ (of these leaves: 2.6 t ha$^{-1}$)
   Phytomass (belowground): 25.5 t ha$^{-1}$ (estimated)
   Net production: litter, 4–6 t ha$^{-1}$ a$^{-1}$
   Wood production not determined

2. Dry monsoon forest in India (24°54'N, 83°E, 140–180 m above sea level)
   Tree species: *Anogeissus, Diospyros, Budenania, Pterocarpus* etc.
   Soil: reddish-brown, lessivated sandy loam
   Phytomass (aboveground): 66.3 t ha$^{-1}$ (of these leaves: 4.7 t ha$^{-1}$)
   Phytomass (belowground): 20.7 t ha$^{-1}$ (estimated)
   Annual net production:
   Trunks and branches: 4.40 t ha$^{-1}$ a$^{-1}$
   Leaves: 4.75 t ha$^{-1}$ a$^{-1}$
   Undergrowth: 0.35 t ha$^{-1}$ a$^{-1}$
   Roots (estimated): 3.40 t ha$^{-1}$ a$^{-1}$

Deciduous forests are exploited by the population for shifting cultivation, 3–5 years at a time. A secondary forest grows up on the fallow patches within about 10–20 years. Trees apparently do not live more than 100 years.

## Savannas (Trees and Grasses)

As previously mentioned, the only ecosystems to be properly designated as savannas are tropical grasslands in which scattered woody plants exist in competition with the grasses (Fig. 104).

Grass and woody species are ecologically antagonistic plant types, one usually excluding the other. Only in the tropics, where both summer rain and a deep, loamy sand coincide, are they found existing in a state of ecological equilibrium. The cause of the antagonism is in differences in (1) their root systems and (2) their water economy.

1. Grasses possess a very finely branched, intensive root system which densely permeates a small volume of soil. Such a root system is especially suited to fine, sandy soils with an adequate water capacity in regions of summer rain where the ground contains plenty of water during the growing season.
   Woody species, on the other hand, have an extensive root system with coarse roots which extend far into the

**Fig. 104.** *Acacia detinens* savanna in Namibia. The grass layer has dried out after the rainy season. It appears as if there were a forest growing in the distance, although it is actually only the same savanna

soil in every direction, thus penetrating a much larger soil volume, but less densely. This type of root system is well suited to stony soil where the water is not uniformly distributed, not only in summer rain regions, but also in the winter rain regions where the water which has seeped down to great depths has to be drawn up again by the roots in summer. For this reason, grasses are unsuited to winter rain regions.

2. As far as the water economy is concerned, it is characteristic of grasses that given sufficient water they transpire very strongly and photosynthesis proceeds very intensively. Within a short period their production is very large. At the end of the rainy season when water becomes scarce, transpiration does not slow down, but continues until the leaves, and usually the entire aerial shoot system, dies. Only the root system and the terminal growing point of the shoot survive. The meristem tissue is protected by many layers of dried-out leaf sheaths and is capable of surviving long periods of drought, even though the soil itself is practically desiccated. Growth recommences after the first rain.

The water economy of woody plants, on the other hand, with their large system of branches and many leaves is well regulated. At the first indication of water scarcity, the stomata close and transpiration is radically reduced. If the lack of water becomes acute, then the leaves are shed; during the dry season, only the branches and buds remain. Although well protected against loss of water, these remaining parts have been shown to exhibit a very small but demonstrable loss of water over the

**Fig. 105.** Schematic presentation of transition from
(**A** and **B**) grassland to (**C**)
shrub savannas and (**D**) tree
savannas. Explanation in
text

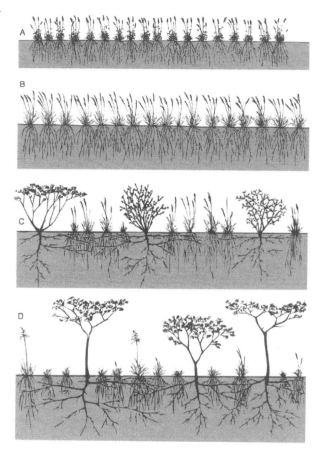

course of several hours. The water reserves in the wood
are insufficient to compensate for water losses over a
lengthy dry period, which means that woody plants are
obliged under such conditions, to take up a certain, albeit small, quantity of water. If there is no available
water in the soil, however, then they dry out and die.

This consideration of these differences helps us to understand the ecological equilibrium of the savannas. As an
example, an area in **Namibia** was chosen with gradually
increasing summer rainfall, uniform relief, and a fine,
sandy soil which takes up all of the rainwater and stores
the larger part of it (see Fig. 105).

This transitory region between zonobiome II and the desert with summer rain is designated as zonoecotone II/III, in which climatic savannas occur with an annual precipitation of 500–300 mm and an 8-month-long dry season.

Where the annual rainfall amounts to only 100 mm (see Fig. 105 A) and the water cannot penetrate very far down into the ground, the roots of the small tufted grasses which permeate the upper, wet soil layers use up all of the stored water and then dry up at the end of the rainy season, with the exception of the root system and the apical meristem. Woody plants cannot survive because the soil offers no available water during the dry season (**semi-desert**). At a rainfall of 200 mm (B) the situation is similar: the soil is wet to a greater depth and the grasses are larger but still use up all of the water (**grassland**). Only when the rainfall reaches 300 mm (C), is some water left in the soil by the grasses at the end of the wet season, and although this is insufficient to keep them green, it is enough to form a **shrub savanna**. If the annual rainfall is 400 mm (D), then the larger amounts of water remaining in the soil at the end of the summer rainy season support solitary trees, and a **tree savanna** is formed. However, it is still the grasses which are the dominating partner, they determine how much water is left over for the woody plants and the amount can vary from year to year.

Only when the rainfall is sufficient for the crowns of the trees to link up to form a canopy, shade of which prevents the proper development of grasses, is the competitive relationship reversed. It is now the woody plants which are the dominant competitors in the savanna woodland or dry tropical-deciduous woodland, and grasses are obliged to adapt themselves to the light conditions prevailing on the ground. Such a labile equilibrium in the savanna is readily disrupted by man when he begins to utilise the land for grazing purposes. Water losses due to transpiration cease when the grass is eaten off, so that more water remains in the soil, to the advantage of the woody plants (mostly *Acacia* species), which can consequently develop luxuriantly and produce many fruits and seeds. They are distributed in the dung of the grazing animals which eat the pods. The seedlings are not exposed to competition from grass roots. The predominantly thorny shrubs grow so densely that thorny shrub land is formed, which is then useless for grazing purposes.

In all grazing areas that are not rationally utilised, there is great danger of such a **bush encroachment,** and it is for this reason that thornbush as a substitute plant community is nowadays more widespread than the cli-

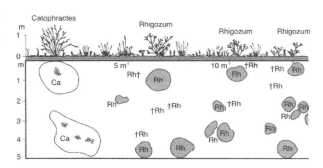

**Fig. 106.** Transect (1 m wide) through a typical patch of vegetation near Voigtsgrund (Namibia). Grasses dry out during the dry season. *Below*: Ground plan of plant cover (without grasses). *Ca*, *Catophractes*; *Rh*, *Rhigozum* († dead)

matic savanna in, for instance, the arid parts of India and in northern Venezuela and the offshore islands (Curacao and others). If the area is more densely populated, and if the woody plants are used for fuel or for making protective thorny hedges around the kraals, then a man-made desert is often produced, which has only a covering of annual grass during the rainy period. The cattle starve during the dry season, with nothing but the straw remnants of grasses as fodder. Such is the situation, for example, in the Sudan. The only remaining natural savannas are in central Argentina at a rainfall of 400–200 mm with *Prosopis* as the woody species.

The transition is of a different nature on stony ground. The grasses are unable to compete with the woody plants and are absent. With decreasing precipitation, the woody plants become smaller and further apart. Each shrub requires a larger root area, and the roots run near the surface since rain moistens only the upper layers of the soil. In approaching zonobiome III, only a few small dwarf shrubs showing xerophyllous adaptations remain (dwarf shrub semi-desert). A special situation is encountered in two-layered soils such as are found in Namibia, where a shrub savanna is found, although annual precipitation is only 185 mm (Fig. 106).

If the soil were deep and sandy, pure grassland would be expected at such low rainfall. In fact, the soil profile reveals that beneath the 100–200 mm of sand there is sandstone of the Fish River formation, which is arranged either in thin layers with small cracks or forms thick banks with larger crevices. The upper sandy layer cannot retain all the rainwater and part of it seeps into the cracks in the sandstone. The grasses can utilise the water from the sandy layer and the roots of the shrubs penetrate into the sandstone layer and take up the water from cracks. Water reserves in the cracks of the thin-layered sandstone

are sufficient only for the small *Rhigozum* bush, whereas
the larger *Catophractes* shrub flourishes in the crevices of
the thicker-banked sandstone. The distribution of bushes
reflects the structure of the sandstone even in places
where the covering layer of sand is missing. Bushes compete with one another and although both types can germinate in the larger crevices, the smaller bushes are, in time,
ousted by the larger ones and only their dead remains are
left. There is no competition between grasses and woody
plants in this case.

In zonoecotone II/III, zonal savannas occur in place of
deciduous forests where the annual precipitation is too
low for the latter. In zonobiome II, however, savannas are
found wherever (in spite of sufficient precipitation) the
soil does not contain enough water for the survival of a
forest during the dry season. On the other hand, too
much water during the rainy season results in a water-
logged soil and excludes woody plants. In such a case, a
pure grassland develops which can dry out during the dry
season and which is typical of the parklands. This is illustrated in Fig. 100 A–D, where A demonstrates a slightly
hilly area of the northern Kalahari with deep sand. Relatively little water is retained at field capacity, so that a
large proportion of the rain seeps through the soil and the
retained water alone is only sufficient for savanna vegetation. In places, laterite pans or layers (black lines in
Fig. 100) are present as barriers, thereby preventing water
seepage. If the layer is relatively deep (left), a dense
wooded grove or dry forest develops on the moist sand. If
the layer is not so deep in a lowland area (centre), the
ground above it is waterlogged, only allowing the growth
of grass, with the exception of the termite mound (T)
which is more adequately drained and suitable for the
growth of trees.

A sandy soil with an uninterrupted laterite layer which
forms a basin is shown in Fig. 100 B (left); the basin collects water by lateral seepage. The soil is well-aerated and
is wet deep underground, providing favourable conditions
for the zonal deciduous forest. In the centre is a lowland
with grassland which is flooded during the rainy season.
On the somewhat higher level on the right, on an alkaline
clay soil, the grassland is interspersed with species of
woody plants adapted to heavy soils (*Mopane, Balanites,*
flute acacia). In contrast to this, Fig. 100 C shows the uninterrupted laterite layer at a uniform distance below the
surface. The soil above it is saturated with water during
the rainy season (heavily shaded) and covered with grassland. The soil is somewhat better drained on the smaller
elevations, which are covered with tree savanna (left) or

**Fig. 107.** Repeated drying cycles of the B-horizon (in I) and loss of the topsoil hardens the plinthite layer, which is rich in iron-oxide, to "iron stone" (laterite; after SCHULTZ 1990, from THOMAS 1974). The hard laterite layer may "cement" the runoff surfaces and so lead to formation of steps (II) on mountain slopes and in formation of table mountains (mesas)

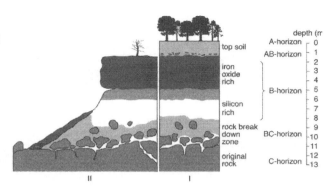

with wooded groves if there is sufficient root space. To the right is a basin with an open groundwater surface and a zone of reeds (R). A relatively dry sandy region with savanna vegetation is shown in Fig. 100 D. Woody plants dry out down to the soil during the drought season, and grow again from the base of the stem or sprout from the roots during the rainy season. Depending on the availability of water, laterite layers at various depths below higher ground support the development of wooded groves or savanna forests. Valleys are characterised by vleys with swamp vegetation which may dry out during the drought season, and sparse tree growth on the edge of the laterite layer along laterally flowing water. This illustrates how the presence of forest and savanna is dependent on the position of the laterite layers, and how the development of parklands depends on water-saturated soils.

The hard layer of the B horizon often becomes exposed through erosion. This leads to the formation of **table mountains** (mesas). At the edge of these mountains the formation of steps leads to steep precipices, the development of which is shown in Fig. 107. The presence of savannas on nutrient-poor soils will be discussed later (p. 185).

Other factors favouring the development of savanna vegetation are **fire** (see p. 102), herds of large herbivores and various human activities. In zonobiome II, fire was already an effective natural factor before the appearance of man. The rainy season usually commences with storms. The abundant dry grass present at this time of year is easily ignited by lightning. Large numbers of pyrophytes (fire-resistant woody plants) attest to the frequency of such fires. These trees or shrubs often possess thick bark which becomes scorched while the cambium remains protected. Many shrubs have dormant buds located below the soil, which sprout when the parts of the plant above ground are burned off. A number of plants have lignified underground storage organs (lignotubers) which allow for rapid regeneration.

In prehistoric times, primitive man almost certainly started **grass fires** in order to protect his settlements from the effects of fire caused by lightning. The tall grass of the wetter zones permits fire to spread much more rapidly and with a greater ferocity than elsewhere. Nowadays, it is common practice to burn the grass in the dry season to facilitate big-game hunting, to destroy vermin (but also snakes etc.). After a grass fire, the grasses sprout earlier, which is initially advantageous for grazing. Grass fires only penetrate dry forests with an undergrowth of grass, but they can also push back the limits of the wet forests. Above all, they prevent the forest from winning back ground that has been cleared and turned into grassland.

In the north of Venezuela fire also plays an important role, besides water supply and distribution of nutrients, in maintaining the equilibrium between *Byrsonima crassifolia* (with a very low density) and the perennial C4 grasses *Trachypogon vestitus* and *Axonopus canescens* (Fig. 112). However, the equilibrium between those two grasses is very labile, as shown by INCHASTU (1995) in transplantation experiments. Only under absolute protection from fire is *Trachypogon* gradually replaced by *Axonopus*.

A very important factor for savannas is **grazing** by large animals (ANDERSON et al. 1973). Growth of young trees is inhibited by their grazing, stripping of bark, breaking branches and trampling. Elephants are especially destructive to forests. They pull out trees by the roots or strip off the bark. Elephant paths cause clearings, allowing grass fires to penetrate into the forest. On average, one elephant is capable of destroying four trees per day. The loss of trees in the miombo forests can be as high as 12.5% annually. The number of elephants in the national parks is increasing rapidly. Murchison Park, on Lake Albert is becoming increasingly deforested by elephants. In Serengeti Park, however, the damage caused by wildlife and the regeneration of the vegetation appear to be in equilibrium. It is interesting to note that many woody plants of the savannas of Africa, with their high wildlife populations, are thorny. However, this is not true for South America, where wildlife is less abundant. This indicates the selection of species protected from grazing by wildlife. Wildlife paths influence the vegetation indirectly by exposing surfaces to channel erosion. This is especially true for hippopotamuses, which climb up the banks of the rivers at night to graze on the grass. The erosion channels may

All influences such as fire, grazing or deforestation, either as a result of shifting cultivation or for collecting fire wood, destroy forests.

lead to drainage of a wet grass area, making the advance of woody plants possible. A summary of the complex effects of big game was presented by Cumming (1982).

The influence of man is even greater, whether he functions as animal breeder or farmer. The use of the savanna north of the equator for grazing began at least 7000 years ago. Forests remain only as small relics in this region, and it may be assumed that a large portion of the savannas are of a secondary nature (Hopkins 1974).

In conclusion, the following types of savanna may be distinguished:

1. Fossil savannas, which arose early under other circumstances, such as in the range of zonobiome I.
2. Climatic savannas in the range of zonoecotone II/III, with an annual precipitation of less than 500 mm.
3. Edaphic savannas which are determined by the soil conditions in zonobiome II
    a) On soils in which the water balance is altered by the presence of impermeable barriers (laterite layers, clay layers, dense silt or sand layers), and is less favourable for trees than could be expected, based on the amount of rain alone;
    b) on soils which are primarily so deficient in nutrients that they cannot support a forest (p. 185 f.);
    c) within parklands where soils are saturated with water during the rainy season, resulting in palm savannas (p. 179).
4. Secondary savannas caused by fire, the effects of large wildlife species and human interventions.

The determination of the respective type of savanna cannot be achieved by superficial observation. Instead, thorough investigation is necessary.

## Parkland

Parkland usually develops in very flat country of zonobiome II. Its formation is due to the minute differences in relief, which are inconspicuous during the dry season. During the heavy rains of summer, all the slightly deeper depressions are flooded, the water taking months to drain away. This biotope is covered by grassland; the soils are grey, whereas the slightly more elevated areas which are not flooded and to which the woody plants are confined, have a thick layer of red, sandy loam. Rivers begin here on the watershed in barely perceptible grass-covered narrow strips which gradually unite and become deeper and

**Fig. 108.** "Termite savanna", periodically flooded grassland with trees on old termite mounds, in northwestern Kenya (Photo. E. WALTER)

deeper, with an increasing gradient, to form streams and river beds (easily recognisable from the air).

A peculiar variant is seen in the **termite savanna**, which consists of large grassy hollows from which the deserted termite heaps rise like islands. Since these mounds are never submerged, they are covered with trees, so that the scenery is a mosaic of different plant societies (Fig. 108), and is, therefore, not a true savanna.

Deeper hollows with black clay soils, designated as "**mbuga**", are a special type of amphibiome with an alternately wet and dry soil, and a hard layer of iron concretion at a depth of 50 cm. Since the potential evaporation far exceeds the 1000 mm of yearly rainfall, the clay soil dries out in August to December to a depth of 50 cm and is split polygonally by deep cracks. Biotopes of this kind are unsuitable for trees, which only thrive where the water table does not rise above a depth of 3 m. At this depth there is a laterite layer, to which the roots of the trees extend.

In contrast to the termite savanna, the palm savanna is a homogeneous plant community. Palms are woody monocotyledonous plants, with a bundle-like root system with uniform, almost branchless roots which extend radially, so that palms are situated individually in the grassland. They tolerate occasional flooding. Although there is no available information on the subject, the soils of palm savannas probably dry out as much as those of pure grasslands (see "Palmares", p. 184), but no research has been done on competition between palms and grasses.

Trees usually stand far apart from each other as isolated individual trees in very open savannas. BELSKY & CANHAM (1994) compared this situation with gaps in forests (Fig. 109). Both are examples of idiotopes. Areas in-

**Table 14.** Comparison of some processes related to gaps in closed forests and to individual trees in grassland (comparison of idiotopes)

| | Gap | Isolated tree in savanna |
|---|---|---|
| Development | Usually suddenly (episodic storm event) | Slow (germination, establishment of seedling) |
| Growth | Very rarely because of fallen branches or stems of neighbouring trees | Gradual by enlarging the crown |
| Disappearance | Usually fast by growth of neighbouring trees | Could be very sudden because of tree death |
| Period of time | Short (5–30 years) | Long (life time of tree, usually well over 50 years) |
| Dynamics of resources | Usually only short additional release of nutrient | Usually continuous preference by wild animals and supply of external detritus |
| Secondary succession | Only in larger gaps | Very rarely recognisable |
| Ecological effect on the environment | Short (5–20 m) | Larger (50–100 m) |

(after BELSKY & CANHAM 1994)

fluenced by a fallen tree and the dynamics of crown closure, on the one hand, and the dynamics in a grassland dominated by individual trees and their development to a group of trees or to grassland without trees, on the other hand, are compared in Table 14 as contrasting structural elements.

## Examples of Larger Savanna Regions

Expansive savanna-like vegetation communities are found not only in Africa north and south of the equator, but also in South America on the Orinoco, in central and eastern Brazil and in the Chaco region. Vegetation changes drastically along the gradient from perhumid tropical rain forests to extremely dry tropical deserts; but also the significance of the relevant life forms is very different. ELLENBERG (1975) demonstrated this in a schematic principally applying to the lower regions of the Andes, but also applicable in principle to other regions with similar gradients (Fig. 110). Some examples of expansive savannas are explained in the following.

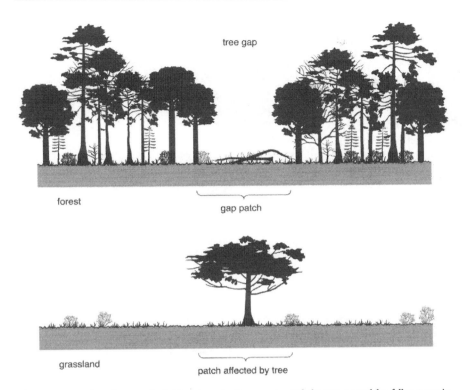

**Fig. 109.** Comparison between individual trees in the savanna and the gap caused by fallen trees in the rain forest shows remarkable parallels in the expression of "island biotopes" (after BELSKY & CANHAM 1994)

## a The Llanos on the Orinoco

The llanos, on the left bank of the Orinoco in Venezuela 100 m above sea level, occupy a basin 400 km in width that was left by a Tertiary sea, and extend for a distance of 1000 km into Columbia. The inflowing rivers have filled the basin with weathering products from the Andes. In the central Llanos around Calabozo (see diagram in Fig. 90), the climate is very typical of zonobiome II: annual precipitation amounts to more than 1300 mm, the rainy period lasts 7 months, and the dry period lasts 5 months. With a climate of this type, wet-deciduous forests would be expected, but they are only present as "**matas**" or small, scattered woods. Apart from this, the llanos bordering the river are pure grasslands, which are flooded in the rainy season, as is usual in zonobiome II (with

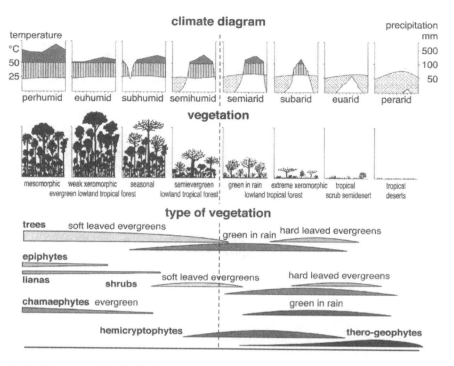

**Fig. 110.** Schematic of the distribution of vegetation in the lowlands of the Andes together with the climate gradient (after ELLENBERG 1975)

trees only on the banks, as gallery forests). Apart from this, the area is covered with grass about 50 cm in height, scattered with small trees (*Curatella, Byrsonima, Bowdichia*) and is in fact a true savanna. Since this savanna cannot be climatic in origin (the rainfall is too high), edaphic factors such as soil conditions must be responsible.

The assumption has often been made that this is an anthropogenic savanna resulting from the use of fire, but this explanation is too simple and uncritical since the savanna existed long before the arrival of the white man. It was neither cultivated nor used as grazing land by the Indians. Fires caused by lightning are common occurrences in grasslands and although the Indians probably often burned down the dry grass, this was only possible because natural grassland already existed. Fire has certainly played its part in forming the savanna, insofar as only fire-resistant woody species could survive in the grassland and on the fringe of the matas, but it is by no means the

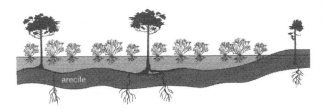

3,85 m     Ground water after the end of the rainy season

2 m

5,85 m     Ground water before the beginning of the rainy season

**Fig. 111.** Scheme to explain the situation in the llanos north of the Orinoco. The changing water table below the arecife can only be reached by plants with deep roots (after WALTER 1990)

primary reason for the existence of these enormous grassy expanses. In the central Llanos it has been shown that at a time when the groundwater table in the basin was very high in this region, a laterite layer was formed which was cemented by ferric hydroxide. This is called "**arecife**" (Fig. 111). Arecife runs beneath the surface at varying depths (mainly between 30 and 80 cm), rarely below 150 cm, and can sometimes also be seen on the surface because of erosion. The statement that arecife is impermeable to water cannot be correct in this case, since the 750 mm of rain falling during only 3 months of the summer rainy season cannot possibly be absorbed by the soil overlying the arecife. In such a completely flat region, flooding would inevitably result, but this does not occur. The reddish colour of the soil is another factor which speaks against prolonged water-logging. The groundwater table beneath the arecife has in fact been found to rise from –5.75 to –3.85 m, a rise of almost 2 m, by the end of the rainy season. Assuming a pore volume of about 50% for the alluvial deposits, this would mean that about 300 mm is retained by the soil above the arecife and 1000 mm seeps through. Arecife laid bare by erosion on the river banks shows quite clearly that there are irregular channels penetrating the hard layer in places. This suggests that grasses can take root in the fine soil above the arecife and use up the 300 mm of stored water in the course of their development. Woody plants, however, grow wherever their roots can find a way through the arecife layer to the damp, underlying rock layers where adequate water supplies are available. Groups of trees can grow wherever the channels through the arecife are large enough or where a number occur close together. However, small woodlands exist only in isolated patches where the arecife is either entirely absent or extremely deep. In such places deciduous forests are found which are, in fact, the type of vegetation corresponding to the regional climate. These savannas can thus be considered as a stable, natural plant community in which the distribution of the trees is a reflection of the arecife structure. The following facts support this interpretation:

1. Wherever the arecife is at the surface, grass cover is completely wanting, but solitary trees grow on it at rather large intervals. The roots must in such cases reach down below the arecife through existing channels.

2. *Curatella* remains green throughout the dry season in contrast to other woody plants in the typical savannas. This indicates that the water supply is sufficient throughout the year. Transpiration measurements have demonstrated that a single tree transpires approximately 10 l per day during the dry period. Since the topsoil is dry at this time the water must be provided from the layer below the arecife. The same holds true for other species of woody plants.

3. Small woods (matas) grow in places where arecife is locally absent and their roots can penetrate unhindered into the ground.

Final proof would be afforded only by excavating the roots over larger areas, which would be difficult to do. The use of dynamite on the arecife would certainly favour woodland expansion. Scattered over the savannas of the Llanos are slight depressions into which the water drains after a heavy downpour (1961: 38 mm in 20 min) and in which sediments of grey clay accumulates. The water in the depressions reaches a depth of about 300 mm during the rainy season, but toward the end of the dry season, the grey soil dries out completely.

This **alternating wetness and dryness** is well tolerated by certain grasses (for example *Leersia, Oryza, Paspalum*), but not by tree species other than palm trees. This results in the formation of "palmares," i.e. **palm savannas** (also widespread in Africa), which are grasslands containing the palm *Copernicia tectorum*. Such areas often burn, but the palms are resistant to fire (as are the tree ferns) because they do not possess easily damaged cambium. The dead leaves sheathing the trunk are burned and the outermost vascular bundles are charred, but the resulting layer of carbon acts as an insulation against later fires. The apical meristem, surrounded by young leaves, survives. Old leaves completely missing from the trunk are a sign that the palm savanna has been burned recently. If the leaves sheath the trunk down to the ground, then the palm has

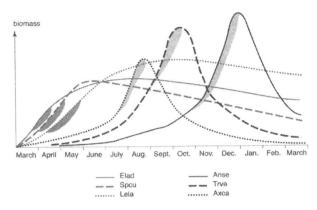

**Fig. 112.** Annual rhythm of green biomass of six dominant grass species of the Venezuelan llanos after the usual March fire shows a tight temporal niching of grass species (*hatched* flowering period) (after SARMIENTO 1996). *Anse = Andropogon semiberbis; Axca = Axonopus canescens; Elad = Elyonurus adustus; Lela = Leptocoryphium lanatum; Spcu = Sporobolus cubensis; Trve = Trachypogon vestitus*

not been exposed to fire at all. If only the lower portion of the trunk is bare, this means that the palm has had a chance to grow for several years since the last fire.

Some of the water must drain away from the areas on which palms grow because otherwise, with a rainfall of 1300–1500 mm and a potential evaporation of 2428 mm (i.e. a negative hydrological water balance), the soil would become brackish. On permanently wet areas, the palm *Mauritia minor* is found. Black, acid, peaty soils are formed on which a few grasses, *Rhynchospora, Jussiaea, Eriocaulon*, and the insectivorous *Drosera* species (sundews) are found. Areas of this kind have to be considered as amphibiomes, a special form of helobiome, just like the alternately wet and dry grassland mentioned above.

Today, burning takes place annually to improve grazing areas which, of course, synchronises different species of grasses in their development demonstrating clearly the periodic niching of biomass production (Fig. 112).

Further east, the llanos are succeeded by a plain with sandy deposits from the Orinoco, which at one time changed its course here to the north, flowing through the Unare lowlands and emptying into the Caribbean Sea. The often white quartz sands are weathered products of the quartzite sandstones of the Guayana table mountains, and correspond to those of the Brazilian shield, which are also low in nutrients. Similarly leached quartz soils occur on other areas of the old Gondwana shield. These savannas and to some extent pure grasslands, were probably induced by the same factors as the Campos Cerrados, which will be discussed next.

## b Campos Cerrados

The Campos Cerrados is a region with a savanna-like vegetation covering an area of 2 million km² in central Brazil (EITEN 1982). The 4–9 m high trees cover between 3 and 30% of the total surface. The climate is characterised by a 5-month-long dry season and an annual precipitation of 1100–2000 mm. RAWITSCHER (1948) presented the first information on the water balance of these savannas and demonstrated that the deep soil is constantly wet at a depth of only 2 m. This provides the deep-rooting species of woody plants with plentiful water, so that they remain evergreen and transpire heavily during the dry season. Only the grasses and shallow rooting species dry out or shed their leaves during the dry season. The soils are weathered products of the granite and sandstone of the Brazilian shield and are very poor in nutrients, especially phosphorus (but also potassium, zinc and boron). This was shown by the application of mineral fertiliser to such crops as cotton, corn and soybean. That fact that the nutrient deficiency, rather than the water factor, is responsible for inhibiting the growth of the zonal deciduous forest is demonstrated by the observation that a semi-evergreen forest may be found on a basaltic soil in the vicinity of São Paulo. The Campos Cerrados is burned regularly and the presence of a large number of pyrophytes is an indication that fire has been a natural factor in this region for ages. Although they are not the actual reason for the lack of forest vegetation, fires do reduce the density of the wooded stands (COUTINHO 1982).

## c The Chaco Region

The Chaco region is situated in the westernmost part of zonobiome II in South America, an expansive plain between the Brazilian shield in the east and the Andean foothills in the west, which spreads across southern Bolivia, most of Paraguay and far into western Argentina, thereby covering an area of 1500 km from north to south with an average width of 750 km (HUECK 1966). The central region of the plain is only 100 m above sea level.

During the heavy summer rains (annual precipitation: 900–1200 mm), large portions of the plain are flooded, especially in the east. Here, a parkland with forests, expansive periodically flooded grasslands, palm savannas and swamps are found. The western section, in Argentina, is heavily overgrown with shrubs, and salt pans with the halophytes *Allenrolfea* and *Heterostachys* are also present. The southern Chaco is succeeded by the pampas. The relief is quite level and there are water-impermeable layers in the soil. The vegetation is primarily a *Prosopis* savanna with a grass cover of *Elionurus muticus* and *Spartina argentinensis*. Predominant tree species of the Chaco forests are tannin-rich Quebracho species, *Aspidosperma quebracho-blanco* (for example *Schinopsis quebracho-colo-rado* and *S. balansae*). *Trithinax campestris* is a common palm species, while *Copernicia alba* is typical in wet depressions.

The mammalian fauna is relatively sparse. *Myrmecophaga tridactyla* and *Tamandua tetradactyla* are termite eaters. The carnivores are represented by the jaguar (*Leo onca*), the puma (*Felis concolor*) and a number of smaller species. Rodents are numerous. The trees are inhabited by the tree sloth (*Bradypus boliviensis*), three species of monkey (*Cebidea*), the tree porcupine (*Coenda spinosus*), the mustelid (*Eira barbara*), a number of insectivores and fruit or nectar-eating bats, as well as the vampire bat (*Desmodus rotundus*). The only bird species to be mentioned here is the large flightless *Rhea americana*. Reptiles are represented by two rare cayman species, three turtle species, several poisonous snakes (25 snake species in all) and various lizards. Up to 30 species of anurans have also been identified. Also present is a large number of invertebrates. Research on the ecosystem is still lacking. The main influences of humans are deforestation and grazing, which may lead to brush overgrowth. A brief summary with references by BUCHER (1982) is available.

## d Savannas and Parklands of East Africa

This huge area at the base of the large volcanoes with the giant crater Ngoro-Ngoro, the East African Rift Valley and the expansive Serengeti region is well known, especially because of its abundant wildlife which is supported by the nutrient-rich volcanic soils and the resulting good quality fodder from the abundant vegetation. This equatorial region, with its diurnal climate and monsoon climate has two rainy seasons, a long one and a short one. These are usually only separated by a short dry season and have the same effect as a summer rainy season, so that with an annual precipitation of 800 mm, savannas and parklands similar to those in zonobiome II may be found here. Forest clearing, annual fires and overgrazing have affected the plant cover substantially, resulting in various stages of degradation. Although this is a typical tree savanna, it is often referred to as an "orchard steppe". Where the climate is drier, and in dry stony regions, large candelabra euphorbia and *Aloe* species are found.

### e Vegetation of Zonobiome II in Australia

This vegetation type exists in only a few small relicts of deciduous forest in north-eastern Australia, with Indo-Malaysian floral elements and some insignificant deciduous species of *Eucalyptus* in northern Australia. Parklands with palms and evergreen *Eucalyptus* species are distributed within the range of zonobiome II. Somewhat more to the south where the annual precipitation is lower, savannas occur with a grass cover of *Heteropogon contortus* and evergreen *Eucalyptus* species. The term "savanna" is not used in the extensive vegetation monography of Beadle (1981). In contrast, the Australian researchers Walker & Gillison (1982) designate as savannas all sparse forests in which grasses and the herbaceous layer comprise more than 2% of the ground cover. According to this criterion, most sparse *Eucalyptus* forests would have to be considered savannas.

##  Ecosystem Research

Grasses and trees are the components of the savanna. The number of grass species is relatively low although their biomass is more important. However, for Leguminosae it is the opposite (Fig. 113). There are other, rarer species (about 25%), however, their biomass is only 1.5%.

The following two savanna ecosystems were investigated ecologically: the lamto savanna in West Africa, a relict savanna in a rainforest region, and the Nylsvley savanna in South Africa, bordering on the Kalahari in the west.

### a The Lamto Savanna

This relict savanna is situated in the Guinea forest zone (in the Ivory Coast) at 5°W and 6°N, and is, therefore, still within zonobiome I. It is burned annually to prevent the advance of the neighbouring rainforest, regardless of whether the soil conditions are capable of accommodating it or not. The mean annual precipitation is 1300 mm and the climate diagram indicates a dry period of only 1 month in August. The weather varies greatly from year to year, however, and the annual rainfall fluctuates between 900 and 1700 mm. The higher levels of the relief are occupied by tree or shrub savannas on red savanna soil with laterite concretions; on the lower levels, however, palm savannas are found on soils saturated with water not far below the surface.

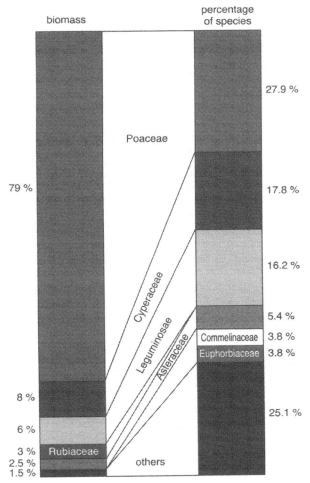

biomass

percentage
of species

**Fig. 113.** Relative biomass
and percentage species
distribution of the most
important plant families in
African savannas (after
MÜLLER 1991)

27.9 %

Poaceae

79 %

17.8 %

16.2 %

5.4 %

Commelinaceae  3.8 %

Euphorbiaceae  3.8 %

Cyperaceae

Leguminosae

Asteraceae

8 %

25.1 %

6 %

3 %   Rubiaceae

2.5 %

1.5 %

others

The various plant communities were studied by ME-
NAULT & CESAR (1982) (see Table 15). Consumers and decomposers in these savannas were recorded by LAMOTTE
(1975). Large animals appear only sporadically; the zoomass (expressed as kg ha$^{-1}$) of birds is 0.2–0.5, of the
12 rodent species 1.2, of earthworms 0.4–0.6. It was not
possible to determine the mass of the grass, humus and
wood-eating termites or of other invertebrates. Soil respiration, which reflects activity of micro-organisms, was cal-

**Table 15.** Nominal values for ecosystems (extreme values) of a low shrub savanna (first number) and a dense tree savanna (second number)

| | | |
|---|---|---|
| Number of woody plants per ha | 120 | 800 |
| Area covered by woody plants | 7% | 45% |
| Leaf area index | 0.1 | 1 |
| Phytomass (aboveground) (t ha$^{-1}$) | 7.4 | 54.2 |
| Phytomass (belowground) (t ha$^{-1}$) | 3.6 | 26.6 |
| Net wood production (t ha$^{-1}$ a$^{-1}$) | | |
| Aboveground | 0.12 | 0.76 |
| Belowground | 0.05 | 0.37 |
| Net production of leaves and green shoots | 0.43 | 5.53 |
| Net production of the grass layer (t ha$^{-1}$ a$^{-1}$) | | |
| Aboveground | 14.9 | 14.5 |
| Belowground | 19.0 | 12.2 |

culated at $8\,t\,ha^{-1}\,a^{-1}$ $CO_2$. An attempt to determine the energy flow of decomposition (LAMOTTE 1982) yielded the following results:

1. Annual fires result in the mineralisation of approximately one third of primary production.
2. Probably less than 1% of the primary production is eaten by consumers. Decomposition by detritus eaters, including the earthworms as the most important group, is of limited importance.
3. 80% of the primary production is decomposed by micro-organisms, making the representation of the energy flow in the form of a pyramid quite doubtful. Therefore, it is confirmed that the long cycle by way of the consumers is quantitatively insignificant (see p. 90)

A large amount of faunistic information on individual animal groups of the savannas is presented by BOULIÈRE (1983).

## b The Nylsvley Savanna

This region lies at 24 °S, north of Johannesburg in the Nylsvley Nature Reserve, and comprises 745 ha, 130 ha of which are stony soils (HUNTLEY & MORRIS 1978; HUNTLEY & WALKER 1982). The climatic conditions are illustrated in the climate diagram in Fig. 114. Here, we have a relatively dry tropical climate with summer rains (zono-

**Mosdene** 1097 m, 609.5 mm
**Nylstroom** 1143 m, 18.3°

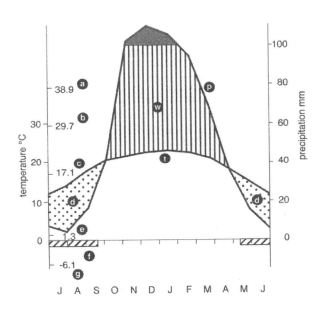

**Fig. 114.** Combined climate diagram of Mosdene and Nylstrom in South Africa near the research area. *a* Absolute maximum; *b* mean daily maximum of warmest month; *c* mean daily temperature fluctuation; *d* arid season; *e* mean daily minimum of the coldest month (1.3 °C); *f* month with possible frost; *g* absolute minimum; *p* rain curve; *t* temperature curve; *w* humid season (from WALTER 1990, after HUNTLEY & MORRIS 1978)

biome II), an annual precipitation of 610 mm and rare frosts down to −6 °C in the months of May to September.

The nutrient-poor soils are sandy latosols (pH = 4), derived from the weathering of bedrock of the Waterberg series (B-horizon at a depth of 30–130 cm and a clay content of 6–15%). The vegetation constitutes a *Burkea africana-Eragrostis pallens* tree savanna with trees of up to 14 m in height. Besides *Burkea*, other dominant deciduous trees are *Terminalia sericea* and *Combretum molle*. Among the brush species are *Ochna pulchra* and *Grewia flavescens*. The crowns of these plants cover 27.5% (20–60%) of the surface. In areas which were inhabited by natives until 50 years ago, the soil is more compacted, more abundant in the elements N, P and K and bears a secondary growth consisting of an *Acacia tortilis-A. nilotica-Dichrostachys cinerea* thornbush savanna (covering 10%) with *Eragrostis lehmanniana* as the dominant grass species. The aboveground phytomass of the woody producers in the *Burkea* tree savanna amounts to 16.3 t ha$^{-1}$, 14.9 t ha$^{-1}$ of which is contained in trunks and branches, 0.3 t ha$^{-1}$ in twigs and

1.1 t ha$^{-1}$ in leaves. Another 1.9 t ha$^{-1}$ are present in the dead wood. The leaf area index is 0.8.

Herbaceous plants are of no special importance in the grass layer. The grasses are less dense beneath the tree crowns than in between them. The phytomass of the grass layer fluctuates significantly within a small area, and varies from year to year according to the amount of precipitation. Measurements over 3 years have revealed maximum values of 235 g m$^{-2}$ between the trees and 62 g m$^{-2}$ under tree crowns and the minimum values were 141 and 16 g m$^{-2}$ respectively. The underground phytomass is calculated at 15.5 t ha$^{-1}$. Between the trees, half of this mass is attributed to the roots of the grasses, 75% of which is located in the upper 20 cm of the soil. In the summer, 13% of the roots are dead, compared to 30% in the winter. Between 7 April and 14 November 1977, the amount of litter in the *Burkea* savanna totalled 160 g m$^{-2}$ (84.8% leaf litter, 9.4% twigs, 5.5% fruits and seeds, 0.3% bark and bud scales); 35% of the litter originated from each *Burkea* and *Ochna*. The average annual litter production is estimated to be 170 g m$^{-2}$. The total litter on the ground on 18 October 1976 was 1853 g m$^{-2}$ and decreased to 1342 g ha$^{-1}$ by 12 July 1977.

Micro-organisms were counted by standard plate counts and showed a strikingly large number of *Actinomycetes*. Activity of soil organisms was established by ATP determinations and measurements of soil respiration. The daily $CO_2$ emission for a 24-h period was 1866 mg m$^{-2}$ $CO_2$ (minimum in August, 226 mg and maximum in January, 4367 mg). For further information on the South African savanna see HUNTLEY & WALKER (1982). The biomass of various *Burkea* savanna was determined by RUTHERFORD (1982).

### c Fauna

Both the vertebrate and invertebrate fauna of the *Burkea* tree savanna are notably different from those of the *Acacia* thornbush savanna. In all, 18 species of amphibians occur in the Nyl River valley, 11 of which are found in the research area. The toad, *Bufo garmani*, and the frogs, *Breviceps mosambicus* and *Kassina senegalensis*, may be found far from the nearest source of water. The reptiles are represented by 3 species of turtles, 19 lizards and 26 snakes.

Of the 325 different bird species, 197 are found permanently within the National Park; 120 species inhabit the research area (14 raptors, 71 insectivorous species, 4 baccivorous (berry-eating) species, 10 granivorous species and 26 omnivorous species). In the park 46 of the 62 mammalian species are found in the research area. The most numerous group is that of the rodents. One species each of porcupine, warthog and jackal and two species of monkey inhabit the area. The most important artiodactyls are the koodoo (*Tragelaphus strepsiceros*), the impala (*Aepyceros melampus*), the duiker (*Sylvicapra grimmia*) and the steenbok (*Raphicerus campestris*).

Since it is so difficult to determine the number of individuals or the living zoomass, this was achieved rarely and then only as an approximation. The number of snakes was estimated at 3 ha$^{-1}$, and the most frequent reptile, the gecko (*Lygodactylus carpensis*), at 195–262 ha$^{-1}$. The population of the common lizard, *Ichnotropis capensis*, was 7–11 animals ha$^{-1}$. The living zoomass of birds on 100 ha of the *Burkea* savanna was 40 kg, which is reduced by 25–30% when the migratory species depart for the winter. Mammals were captured so seldom and so sporadically that the results are of little meaning. Captures of *Dendromys melantois*, for example, averaged approximately 5 (0–15) animals ha$^{-1}$, and of other rodents only 2 ha$^{-1}$. The following mean values were given per 100 ha for the artiodactyls: impala 13, koodoo 2, warthog 1, duiker 2, steenbok 1 to 2 (reedbok, rare).

According to the previous owner of Nysvley, grazing was only allowed in the area between the months of January to April during the last 40 years. Otherwise, losses would occur due to the poisonous geophytic relative of the Euphorbiaceans, *Dichapetalum cymosum*. The biomass of the cattle during those 4 months was approximately 150 kg ha$^{-1}$. In 1975, effects of overgrazing became evident, so that the herd was reduced to almost half in the following year.

The number of invertebrates is so large that it was necessary to limit the study to certain groups which play important roles in the ecosystem: wood-eating Coleopterans, Lepidopterans, social insects, root-eaters and spiders. The zoomass of the invertebrates in dry weight averaged 135 g ha$^{-1}$ (minimum in August = 60 g ha$^{-1}$, maximum in March = 300 g ha$^{-1}$) on woody plants. The dry weight of the insect zoomass is higher in the grass layer. Occasionally, large numbers of caterpillars (*Spodoptera exempla*) or beetle larvae (*Astylus atromaculatus*) were seen on the grass species *Cenchrus ciliaris*.

During the warmer season, 77% of the dung is removed within a day by dung beetles (*Coprinae, Aphodiinae*), which dig below the deposit, while dung rolling beetles (*Pachilomera* spp.) distribute the dung over a large area. These coprophages lead to the following group.

The following figures give the dry mass in kg ha$^{-1}$ for the *Burkea* savanna; the values in parentheses are for the *Acacia* thorn savanna: Acridoidea 0.76 (2.32), other Orthopterans 0.06 (0.02), Lepidopterans 0.05 (0.03), Hemipterans 0.08 (0.08), others 0.05 (0.05), thus total insect mass 1.00 (2.50) kg ha$^{-1}$.

The decomposers are the small saprophagous animals of the soil and within the litter layer, which feed on dead plant and animal remains, thereby breaking them down further in size. Finally, protozoa, fungi and bacteria provide for their complete mineralisation. The most important saprophages are the termites. Oligochaetes, Myriapods and Isopods are less significant. Acarids and Collembola eat bacteria and fungi. Of the 15 termite species, the most common are *Aganotermes oryctes, Microtermes albopartitus, Cubitermes pretorianus* and *Microcerotermes parvum*. Of these 15 species, 4 are humus-eaters, while others eat dead wood or grass litter. More detailed studies on these termites are not available. An average of 2540 termites were found under $1 \text{ m}^2$ ground surface (maximum: 8204 in November; minimum: 596 in July).

The fauna of other savanna regions in Africa is particularly rich in large mammals compared to the savannas and grasslands on other continents. However, it must be assumed that some thousands of years ago the number of large mammals in these continents was considerably higher. For larger herbivores the following functional groups can be listed:

- grass eaters (grazers)
- leaf eaters (leaves from shrubs and trees; browsers)
- grain eaters (seed eaters; granivores)
- nectar eaters (nectivores)
- fruit eaters (frugivores)

## Tropical Hydrobiomes in Zonobiomes I and II

High rainfall combined with relatively low potential evaporation accounts for the large surplus of water in the wet tropics. San Carlos de Rio Negro in southern Venezuela, for example, with a rainfall of 3521 mm, has a potential evaporation of only 520 mm. In flat, poorly drained areas, therefore, extensive swamps have developed. In Uganda, such swampy areas cover 12,800 $\text{km}^2$, about 6% of the entire country. The drainage basins of the river systems are not separated by watersheds, but are connected with one another by a network of swamps. The flight from Livingstone to Nairobi provides excellent views of the large Lukango swamps as well as those surrounding Lakes Kampolombo and Bangweulu. The largest swamp area of all is formed by the White Nile in southern Sudan, which, together with its left tributary the Bar-el-Ghasal, fills an enormous basin lying 400 m above sea level. This region is known as the "Sudd" and extends 600 km from north to south and from east to west at its widest and longest points. The total area covered is estimated at 150,000 km$^2$ and varies according to the water level. Half of the water of the Nile is lost by evaporation in the Sudd region. Seen from the air, the water appears to be dotted with small islands barely extending above the surface, but these are in fact, swimming islands or floating lawns formed by the shoots of *Vossia* grass and *Papyrus*. There are also grassy rafts of the South American *Eichhornia* and *Pistia*. From the air, it is possible to distinguish free waterways and small stretches of water. When the water level sinks, part of the land emerges, and grassland consisting of tall *Hyparrhenia rufa* and *Setaria incrassata* develops. The wettest parts are covered with *Echinochloa* species, *Vetiveria* and *Phragmites* (reed).

It was previously assumed that the **"Great Pantanal"** in the Mato Grosso in Brazil, bordering Bolivia and Paraguay, was a similar type of swamp, from which the southern tributaries of the Amazon and the right tributaries of the upper Paraná arose. However, this region is flooded only during the rainy season and is used as grazing land in the dry season, although ring-like lakes bordered with woodland remain. Swamp areas and watery basins are also widespread in the rest of the wet tropics. The aquatic vegetation consists of some cosmopolitan and pantropical species, but each region also has its own floristic peculiarities.

## Mangroves as Halo-Helobiomes in Zonobiomes I and II

Approaching a tropical coast with its protective coral reefs from the sea, one's attention is caught by partially submerged mangroves. At high tide, the crowns of the trees barely extend above the surface of the saltwater, and only at low tide are the lower portions of the trunks and the stilt roots and pneumatophores visible. Such forests are found in the tidal zone in saltwater, where the salt concentration is about 35 g l$^{-1}$, which corresponds to an osmotic potential of $-2.5$ MPa.

More than 30 woody mangrove species are known. The best developed mangroves are found near the equator in Indonesia, New Guinea, and the Philippines, however, with increasing latitude, the number of species decreases until only one species of *Avicennia* remains. The last outposts are to be found at 30°N and 33°S in East Africa, 37–38°S in Australia and New Zealand, 29°S in Brazil, and 32°N

Mangroves are an azonal vegetation confined to the saltwater of tidal regions. Mangroves grow on very finely grained soils, protected from surf and free of frost.

Distinction must be made between the species-rich eastern mangroves on the coasts of the Indian Ocean as well as the west coasts of the Pacific Ocean and the western mangroves of the coasts of the Americas and the east coasts of the Atlantic Ocean which are species-poorer.

**Fig. 115.** Inner mangrove
zone with *Ceriops tagal*
(Rhizophoraceae) north of
Mombassa (Kenya) (Photo:
S.-W. BRECKLE)

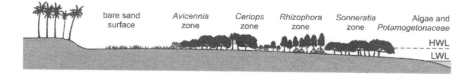

**Fig. 116.** Zonation of the
East African coastal man-
groves (from WALTER &
STEINER 1936). *HWL* High
water level; *LWL* low water
level

on the Bermuda islands. Thus, although the mangrove is
at its best in equatorial regions, it also extends throughout
the tropical and subtropical zones almost as far as the
winter rain regions or the warm-temperate zone (CHAP-
MAN 1976).

The most important genera of mangrove are *Rhizo-
phora*, with stilt roots (Fig. 115) and viviparous seedlings,
and the non-viviparous *Avicennia*, the white mangrove,
which has thin pneumatophores growing up out of the
ground. *Laguncularia* is a western mangrove. *Conocarpus*
grows only where the salt concentration is low. In the
eastern mangrove vegetation, species of the genera *Bru-
guiera* and *Ceriops* (both viviparous, with knee-like roots)
occur, as well as *Sonneratia* (non-viviparous, with thick
pneumatophores) and species of *Xylocarpus, Aegiceras,*
and *Lumnitzera* and others. Different mangrove species
usually grow in distinct zones, and only seldom are they
mixed. This zonation is dependent upon the tides, since
the nearer a species grows to the outer edge of the mangrove, the longer and deeper does it stand in saltwater
(Fig. 116).

The tidal range (the difference in depth between high
and low tides) not only differs from place to place on the

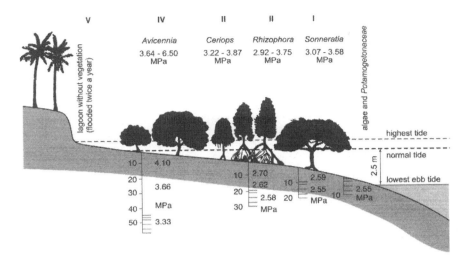

**Fig. 117.** Osmotic potential of the cell sap in MPa (negative values; lowest and highest) of the leaves of mangrove species and of the soil solutions at various depths (in cm). Coastal mangroves of eastern Africa, arid type

coast but also varies periodically, according to the phase of the moon and position of the sun. It is at its maximum at new and full moon (spring tide) and at a minimum midway between the two (neap tides). The spring tides are at their highest twice a year, when day and night are equally long (equinoctial spring tides). In addition to the **coastal mangroves** growing on flat shores, often in a belt many kilometres wide and with no influx of fresh water from the land, there are also the **estuary mangroves**, extensive in the delta regions of rivers, as well as the less common **reef mangroves** growing on dead coral reefs protruding above the surface of the sea. Coastal mangroves have been investigated in great detail in Australia and in eastern Africa, with special regard to salt relationships (Fig. 117).

The East African coast in the neighbourhood of Tanga has a relatively dry monsoon climate. Potential evaporation is equal to or greater than annual rainfall. Apart from a short dry season there is also a pronounced period of drought. This is responsible for the higher salt concentration of the soil within the tidal region, which is greater the further inland and the shorter the period of time during which the ground is flooded. The most extreme conditions in the mangrove zone are to be encountered at the

inland margin, which is reached only by the equinoctial
spring tides. Here, the saltwater in the soil is strongly
concentrated by evaporation during drought, whereas during the rainy season the soil may be completely leached.
Since no plant can tolerate such drastic variations in salt
concentration, these areas are devoid of vegetation. They
occur wherever the climate prevailing at the inner boundary of the mangroves is such that a period of drought occurs. In northern Venezuela, however, small clumps of
salt-sensitive plants such as columnar cacti and *Opuntia*
or bromeliads do, in fact, grow under such conditions. It
is known, though, that bromeliads take up water through
their leaves and do not root in the soil. Cacti invariably
grow on small sand heaps from which the salt has been
washed away in the rainy season. They obtain their water
by means of shallow roots spreading in the sand and are
therefore unaffected by the underlying salty soil. The tissues of both cacti and bromeliads are free of salt and they
are not halophytes – a further example of the observation
that obvious soil properties may not always indicate the
ecological conditions under which plants grow and require careful analysis.

In very humid regions, on the other hand, the exposed
areas are continually being washed by rainwater so that
the salt concentration of the water in the soil must decrease landward. Moving upriver this is also true of the
estuary mangroves which, via a brackish zone where the
fern *Acrostichum*, the *Nipa* palm, *Acanthus ilicifolius* and
many other species are found, give way to a freshwater
community without the interpolation of a barren zone
(Figs. 116–118). In spite of the fact that mangroves are an
azonal vegetation, their zonation is determined by the climate and they are, therefore, different in the humid zonobiome I than in climates with pronounced periods of
drought (Fig. 118). In this respect, zonation of mangroves
between zonobiome I and II or even III is fundamentally
different.

Plants rooting in saline soils take up a certain amount
of salt and store it in the cell sap. In the case of mangroves, the salt concentration of the cell sap in their succulent leaves is roughly equal to that of the soil. Apart
from this, non-electrolytes are present in a concentration
usual for tropical species. Figure 117 illustrates the typical
zonation and the potential osmotic pressure in the soil
and the mangrove leaves, and Fig. 118 shows the differences between mangroves in arid and humid regions.

Zonation in mangroves results from competition between the various species. Investigations carried out in
eastern Africa suggest that the salt factor is decisive in de-

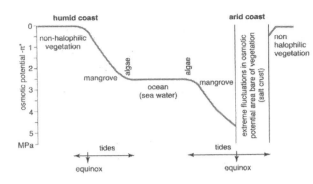

**Fig. 118.** Scheme showing the osmotic potential ($-\pi^*$; MPa) due to salt in the soil and the mangrove formations on humid and arid coasts

termining this zonation. *Avicennia* is the weakest competitor but has the highest salt resistance, so that stunted individuals of this species constitute the landward limit of mangroves. *Sonneratia* is the strongest competitor but seems not to tolerate a salt concentration above that of seawater and is therefore confined to the outer fringes. In continuously humid areas the situation is more complicated: *Avicennia* appears to be confined to sandy ground, whereas *Sonneratia* prefers silty soil. In such regions, type of soil, soil aeration, duration of inundation, water movement, decrease in salinity, and variations in salt concentration seem to be of greater significance.

The **salt economy** of mangroves presents an interesting problem. Mangroves are unable to take up seawater in its original form because the salts which would be left behind after the loss of water by transpiration would soon lead to the formation of a saturated salt solution in the leaves. It has recently been shown by direct measurement that osmotic potentials from $-3.5$ to $-5.5$ MPa can be produced in mangrove leaves. This is lower than the osmotic potential of the soil solutions. This potential difference is transmitted to the roots by the cohesion tension in the conducting vessels. The roots act, at the same time, as an ultra-filter permitting only the passage of almost pure water which is then transported to the leaves. Only the very small amount of salt necessary for the osmotic potential enters plants, and salt is stored in vacuolar solution in the leaf cells.

The mechanisms whereby the salt concentration is regulated have not been completely elucidated. Some of the salt from the old, senescing leaves can be transported to the young, developing leaves or excess salt can be eliminated when the old leaves drop. *Avicennia* has salt glands situated on the underside of its leaves for regulating the salt concentration. This species can excrete a 4.1% salt solution which is more concentrated than seawater; the proportion of NaCl to KCl is the same as in seawater (90 and 4%). No excretion takes place at night but it reaches a maximum at midday. In 24 h, 0.2–0.35 mg of salt per 10 cm$^2$ leaf area is excreted. In the dry season the salt crystallises on the underside of the leaves during the day and at night, when the humidity is greater, the salt dissolves and drips away. It is interesting to note that the viviparous seedlings are almost free of salt and have an osmotic potential of only –1.3 to –1.8 MPa. This means that water must somehow reach them with the help of glandular tissue in the cotyledons. However, as soon as the seedlings drop off and take root in the salty ground, their salt concentration increases, and the osmotic potential attains the normal level. The radicle appears, initially at least, to be permeable to salt.

The function of pneumatophores has been clarified. They are equipped with lenticels with minute openings which permit the entry of air, but not water. When the pneumatophores are completely submerged in water, their intercellular oxygen is used up in respiration and a negative pressure develops because carbon dioxide, being very soluble, escapes into the water. As soon as the roots emerge from the water, a pressure compensation takes place, whereby air (and oxygen) is drawn in. The oxygen content of the intercellular spaces in such roots, therefore, varies periodically between 10 and 20%.

Together with their fauna of numerous small fiddler crabs and the mangrove fish (*Periophthalmus*) which can be seen crawling out of water and up the trees, mangroves present a highly interesting ecosystem belonging to neither the sea nor the land. Mangroves are very endangered in many places because of the use of their wood for charcoal, and the areas for breeding crabs.

## Shore Formations – Psammobiomes

Sandy shore formations of tropical coasts offer fewer peculiarities. Beyond the barren zone resulting from exposure to the force of the breakers, sand plants with long runners can be found, among them the widespread *Ipomoea pes-caprae* and the halophytic *Sesuvium portulacastrum*, *Batis maritima* and *Sporobolus virginicus*. Still further inland, beyond the influence of saltwater, the sand in the tropics very soon becomes covered by shrubs and trees, the floating fruits of which can be seen in the tide-mark on all tropical shores. Typical representatives are *Terminalia catappa* and coconut palms, although nowadays palms are nearly all planted by man. *Barringtonia, Calophyllum, Hibiscus tiliaceus* and *Pandanus* are characteristic of the Eastern oceans, and *Coccoloba uvifera* (Polygonaceae), *Chrysobalanus icaco*, and the poisonous *Hippomane manicinella* (Euphorbiaceae) of the western oceans.

No large dune areas occur in the tropics, with the exception of the northern coast of Venezuela, near Coro, where a semi-desert type of climate prevails. As a result of the continual trade winds blowing in from the northeast or east-northeast, large quantities of sand drift landward from the beach and are trapped by *Prosopis juliflora*. The dunes, thus formed, continue to grow in the direction of the wind and are soon covered by *Prosopis* bushes. In this manner, a series of dune ridges is established, running parallel to one another and to the wind direction and reaching a considerable height. Migrating dunes, or barchans, are found in one part of the dune region, probably resulting from wood-clearing. They join up to form ridges at right angles to the wind direction.

## 10 Orobiome II – Tropical Mountains with an Annual Temperature Periodicity

Whereas a brief dry period in the alpine belt has no effect on the water supply of the plants in orobiome I, the dry period in ZB II, even at considerable heights, has very obvious consequences, depending upon its duration. Precipitation increases in the montane belt and cloud cover reduces the number of hours of sunshine to such an extent that an evergreen montane forest develops. Cloud forests may even be found above this, in the trade wind or monsoon region (Fig. 91). In the monsoon area of India even smaller mountains have a large effect on precipitation, at least on the distribution over the whole year (Fig. 119).

The entire sequence of the altitudinal belts of orobiome II is best seen in the southern slopes of the eastern Himalayas, by following the extremely humid Sikkim profile from Darjeeling toward the north. The forest belts are not easy to distinguish from one another, however, and a further complication is caused by progressive replacement of Paleotropical floral elements by Holarctic elements. The foot of the mountains is occupied by wet-deciduous forest with *Shorea robusta* or with bamboos and palms on wet ground. An evergreen tropical-montane forest (*Schima, Castanopsis*) with tree ferns begins at an altitude of about 900 m above sea level. In its higher regions, Holarctic gen-

**Fig. 119.** Increase in quantity of precipitation in the monsoon region of India. Climate diagram of Bombay at sea level and two meteorological stations in nearby mountains. At the higher station, 1380 m above sea level, almost 3000 mm of rain falls in July. The rainy season is increased by only 1 month, although the annual precipitation exceeds 6000 mm

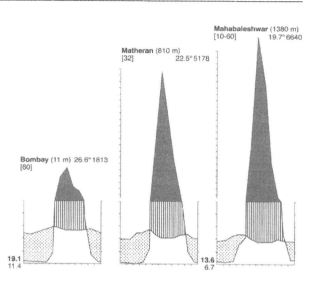

era (*Quercus, Acer, Juglans* and *Vaccinium*) are well represented. Above this is a cloud forest with Hymenophyllaceae and mosses. The proportion of Holoarctic genera (e.g. *Betula, Alnus, Prunus, Sorbus*) increases with increasing altitude. The frost limit lies between 1800 and 2000 m above sea level. In the belt immediately above this, there are large numbers of tall *Rhododendron* and *Arundinaria* species, which are replaced further up by conifers, e.g. *Tsuga, Taxus*). An *Abies densa* fir forest containing broadleaf species occurs between 3000 and 3900 m above sea level. The tree line is composed of *Abies* and *Juniperus*.

The subalpine belt is again characterised by tall *Rhododendron* species, which gradually diminish in size in the alpine belt, with its flower-filled meadows, until at 5400 m above sea level, *Rhododendron nivale* is merely a tiny shrub. This orobiome system is especially complicated in the Himalayan Mountain range (TROLL 1967, MEUSEL & SCHUBERT et al. 1971, MIEHE in WALTER & BRECKLE 1994). From 5100 m above sea level upward, the plants are mainly hemispherical and cushion-like in form (e.g. *Arenaria, Saussurea, Astragalus, Saxifraga*). The snow line is reached at 5700 m above sea level.

In the **Andes**, the order of the altitudinal belts on the eastern slopes differs from that on the western slopes, and is different again in the mountainous valleys of the interior. A brief schematic survey was published by ELLENBERG (1975). The high plateau of the **Altiplano** has undergone anthropogenic changes due to human settlements and grazing of llama herds. In accordance with the climate, the altitudinal belts on the western slopes become increasingly xerophytic toward the south. In addition, the deciduous-forest belt, in leaf only during the rainy season, extends further and further upward and the evergreens become increasingly sclerophyllous with smaller leaves.

The occurrence of a warm season pushes the forest limit up to 4000 m above sea level, with scattered stands of *Polylepis* even at 4500 (4900) m (see p. 152). The páramo is replaced first by puna of the wet-type with cushion-like plants, then further south by dry puna with tussocks of xerophytic grasses (e.g. *Festuca orthophylla, Stipa ichu*), and finally by desert-like puna with *salares* (saltpans) in orobiome III (CHONG-DIAZ 1988). Soils of the alpine belt change correspondingly, from peat soils to chestnut-brown soils and serozems and even solonetz and solonchak. The puna of north-western Argentina (between 22 to 24.5 °S has been investigated in detail, including microclimatic studies (RUTHSATZ 1977).

## Man in the Savanna

Large areas of the savanna are nowadays used for grazing cattle herds. Even in previous times, herding nomads used the savanna over large stretches and with their large herds were in competition for fodder resources with the wild animals. In Neotropic savannas African grasses have been introduced in large areas and have thus considerably reduced the original biodiversity. However, productivity has been increased in parts by the use of large areas for grazing cattle (SOLBRIG et al. 1996). Improvement of the growth of new green grass is supposed to be achieved by frequent grass and bush fires, often intentional before the start of the rainy season. However, in the long term this is bound to lead to increasing depletion of nutrients in the soil.

## Zonoecotone II/III

### a Sahel Zone

This zonoecotone includes the open, climatic savannas, such as those of Namibia, which have already been discussed (see p. 105). Similar conditions prevail in the Sahel zone, south of the Sahara, where the transition to the summer rain region of the Sudan (zonobiome II) occurs. The Sahel zone has become completely degraded by overpopulation and overgrazing as a result of the typical recurring years of drought in this zone. Due to the limited number of natural water sources in this area, it was only capable of supporting a very small population and a correspondingly small number of cattle until foreign aid programs attempted to develop the land by drilling a large number of wells. This made water available for larger herds of cattle and also resulted in a rise in the human population as long as the annual precipitation remained above the long-term average. The occurrence of several years of drought led to a catastrophe. Although enough water was available for both humans and animals, the parched pastures no longer yielded grass. Cattle died of starvation and humans were forced to leave the area or were supported by foreign aid projects. The pastures, however, were irreparably damaged and were converted into a "man-made desert". In Namibia with its similar climate, several successive years of drought also proved disastrous, although the small number of farmers is able to survive by reducing the size of their herds in time. The economy recovers quickly after a few good years of rain.

## b Thar or Sind Desert

A further zonoecotone II/III is situated in the border region of India and Pakistan in the Thar or Sind desert. This homogeneous arid region between the Aravalli Mountains in the east and the Baluchistanian Heights in the west is also referred to as the Great Indian Desert (Fig. 120). The aridity increases from east to west.

Reference in the literature to a Saharo-Sindic desert zone is incorrect. The Sahara, a rainless region (or one with scant winter rain) is floristically mainly part of the Holarctic, and extends eastward into the Egyptian-Arabian desert as far as Mesopotamia. The Sind Desert, however, is the final arid extension of the Indian monsoon region and floristically it must be considered part of the Paleotropics. Climatically, the Indian Thar Desert is a zonoecotone II/III which may be compared to the "Sahel", the transitional region between the Sudan and the southern Sahara. Both receive light summer rains although the Indian region lies north of the tropic of Cancer and the mean annual temperature is 2–3 °C lower than in the Sahel, meaning that frosts may occur between the months of December and February (Fig. 120). Only the Indus valley has an annual rainfall of less than 100 mm and would,

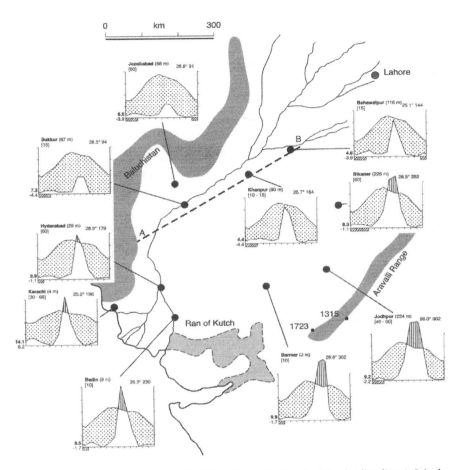

**Fig. 120.** Climate diagram map of the Sind-Thar Desert. Northwest of the dividing line *A–B* is the very arid region

therefore, qualify as a desert. Because of the Indus and its tributaries, however, it is a water-rich irrigation region.

On the other hand, the "Great Indian Desert" is a "man-made desert". The area was settled 4000 years ago, but since the campaigns of Alexander the Great, the population has increased to an extent that overgrazing, wood clearing and to some degree farming, have completely degraded the countryside (MANN 1977). In its natural state, the region was a *Prosopis* savanna with an annual precipitation of 400–150 mm on a deep reddish brown sandy sa-

vanna soil, as can be seen today in the area around Jodhpur (RODIN et al. 1977). Here, the following thorny shrubs
are found: *Prosopis cineraria, Ziziphus nummularia, Capparis decidua* (= *C. aphylla*) and others. With an annual
precipitation of 500–600 mm, *Prosopis* grows to a height
of 8 m, forming stands of 150–200 individuals per hectare.
With less precipitation, 300–400 mm, it only grows to 5–
6 m in stands of 50–100 individuals per hectare and with
only 200 mm it only grows to 3–4 m in stands of 25–30
plants per hectare. Decreasing amounts of rain also result
in the replacement of high grass species (*Lasiurus, Des-
mostachys*) by lower growing species (*Aristida*) (GAUSSEN
et al. 1972). The conditions are, therefore, similar to those
in south-west Africa (p. 172 and Fig. 105). In regions with
over 250 mm annual precipitation, the savannas are used
for grazing and are degraded as a result of overstocking
with cattle, so that the annual grass species *Aristida ad-
scensionis* becomes a weedy dominant in the pastures.

The soils are extremely sandy in the Bikaner District
(Fig. 120). As a result of overgrazing in the vicinity of the
villages, moving barchans (dunes void of vegetation) give
the impression of an extreme desert (Fig. 121). Actually,
however, the water content in the sand of such unvegetated dunes is much higher than of overgrown dunes as is
illustrated by the following figures (Table 16) from an area
with an annual precipitation of 260 mm.

This difference is explained by the fact that growth of
*Prosopis* requires approximately 220 mm of water for transpiration. The often-planted grass species, *Pennisetum ty-
phoides*, requires 160–180 mm. The inhabitants take advantage of the water content of the sand in the unvegetated dunes by planting watermelons every 2×2 m. They
use branches to prevent the sand from blowing away. No
information is available on the natural vegetation of the
driest parts of the Sind Desert in the Indus Valley. This irrigated region is densely populated and there are no areas
with natural vegetation to be found. Due to inappropriate
irrigation practices, the water table is rising resulting in a
secondarily increased salt concentration in the moist soil.
This has caused the loss of 40,000 ha of agricultural land
annually and means that the food production will not be
able to keep pace with the growing population. Rehabilita-
tion of the brackish soil, for example by lowering the
water table, would be an extremely costly endeavour in
this level landscape.

Natural saline soils are widely distributed in the southern part of the Thar Desert on the Gulf of Kutch. Mangroves grow in the tidal zone, followed further inland by
salt marshes with *Salicorina, Suaeda, Atriplex* and the salt

**Table 16.** Water content (in mm) in the sand of unvegetated (I) and vegetated (II) dunes near Jaisalmer

|  | March | | June | | September | | January | |
|---|---|---|---|---|---|---|---|---|
| Depth (in cm) | I | II | I | II | I | II | I | II |
| 0–105 | 41 | 10 | 33 | 17 | 45 | 10 | 34 | 7 |
| 0–210 | 106 | 39 | 94 | 48 | 120 | 33 | 105 | 28 |

(after MANN 1976)

**Fig. 121.** Barchans that have started to move recently in the region between Jaisalmer and Jodhpur with single specimens of *Prosopis*, *Acacia* and *Calotropis* (Photo: M. P. PETROV)

grass species *Urochondra*. In the region of Ran of Kutch, with its high water table, nearly sterile salty clay soils are found with halophytes (*Haloxylon salicornicum*, *Aeluropus*, *Sporobulus*), and *Cenchrus* spp. and *Cyperus rotundus*, or scattered woody plants on good locations (BLASCO 1977).

## c Caatinga

A difficult region to categorise ecologically, is the arid "Polygono da Sêca" in the Caatinga of north-eastern Brazil. It is characterised by a precipitation which may vary extremely from year to year. In the driest location in Cabaceira, for example, the years 1940–1946 with plentiful rainfall (664–150 mm) were followed by the drought years of 1948–1958 with precipitation below 80 mm (1952 only 24 mm, 1958 only 22 mm), excepting the years 1954 with 170 mm and 1955 with 187 mm. In such an unreliable cli-

**Fig. 122.** *Adenium socotranum* (Apocynaceae) with a trunk diameter of 2 m on western Socotra (Photo: F. KOSSMAT)

mate, large succulent columnar cacti and large thorny bromeliads which grow close to the ground survive best, as well as water storing trees (*Ceiba* and others) or deciduous shrubs which are leafless for long periods. This region is difficult to use agriculturally and is only sparsely settled since the drought periods are unpredictable and force the inhabitants to leave the area. Similar conditions are found in the trade wind deserts on the Venezuelan-Columbian border on the north coast of South America and on the Galapagos Islands. These dry regions also experience years with very high precipitation.

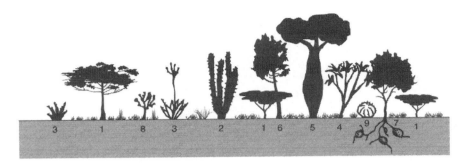

**Fig. 123.** Characteristic plant types in the thorn-succulent savanna (after TROLL 1960). *1* Thorny pinnate leaved umbrella trees (*Acacia* type); *2* stem succulent candle or candelabra trees (cacti type); *3* succulent and thorny leaved tufted plants (*Aloe* types); *4* succulent and thorny leaved tufted plants (*Dracaena* type); *5* barrel-trunk sappy wooded trees with deciduous leaves (*Adansonia* type); *6* sclerophyllic trees with thorns (*Balanites* type); *7* deciduous trees with woody roots or lignotubers; *8* sclerophyllic shrubs and small trees (*Capparis* type); *9* low growing stem succulents (*Stapelia* type); grasses in between

## d Tropical East Africa

Finally, and still belonging to the Paleotropic realm are the extensive arid regions in the tropical parts of eastern Africa, as well as a small area in the rain shadow between the Pare and West Usambara mountains where very odd succulents are found (*Adenia globosa*, the boulderlike *Pyrenacantha*, *Euphorbia tirucalli*, *Caralluma*, *Cissus quadrangularis*, *Sansevieria* and others). This is probably the driest region near the equator, with an annual mean temperature of 28 °C and a rainfall of only 100–200 mm. In northern Kenya, western Ethiopia, Somalia and on Socotra, there are even more extensive arid regions where *Adenium socotranum* (Apocynaceae), a plant with bizarre succulent stems achieving diameters of up to 2 m, is found (Fig. 122), and *Dracaena cinnabari* with a trunk diameter of 1.6 m. The different life-forms of the thorn succulent savanna are shown schematically in Fig. 123. Most of these life-forms should be seen as adaptations to long dry periods, however, they were not able to penetrate into the actual deserts of zonobiome III.

## e South-West Madagascar

Madagascar with its unique flora and fauna, has rain forest of zonobiome I with annual rainfall of up to 2000 mm on its east coast. However, the largest part of the island has a climate of summer rains with deciduous forest. The tree and shrub flora of Madagascar was also unique with about 94% of endemic species. Whilst there used to be a high floristic biodiversity on Madagascar, now many of the forests and savannas have been deforested and large areas are degraded. Enormous grass regions are burnt annually, allegedly in order to gain grazing areas for at least 10 million zebus and in dryer parts goats are kept. The driest south-eastern corner of Madagascar is distinguished by baobab trees and plants of the columnar cactus-like family Didiereaceae (4 genera with 11 species), which only occurs here. With an irregular annual rainfall of 350 mm a thornbush succulent semi-desert developed. Many succulents of the genera *Euphorbia, Aloe, Kalanchoe* and *Crassula* occur, also bottle trees of the genera *Adansonia, Moringa* and *Pachypodium*. Other species have very small leaves, are thorny or even leafless. Poikilohydric vascular plants and ferns also occur. Many of these species are endemic.

## Questions

1. How is the vegetation of evergreen and deciduous tree species interlinked during the transition from humid to semi-arid tropical regions?
2. How does the zonation of mangroves differ in humid and arid regions and what are the salinity characteristics?
3. Which factors control the equilibrium between grasses and trees in the savanna?
4. Which types of savannas exist and which environmental factors determine their characteristics?
5. Because of overgrazing and too many fires savannas become covered with bush. What are the mechanisms of competition?
6. What is the role of impermeable layers in the soil in the expression of vegetation mosaics and how are these layers formed?
7. What is laterite?
8. Why do nutrient-poor quartz soils often occur in regions of zonobiome II?
9. What is the most important ecological factor in park savannas: flooding during the rainy period or the long drought during the dry period?
10. What are the vegetation types in orobiome II above the forest line?

# III Zonobiome of Hot Deserts (Zonobiome of Subtropical Arid Climates)

## Climatic Subzonobiomes

Deserts cover more than 35% of the earth's land surface. The cold winter period, which is typical of the arid regions of the temperate zone, is lacking in the subtropical desert zone (see Chap. VII). The term desert is a relative one. Contrasted with the humid eastern part of North America, the south-west looks like a desert, although Tucson (Arizona) has an annual rainfall of 300 mm. On the other hand, the Mediterranean coast of Egypt, with barely 150 mm of rainfall, is not considered to be a desert by an Egyptian from Cairo. In general, a hot region is termed desert when the annual rainfall is less than 200 mm and the potential evaporation more than 2000 mm (up to 5000 mm in the central Sahara).

The sparse precipitation of arid regions falls at different times of the year, thus providing a basis for a subdivision of zonobiome III into subzonobiomes (sZB), as follows:

1. sZB with two rainy seasons (Sonoran Desert, Karroo)
2. sZB with a winter rainy season (northern Sahara, Mohave Desert, Middle-Eastern deserts)
3. sZB with a summer rainy season (southern Sahara, inner Namib, Atacama)
4. sZB with sparse rainfall occurring at any time of year (central Australia)
5. sZB of the coastal deserts with almost no rainfall but much fog (north Chilean-Peruvian desert, outer Namib)
6. sZB of the rainless deserts devoid of vegetation (central Sahara)

Figure 124 shows the climate diagrams of the various subzonobiomes, with the exception of sZB 5 (since fog is very difficult to measure as precipitation and thus does not show up in the diagrams (see Fig. 136). A very distinctive feature of all arid regions is the large variability in the amount of rain falling in different years. This means that average figures are of little value. Although the years in

Deserts are arid regions where potential evaporation is very much higher than annual precipitation. These regions can be further subdivided into semi-arid, arid and extremely arid. In **zonobiome III** "hot deserts" occur, in zonobiome VII "winter cold deserts".

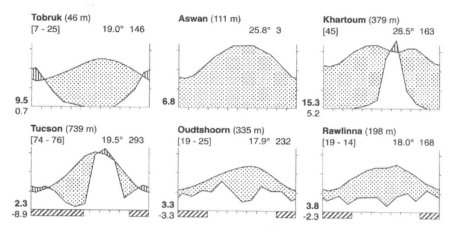

**Fig. 124.** Climate diagrams of desert stations. *Above* From northern Africa, with winter rain, no rain and summer rain; *below* with two rainy seasons (Sonoran Desert and the Karroo) and with rain that may fall at any season (Rawlinna, Australia). Compare also Fig. 134

**Fig. 125.** Curve showing variability in annual precipitation near Cairo, from 1906 to 1953 (from WALTER 1973)

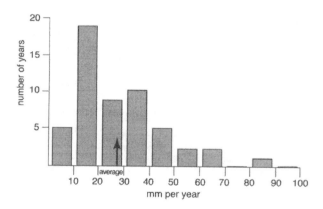

which the rainfall is below average are most frequent, the water reserves of the soil can be replenished for decades in the few years in which precipitation is large.

The variability curve for Cairo (winter rain region) is shown in Fig. 125. The curve for Mulka, the most arid station in central Australia, is similar except that the mean value is 100 mm and the extreme values are 18 and 344 mm. In Swakopmund (outer Namib) the mean precipitation is 15 mm, the lowest and highest values being zero and 140 mm. Each one is a skewed frequency distribution curve, so it would be much more sensible to show the median value.

Ecological conditions differ so much from year to year that an accurate picture of desert ecosystems can only be formed on the basis of long-term observations and each desert has to be considered individually. The few features that they have in common will be discussed first. In all deserts (except in the fog variety), the air is very dry. Both incoming and outgoing radiation are extremely intense, which means that the daily temperature fluctuations are large. In the rainy season, however, the extremes are greatly reduced.

## Soils and Their Water Content

Desert soils are not soils in the true sense of the word, but rather lithosols (syrozems), consisting of the erosion products of the underlying rock, modified by the action of wind and water. Therefore, the properties of the frequently loose bedrock are decisive. In other words, we cannot speak of climatic soils, but only of soil texture. Furthermore, instead of euclimatopes supporting a vegetation typical of the climatic zone, there are pedobiomes (e.g. lithobiomes, psammobiomes, halobiomes). The water supply of plants in arid regions depends upon the substrate, including the soil texture (particle size). The quantity of rain is only of indirect importance; the amount of water remaining in the soil, and thus available to plants, is far more important. Part of the rainwater runs off and a further portion evaporates (Fig. 126). How much of the water remains in the soil and is thus available to the plants, is determined by the texture of the soil. In humid regions, sandy soils are dry because they retain only small amounts of rainwater, whereas clay soils are wet. The reverse is true for arid regions.

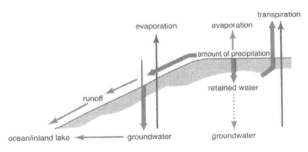

**Fig. 126.** Diagram showing the fate of rain in arid regions. Water retained in the soil is important to plants. The runoff water seeps down to the groundwater in the dry valleys and is only seldom reached by roots

**Fig. 127.** Schematic representation of water retention in various kinds of soils after a rainfall of 50 mm in arid regions. *h–h* Lower limit of moistened soil; *e–e* lower level to which the soil dries out again. The clay soil retains 50% or less; the sandy soil, 90%; and the stony soil, 100% (*hatched*)

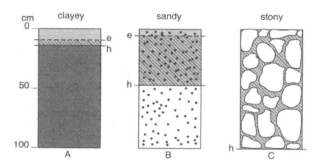

On flat ground in arid regions, water does not sink down to great depths and thus does not reach the groundwater. Only the upper soil layers are damp and the depth to which the water penetrates depends upon the field capacity of the soil and rainfall. Let us assume that 50 mm of rain falls upon a dry desert soil and that it completely soaks into the ground. If the soil is sandy, then the upper 50 cm is wetted to field capacity. If the soil is a finely granulated clay with a field capacity five times as large, the water can only penetrate to a depth of 10 cm. On rocky ground with small cracks, the water goes down much further, possibly 100 cm (Fig. 127).

Evaporation follows the rain. If the upper 5 cm of a clay soil dries out, then 50% of the water originally entering the soil is lost. Sandy soils do not dry out so much and even if the upper 5 cm were to dry out, this would involve a loss of only 10% of the water. On rocky ground there is almost no evaporation and nearly all of the water is retained. This means that, contrary to the situation in humid regions, clay soils form the driest habitats whereas sandy soils offer better water supplies. Fissured, rocky ground provides the wettest habitat if there is no runoff from the rock surface and if there is enough fine soil in the cracks to retain the water.

Such considerations have been confirmed by measurements in the Negev Desert. Comparison of areas with identical absolute rainfall revealed that loess soil offered plants the equivalent of 35 mm of rainfall in available water; rocky habitats with a relatively high runoff, 50 mm; sandy soils, 90 mm; and dry valleys with a large inflow, 250–500 mm. That sandy soils are favourable habitats for plants in arid regions can be seen from the fact that the same type of vegetation occurs on sand at a lower rainfall than on clay. In the Sudan, *Acacia tortilis* semi-desert is found on sandy soil in a zone which has a rainfall of 50–

250 mm. On clay, however, this species is found only in a zone with a rainfall of 400 mm. The *Acacia mellifera* savanna begins to grow on sandy soil at an annual rainfall of 250–400 mm, but on clay soil at 400–600 mm. In the short-grass prairie region of the Great Plains in western Nebraska, tall grass prairie occurs on sandy soil although, otherwise, it only occurs further east at a higher rainfall. The favourable water supply on rocky ground is often marked in arid regions by the occurrence of trees in the midst of the lower vegetation supported by finely grained soils. If sandy soil or the soil in rocky clefts is wet down to the groundwater, the roots of the plants also grow down this far, thus securing their water supply. The following example is worthy of mention. North of Basra in Mesopotamia, groundwater is present at a depth of 15 m and is constantly replenished from the Tigris and Euphrates via gravel strata. Since, however, the rainfall amounts to only 120 mm annually the upper soil layer alone is damp and the plants are unable to reach down to the groundwater. As a result, there is only a sparse ephemeral vegetation after the winter rains. The native population has dug wells and uses the water for cultivating vegetables, which they plant in furrows and irrigate several times daily since the temperature can reach 50 °C. The soil rapidly turns brackish due to the high evaporation, so that vegetables cannot be planted in the same spot for more than 1 year. Between the vegetable plants, however, *Tamarix articulata* cuttings are planted and rapidly take root. In the second year the furrows are not irrigated, but the soil is still damp down to the groundwater from the previous year. The tamarisk roots can therefore grow deeper and deeper in the ensuing years until they finally reach groundwater, after which their water supply is secure and they can develop into large trees. They are cut down every 25 years for fuel and shoot up again from the stumps; in this way, the farmland is transformed into tamarisk forest. Deserts with deep-lying groundwater can be converted into forest if the soil is irrigated to such an extent in the first year after the trees have been planted that it is wetted down to the groundwater.

This example provides the explanation of the fact that phreatophytes, which are dependent upon groundwater, can reach down to it with their long roots, even through several meters of overlying dry soil. This can only be accomplished, however, if the soil is wet from the surface down to the groundwater, as it is after several years in succession with good rains. Once the groundwater has been reached, however, woody plants can attain their usual life span. It need not necessarily be groundwater

Desert vegetation has larger water reserves at its disposal than is apparent to the observer at first sight.

that the plants seek; often it is merely the ground moisture, i.e. interstitial water stored in the soil. Once this
water has reached a depth of 1 m or more, it remains
there for long periods if there is no vegetation or if only a
few plants have succeeded in tapping it with their roots.
Saline soils are very frequent in the desert, particularly in
depressions and will be dealt with separately (see p. 218).

# 3 Substrate-Dependent Desert Types

Desert biomes can be classified on the basis of soil texture
into the following biogeocene complexes. Since these soils
were first studied in the Sahara they bear the local descriptive names.

### a Rocky Desert (Hamada)

A rocky desert occurs if the parent rock developed during
the course of geological history remains on the surface.
This is only seldom found in arid regions as mountains
often disappear completely under their own rubble because of physical weathering. Larger stones are mainly
found on plateaux of the table mountains (mesas), from
which all the finer products of weathering have been
blown away and where the exposed rocks have undergone
severe wind erosion due to sandblasting. The surface is
covered with a pavement formed of hand-sized stones,
darkly stained by desert-varnish (Mn oxides), which lend

**Fig. 128.** Fish River Canyon
in the desert of southern
Namibia (Photo: E. WAL-
TER)

a forbidding aspect to the landscape. Beneath the stony pavement there may be a water-repellent, dusty soil, rich in gypsum and salt if it has originated from marine sediments, which prevents the development of plant cover. Hamada areas are cleft by deep erosion valleys with steep, rubble-covered slopes (Fig. 128). In the cracks and crevices of the rocks, a few xerohalophytic species and a few others can take hold.

### b  Gravel Desert (Serir or Reg)

Gravel deserts arise from heterogeneous, conglomerate parent rock. The cementing substance is readily weathered and removed by the wind and the harder pebbles collect on the surface. Such autochthonous gravel deserts are in contrast to the allochthonous ones, which consist of alluvial deposits of earlier rainy periods and from which, again, the finer material has been blown away. Under the darkly stained gravel layer there may also be a crust, cemented hard with gypsum. It is a particularly monotonous type of desert, slightly undulating, with broad, shallow, sand-filled valleys offering better growth conditions for plants typical of sandy soil and a few xerohalophytes.

### c  Sandy Desert (Erg or Areg)

Sandy deserts are formed in large basin areas by the deposition of sand blown off raised ground and leading to sand dune formation. If there is a prevailing wind direction, then sickle-shaped dunes or barchans are formed, gently sloping on the convex, windward side, and steeply sloping on the concave, lee side. The dunes move in the direction of the wind, but if the wind direction varies periodically, the crest of the dunes alters while the base remains fixed. A thin covering of iron oxide on the sand grains accounts for the bright-red colour of dunes in hot, dry regions. Near the coast, where the air is more humid, the colour changes to yellowish-brown. These mobile, and therefore barren, dunes can store water because the rain sinks in readily and hardly evaporates. Even at an annual rainfall of only 100 mm, a fresh groundwater horizon is present so that water can be obtained by sinking wells or the water comes out between the dunes.

If the sandy covering is not very deep, colonisation by plants is possible (non-halophytes such as dune grasses, *Ziziphus* and others). Adapted perennial species, including shrubs, serve as sand catchers and grow up through the sand that has accumulated around them, thus trapping still more sand. In this manner each plant can form its own dune-hillock (several meters high), called a **nebkha**. These miniature dunes lend a characteristic note to the entire landscape.

### d Dry Valleys (Wadis or Oueds)

Known in south-west Africa as 'riviers' and in America as 'washes' or 'arroyos', they are an important feature in all deserts. They mostly originated in the past during pluvial periods, when the rainfall was higher. The dry valleys commence as scarcely noticeable erosion gullies which then unite to form deep ditches or small valleys and often end in deep canyons. Gravel and sand are deposited by the water as it drains off after a shower. Some of the salt is washed out and the soil is soaked to a considerable depth providing favourable conditions for the growth of halophytic plants (*Tamarix, Nitraria*). The beds of the larger dry valleys bear no vegetation owing to the redistribution of the soil by occasional floods. Vegetation is confined to the valley sides, which are safe from floods and its degree of luxuriance depends upon the amount of water held in the alluvial deposits. There is often a permanent underground flow of water, and in such cases, dense, often linear, non-halophytic groves are present as extrazonal vegetation. In smaller wadis, water running off is caught in terraces which makes agriculture possible.

### e Pans (Sebkhas, Dayas or Chotts)

Hamada – serir – erg – takyr – sebka: this is often a geomorphic sequence of types of deserts which form and correspond in their structure, a large catena (here because of areas of the landscape interlinked by erosion and accumulation) with a substrate sorted according to the particle size.

They are hollows or larger depressions in which silt or clay particles brought in by the water from the wadis are deposited. If there is subterranean drainage (in karst areas), they do not turn brackish. This is also true of the **takyr**, or delta-like formations at the valley exits from which a part of the water drains off after a particularly heavy rainfall. The heavy clay soils, however, provide unfavourable habitats since the water can scarcely penetrate the soil and the ground rapidly dries out again after a flood. For this reason, mainly algae, lichens and ephemeral species grow on takyr soil. If there is no outflow and all the water evaporates, then salt concentration takes place, and in such salt pans, called halobiomes, compact layers of salt form in the deepest places. At the edges, where the salt concentration is lower, hygrohalophytes take hold. The salt content of the groundwater is often low and a salt layer forms only on the surface. If a thin layer of sand is deposited on the surface of such a salt pan, there can be no capillary rise of water and hence no salt concentration. Plants soon establish themselves on the sandy deposits and serve to trap even more sand so that a hillocky, or **nebkha**, type of landscape is formed around the pans.

## f Oases

Oases are those sites in the desert with a dense vegetation, where water of a low salt concentration reaches the surface, either by means of normal springs or artesian wells. Hygrophylic species can grow here. Such oases are nowadays densely populated, and the natural vegetation has been replaced by cultivated plants or weeds. Oases with abundant water are often fringed by salt pans (chotts), where the excess water collects and evaporates (southern Tunisia, Algeria).

##  Water Supply of Desert Plants

The extreme dryness of arid regions has led to the false assumption among researchers with no personal knowledge of the desert that desert plants possess special physiological properties – a physiological resistance to drought – enabling them to grow under arid conditions. In particular, an allegedly high cell sap concentration is often mentioned in connection with the ability of the plants to take up water even from almost dried-out soils. However, detailed ecophysiological investigations over the past decades have shown that this view is incorrect. The water supply of desert plants is not so poor as would be suggested by the low rainfall. Rainfall measured in millimetres is equivalent to litres of water per square metre of ground surface. In order to judge how much water is available to the plants, the transpiring surface per square metre of ground surface must be calculated.

Although there are many different kinds of deserts, they are all alike in the sparseness of their plant cover.

The character of the landscape in deserts is determined by the naked rock and not by the plants. In order to study the exact relationship between rainfall and density of vegetation, identical life forms must be compared (for example, grasses or trees with similar foliage), and a region must be chosen in which rainfall varies over a relatively short distance and temperature conditions remain more or less constant. Furthermore, euclimatopes should be chosen, where the vegetation has in no way been disturbed by human action.

Suitable regions are to be found in South West Africa, with grass cover and an annual rainfall of 100–500 mm, and south-western Australia, with *Eucalyptus* forests and

Water supply per unit of transpiring surface is more or less the same in arid and in humid regions (annual rainfall of 100–1500 mm).

**Fig. 129.** Organic material production (above ground dry weight in t ha$^{-1}$) of grassland in south-west Africa, in relation to the annual precipitation in mm

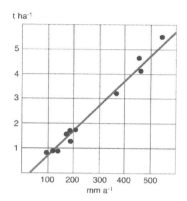

a rainfall of 500–1500 mm. Such investigations have revealed a linear relationship between amount of rainfall and production of plant mass or transpiring area (Fig. 129). This also holds true for the creosote bush desert (*Larrea divaricata*) in south-eastern California as well as for more ephemeral vegetation in arid areas with rainfall up to 100 mm. Grass seedlings require 16–17 mm for the process of germination and are less economical with water than older grasses so the curve rises less steeply.

The drier the region the further apart the plants grow, thus leaving a greater area from which the individuals can take up water. This has been confirmed in northern Africa in olive plantations. The number of trees per hectare decreases with decreasing rainfall until, finally, only 25 trees/ha are left. However, since the individual trees bear approximately the same amount of fruit, it is evident that the water supply is unchanged. In cereal-growing areas, too, it is known that crop-plant density should decrease with decreasing rainfall. To take up water from a larger soil volume, a correspondingly larger root system has to be developed.

The **diffuse vegetation** with an even distribution of perennial (persistent) plants over apparently flat areas changes to a so-called **restricted type of vegetation** ("contracted vegetation"), i.e. plants grow only in the often barely noticeable erosion gullies or depressions, whilst raised areas have no vegetation at all.

Another essential characteristic is that plants reduce the transpiring surface with increasing aridity, but develop a denser root system. It has been shown that, with increasing cell sap concentration, a marked inhibition of shoot growth occurs immediately whilst root growth is at first actually stimulated. In wet regions, the larger part of the phytomass is above ground and in arid regions it is underground. This does not mean that the roots penetrate deeper in dry regions as is usually suggested. Rather, the opposite is true: the root system flattens out. The scantier the rainfall, the less deeply the soil is wetted and beneath the upper layer containing water, there is no water available at all for the plants. Long taproots have been observed only in plants that are dependent upon groundwater. These are, however, special cases from which generalisations should not be made.

In extremely arid regions with a rainfall below 100 mm, a change in the type of plant cover is noticeable and there is no homogeneous vegetation. This is connected with the distribution of water in the soil. In extreme desert regions, apart from the shifting sands, soils usually have a surface crust which can be moistened only with difficulty. The rare, but usually torrential rains, therefore, scarcely penetrate the soil and to a large extent run off the surface. The sandy erosion gullies and depressions therefore receive much more water than the rainfall would suggest and this runoff water penetrates deeply into the soil. In such places, plants develop roots that reach down to water which may be at a depth of several metres. In some places, groundwater even collects in the valleys. With a rainfall of only 25 mm, vegetation is present in all of the valleys in the desert near Cairo-Heluan. Assuming that 40% of the rainwater runs off into the deeper parts of the relief and that these depressions account for only 2% of the total area, then at a rainfall of 25 mm the amount of water available to plants in these biotopes is the same, because of the inflowing water, as if they were growing on level ground with a rainfall of 500 mm. Water loss due to transpiration of the plant cover in such a habitat near Heluan was, in fact, found to be 400 mm. Even in a rainless summer, the cell sap concentration of these plants does not rise, thus confirming that they are well supplied with water. The sandy depressions in the gravel desert along the Cairo-Suez road permanently contain 2.4% water at a depth of only 75 cm (wilting point 0.8%), so they never dry out and are capable of supporting a sparse perennial vegetation. In some erosion gullies, roots descend as far as 5 m, depending upon the depth to which the soil remains damp. Despite the extreme aridity, 200 vascular plant species can be found in the area around Cairo.

The water supply of plants in extreme deserts is therefore not so poor as is usually assumed. However dry the soil surface may appear to be, there is always some water available, at least at certain times, wherever plants are found growing in the desert. Of course, these plants must be able to resist long periods of drought and this they achieve chiefly by means of special morphological adaptations. There is no significant protoplasmic resistance to drought and cell sap concentration is generally low (with the exception of halophytes).

The Berber population of southern Tunisia has exploited the principle of restricted vegetation for countless centuries in order to obtain crops at an annual rainfall of
200 mm or less. Each small gully has a dam to prevent the
water from draining off and date palms, cereals or beans
can be cultivated on the damp soil caught behind the
dam. It has been established that a similar type of runoff
farming was practised in pre-Arabian times by the Naba-
teans in the Negev Desert. The old dams have been reno-
vated in recent times and experiments with various cultivated plants have been successful (EVENARI et al. 1982).

## Ecological Types of Desert Plants

All plants growing in arid regions have been called xerophytes. This is not appropriate, since in every such region
there are habitats, such as the oases, where plants are well
supplied with water. Even species typical of the humid
tropics grow in such habitats. In the rainless Aswan Desert, on an island in the Nile, coconut palms, mangoes, pa-
paya, maté, sweet potatoes, manioc, camphor trees, mahogany trees, coffee, pomegranates and many other species
typical of the Indian monsoon forests are cultivated with
the aid of artificial irrigation. The microclimate in the
dense plantations is less extreme than in the open desert.
In dry valleys with groundwater, plants can grow under natural conditions without suffering any water shortage and
without any sign of adaptation to the dryness. Besides, most
deserts have at least a brief wet season, with the exception of
the rainless central Sahara, the Namib and the Peruvian-
Chilean desert. The species which develop during these
damp periods (*therophytes, ephemerals*), including those
which survive the rest of the time as seeds (*therophytes*)
or in the ground (*geophytes = ephemeroids*), do not exhibit
any particular adaptation to water shortage either.

It would be illogical for an ecologist to draw a distinction between plant species that avoid drought and those
resistant to it. All plant species tolerate drought, some as
seeds (ephemerals), tubers, or bulbs (ephemeroids), some
in a latent condition, like poikilohydric lower plants (algae, lichens), a number of ferns (*Cheilanthes, Notholaena*,
see Fig. 130) or *Selaginella* species, and even flowering
plants, the best known of which is *Myrothamnus flabellifolia* (Rosales; Fig. 130), and some in a state of reduced
activity (xerophytes and succulents).

The term **xerophyte** is used here to describe ecological
groups that require a certain minimum amount of water

**Fig. 130.** *Myrothamnus flabellifolia* (*left*) in latent state (twigs drawn together, leaves folded) and a *Notholaena* species (*right*), two poikilohydric species on mica-schist of the steep escarpment to the Namib Desert in Namibia (Photo: S.-W. BRECKLE)

uptake even in periods of drought, since they possess no adequate water-storage organ. These fall into three subgroups linked by transitional forms:

1. **Malakophyllous xerophytes** are characteristic of semi-arid regions. They have soft leaves which wilt under dry conditions while the cell sap concentration rises steeply. They lose their leaves in lengthy dry periods and only the youngest of the leaves within the hair-covered buds survive. Typical examples are the many Labiatae and Compositae of arid regions and *Cistus* (rock roses) of arid regions.
2. **Sclerophyllic xerophytes** have small, hard leaves and owe their rigidity to mechanical tissue. They are found especially in regions with a long summer drought and are able to reduce their transpiration to a minimum when water is scarce, whereby the cell sap concentration rises only in extreme circumstances. Typical examples are evergreen oaks and *Phillyrea*.
3. **Stenohydric xerophytes** close their stomata in drought conditions at any sign of water shortage and are thus able to avoid a rise in cell sap concentration. As a consequence, gas exchange and thus, photosynthesis are halted, so that plants are in a state of starvation. Leaves of such species do not dry out during the long droughts, but turn yellow and finally fall off. Some non-succulent spurges (*Euphorbia* species) may be cited as examples. Most plants of extreme deserts belong to this group.

In deserts it is more important for plants to survive, and not to produce large quantities of phytomass. There is no competition between the parts of the plants above ground.

Plants achieve survival in the desert with incredible endurance, often as pitiful-looking cripples, and may live for

100 years or more. Although many branches die, enough survive to ensure further growth after future rainfall.

A special group is formed by the **succulents,** water-storing species which use their stored water extremely sparingly in times of drought. Their small water-absorbing roots die, so that no water at all is taken up from the ground during the dry period. Succulents can be divided into the following three groups according to the nature of the organ responsible for storing water during the rainy season:

1. **plants with succulent leaves,** such as *Agave* and *Aloe*, or *Cotyledon, Crassula, Sansevieria.*
2. **plants with succulent stems,** such as cacti and many species of *Euphorbia, Stapelia, Kleinia.*
3. **plants with succulent roots**, that is to say with underground storage organs, such as *Asparagus* species, *Pachypodium,* as well as some Leguminosae with enormous tubers found growing in the sandy regions of the Kalahari.

Von WILLERT (1990) has divided succulents into more exact groups. He differentiates between the types listed in Table 17 on the basis of their development during the seasons.

The concentration of the cell sap in succulents is very low and does not rise even during long periods of drought when large amounts of water have been lost. The water content, calculated on the basis of dry weight, re-

**Table 17.** Life-forms of succulents in the deserts of southern Africa

| |
|---|
| 1  Ephemerals (germination possible after each rain) |
| 2  Annuals<br> • Summer annuals (germination only with summer rains)<br> • Winter annuals (germination only with winter rains) |
| 3  Pauciennes (live only a few years) |
| 4  Perennials (persist for many years)<br> • Geophytes<br>   • Flowers and leaves simultaneously<br>   • Flowers and leaves at different seasons<br> • Root system persisting<br>   • Aboveground flowers only<br> • Aboveground persistent plants<br>   • Except cotyledons no green leaves<br>   • With annual change of leaves (rain-green)<br>   • Evergreen |

(adapted after VON WILLERT et al. 1990)

mains constant since respiration involves the breakdown of organic compounds (sugars, organic acids and others). Many succulents can survive for a year without taking up any water, and in many of them diurnal acid metabolism (CAM = Crassulacean Acid Metabolism) has been demonstrated. Plants with this type of metabolism open their stomata at night when water loss due to transpiration is small and take up $CO_2$. This leads to the formation of organic acids and, thus, to a rise in acidity of the cell sap. The stomata are closed in the daytime and the $CO_2$, which has been bound during the night, can be assimilated in daylight, with an accompanying decrease in acidity. In this manner, the necessary gaseous exchange is effected with a minimum loss of water (DINGER & PATTEN 1974).

For annual succulents the summer annuals are predominantly C4 plants (for example *Zygophyllum simplex*), winter annual CAM plants are, for example, *Opophytum aquosum*.

**Salt plants, or halophytes,** constitute a very important group in many deserts (see p. 56). Their occurrence depends upon the presence of a saline soil rather than upon climate. Their distribution extends beyond the limits of the zonobiome.

## Productivity of Desert Vegetation

If a plant lowers its rates of transpiration and photosynthesis during times of drought by means of a reduction of the active surface, its production is also reduced and may even stop if dryness continues for longer periods. In years with plentiful rainfall, on the other hand, plants develop more luxuriantly, but they cannot use up all of the available water. The surplus is exploited by the ephemerals which develop particularly well under such circumstances and can be considered as fulfilling a buffer role in smoothing out the larger fluctuations in annual precipitation.

In years with meagre rainfall, ephemerals scarcely develop at all or, at best, are represented by dwarf forms. If surface reduction of perennial forms is insufficient to maintain their water balance, the larger part of the plant dies off because the maximum $\pi$ has been exceeded. For survival, it suffices if the shoot meristem of only one branch remains alive to sprout after a rain. All woody plants in the desert bear large numbers of dead branches, indicating past years of drought. Reproduction by means of seeds only takes place after a good rain-year or several successive good years, which seldom happens more than once in a century. Young plants are therefore hardly ever found. Under such circumstances, it is impossible to obtain mean values for production.

The leaf area index for perennial species is less than one, even in exceptionally good years, and only a luxuriant growth of ephemerals in a good rain year can lead to a reasonable production. In the extreme desert near Cairo the production of ephemeral vegetation has been measured after the upper 25 cm of the soil had been soaked by winter rain amounting to 23.4 mm. Of this, 68% was lost by evaporation, and transpiration of the ephemerals during the winter months accounted for 7.3 mm, or 32%, which is the equivalent of 730 kg of water per 100 m². Over the same area, the ephemerals produced 9.384 kg fresh mass, or 0.518 kg dry matter. This gives a transpiration coefficient of 730:0.518 = 1409, which is very high compared with values for central European crops (400 to 700), and is attributable to the low air humidity of the desert. Similar values were found with very little precipitation by SEELY (1978) for annual grasses in the Namib. Zoomass in the desert is extremely small and secondary production is thus almost non-existent. Nevertheless, even in the desert, food chains play a not insignificant regulatory role in the ecosystem (pp. 91 f.).

Special investigations on the productivity of agaves and spherical cacti, such as have been carried out in the western part of the Sonoran Desert in California with its summer drought period, should also be mentioned.

a) NOBEL (1976) gives exact quantitative data (mean values) for *Agave deserti*, which also occurs in the eastern Sonoran Desert. The following data are for plants with an average of 29 leaves: length of leaves 30 cm, surface area 380 cm², fresh weight per leaf 348 g, dry weight 47 g, number of stomata per mm² 30, number of roots per plant 88, root length 46 cm. The roots are spread out flat and radially in order to take optimal advantage of each rainfall.

The stomata open during the rainy season (November to May) for 154–175 days at a soil-water potential of –0.01 MPa. Water uptake ceases when the potential decreases to –0.3 MPa at the beginning of the drought period. The stomata open for the first 8 nights of the drought period and then remain closed. Diurnal acid metabolism (CAM) does not take place until the following rainfall.

Transpiration losses in 1975 were 20.3 kg/plant, corresponding to a rainfall on the densely rooted soil of 26.9 mm, or 35% of the annual precipitation. The transpiration coefficient, i.e. the ratio of the transpired water (in kg) to the amount of dry matter produced, was 25 kg which is quite low and shows a very economical use of water. Per plant, 0.8 kg dry weight was produced annually. Growth is, therefore, very slow and only the older plants blossom once and then die, since in order to produce the large inflorescences all material and water reserves of the plant are exhausted.

This is confirmed by the following observations (NO-BEL 1977a): coming into blossom, the old *Agave* plant had 68 leaves which were 4.1 cm thick as the inflorescence was just becoming visible. After development of the inflorescence, the leaves shrivelled up, faded in colour and were only 1.4 cm thick. In all, the leaves lost 24.9 kg fresh weight and 1.84 kg dry weight during blossoming. Since water uptake is insufficient, 17.8 kg of water was taken from the leaves to supply the inflorescence. The dry weight of the inflorescence was 1.25 kg and 0.59 kg was respired. A plant in blossom produces 65,000 seeds, 85% of which is consumed by animals. Not a single young agave could be located within a 400-m$^2$ area on which 300 agaves were found. Reproduction by seed occurs only in favourable years, otherwise vegetative propagation by runners takes place. These values clearly indicate why agaves are hapaxanthic (monocarpic) species, i.e. plants which only produce blossoms and fruit once and then die.

A further detailed production analysis was carried out in the same area with the spherical cactus *Ferocactus acanthoides* (NOBEL 1977b). Although this is also a species with CAM metabolism, the expenditure required for the blossoms is so low that it blossoms annually. The plant studied was 34 cm tall, 26 cm in circumference and weighed 10.8 kg (with a water content of 8.9 kg). The loss by transpiration in 1 year was 14.8 kg, plus an additional 0.6 kg for the transpiration and growth of the generative organs. $CO_2$ assimilation resulted in the production of 1.6 kg in 1 year, one third of which was respired. The annual growth was measured at 9% and the transpiration coefficient was 70. Although still quite low, it was higher than that of the agave. The behaviour of the stomata was similar to that of the agaves.

## Desert Vegetation of the Various Floristic Realms

At the time when the conquest of the desert by plants took place, during the evolution of a terrestrial vegetation, the floristic realms were already differentiated. Since the plant families, or speaking more generally the taxa of the various floristic realms, differ in their genetic constitution, adaptation to life under arid conditions has also taken different directions in the various floristic realms. Deserts are not only floristically different, the life-forms need not necessarily be similar although convergences do occur.

### a Sahara

Only the northern part of the largest subtropical desert, the northern Sahara-Arabian desert, belongs to the Holarctic realm. In the east, this desert borders the Irano-Turanian and central Asiatic deserts, which have cold winters. The northern limit for productive date palm cultivation forms the border between the two. Chenopodiaceae are especially well represented in the Sahara-Arabian desert, partly on account of the extensive occurrence of saline soils. Succulent species of *Euphorbia* are found only in western Morocco, most of the species are xerophytic dwarf shrubs, some of them broom-like bushes. The only grasses present are xeromorphic with hard leaves: *Stipa tenacissima* and *Lygeum spartum* (transitional zone), *Panicum turgidum, Aristida pungens* and others. Many ephemeral species appear after a good winter rain.

The central part of the enormous Sahara was once, but is no longer, a transitional zone between the northern winter rain regions on the Mediterranean and the southern summer rain regions. This central part is now a desert almost without rain, an extreme desert with very rare periods of precipitation. Despite this a flora exists, even if only a few species. Small, locally limited showers can, in a very limited area, cause the germination of certain annuals, particularly *Zygophyllum simplex*. The landscape is determined by the existing geological layers of rocks and their specific characteristics of weathering (see Fig. 131). Often large blocks or even smaller individual mountains remains as islands in the landscape.

Shrubs such as *Tamarix, Nitraria* and *Ziziphus* are confined to wet habitats and vegetation is restricted to small channels or wadis. Paleotropic elements are numerous, including the species of *Acacia* found in the dry valleys carrying groundwater. The southern Sahara, with the Sahel providing a transition to the summer rain region of the Sudan, belongs to the Paleotropic realm. Grasses with less-hard leaves (*Aristida, Eragrostis,* Panicoideae) are much more common here. There are also far more shrubs (*Acacia, Commiphora, Maerua, Grewia*) as well as herbaceous plants (*Calotropis, Crotalaria, Aerva* and others) which are also typical of the Thar of Sind deserts (see p. 204).

**Fig. 131.** Extreme deserts in the Egyptian Sahara south of Aswan with long-term, mean annual precipitation of 1–2 mm. Rocky outcrops and stony desert (hamada) with some sand dunes (erg) (Photo: S.-W. BRECKLE)

## b  Negev and Sinai

They serve as a bridge from the east of the Sahara to the Arabian deserts. On the peninsula of Sinai, mountain deserts predominate with Irano-Turanian plants growing in the higher regions. The northern Sinai and Negev region is characterised by extensive areas of sand, but sand dunes occur only if there is excessive grazing. Precipitation shows a stark gradient from north to south (see map in Fig. 132).

The north-eastern part of the Sinai peninsula, the Negev Desert leads, via the rift valley which includes the Arawa valley, the Dead Sea and the valley of the Jordan, to the Jordanian desert. Intensive ecological research has been performed in this area for many decades, it is therefore one of the most researched deserts (see WALTER & BRECKLE 1991 a). The Negev Desert is fairly small in area, but is nevertheless very important from a floristic point of view as the transition area between different floristic regions. In this area, various desert vegetations meet over a very short distance, the Mediterranean from the north, the Irano-Turanian from the north-east, the Saharan from the west and south-west and the Arabian from the east. Furthermore, there are Sudanese enclaves, particularly in the lower-lying rift valley, for example with *Salvadora persica, Cordia gharaf, Maerua crassifolia. Cyperus papyrus* occurs also in the Huleh swamps on the upper Jordan where, at the same time, *Nymphea alba* (as a Holarctic plant) reaches its southernmost limit.

**Fig. 132.** Annual precipitation in the region of Israel. Note the substantial increase in gradient from south to north

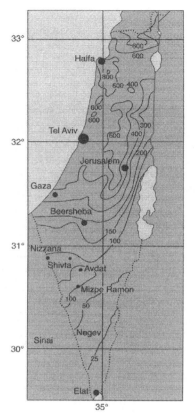

### c Arabian Peninsula

The Arabian desert, at the same latitude as the Sahara, continues the belt of deserts towards the east. On almost all of the peninsula, precipitation is between 15 and 100 mm; in some very steep areas it can occasionally be above 100 mm and in the mountainous areas above 2000 m the precipitation is between 250 and 650 mm. In northern Yemen a very marked sequence of altitudes is recognisable with a rich vegetation, evergreen sclerophyllic woodland and even some tropical genera. The eastern part of the peninsular is the Rub-al-Khali, an enormous area of sand desert. Here, the same differentiation of vegetation occurs as in the Sahara, based on geomorphological conditions. The vegetation is almost exclusively restricted to the groves in the larger wadis, characterised by acacias and mixed at times with several other woody species. In the southern area there is a transition to the acacia-thorn savanna (zonoecotone III/II). At times it also rains in summer (for example in Sana).

## d Sonora

In North America only the deserts in southern California and southern Arizona belong to the subtropical deserts, however, with Holarctic floristic elements. The arid regions in northern Arizona, Utah and Nevada have very cold winters (ZB VII). Several semi-desert to desert regions are Neotropic: the Sonoran Desert (northern Mexico and southern Arizona), although actually in North America, belongs to the Neotropic realm, floristically speaking. Extensive investigations of this desert (or rather semi-desert) have been carried out from the Desert Laboratory in Tucson, Arizona. The vegetation with its tall candelabra cacti, is termed a "cacti forest". By means of a sort of bellows mechanism, these succulents can store so much water that they are capable of surviving for more than a year without any further water uptake (Figs. 33–35). They have shallow root systems, but within 24 h after wetting of the upper soil layer, fine absorbing roots are put out and the water-storing tissue fills up. Apart from the succulent cacti, other ecological types are represented here: winter and summer ephemerals, poikilohydric ferns, malakophyllous half-shrubs (*Encelia*), sclerophyllic species, stenohydric plants and the deciduous *Fouquieria* which develops new leaves after each heavy rain shower although afterwards they rapidly turn yellow due to water shortage. Wide, flat, dry areas are covered with the particularly drought-resistant creosote bush (*Larrea divaricata*), which smells strongly of creosote when its leaves are wetted. This species is also characteristic of the Mohave Desert, which only receives winter rain and is poor in succulents. In the lee of the High Andes, along their eastern foot, a *Larrea* desert stretches more than 2000 km, from northern Argentina into cold Patagonia. The predominating species, *Larrea divaricata*, is probably identical with that found in Arizona (BÖCHER et al. 1972).

## e Australian Deserts

A very different situation is encountered in the arid regions of Australia. The whole of central Australia is arid, however, it has no climatic deserts. Sand-dune regions (the Gibson and Simpson deserts), although not climatically the driest parts of Australia, are desert-like in character as are the gibber plains, bare, stony areas produced by overgrazing. Vegetation in the driest parts, with scanty rainfall at any time of year, is composed of "salt bush" (*Atriplex vesicaria*) and "blue bush" (*Maireana* (*Kochia*) *sedifolia*), both Chenopodiaceae. They occur either in pure populations, or mixed (shrub semi-desert). The soils upon which *Atriplex* grows contain little chloride (about 0.1% dry weight), but since these loamy soils dry out to a considerable extent the concentration of chloride can in fact be very high. The cell sap concentration of *Atriplex* is also correspondingly high (equivalent to -4 to -5 MPa), chlorides accounting for 60-70% of the total; it is in fact an euhalophyte, the growth of which is enhanced by salt. A certain degree of salt excretion is achieved by means of the short-lived vesiculated hairs, which are continuously being replaced. *Atriplex* is a half-shrub which lives for about 12 years. Like most halophytes, it possesses weakly succulent leaves and a root system spreading widely at a depth of about 10-20 cm (above a chalk layer). Therefore, the bushes grow rather far apart.

In contrast, *Maireana sedifolia* is said to be long-lived. Its root system not only penetrates to a depth of 3-4 m into cracks in the chalk layer, but also spreads equally far laterally. This species grows wherever rainwater percolates to greater depths, such as on a light or stony soil. The cell sap concentration of this species is only half that of *Atriplex* and the part played by chloride is also smaller (about 20-40%). This species is thus, probably a facultative halophyte. It can attain dominance if the climate becomes more humid. In the salt bush region there are scattered sand dunes or sandy areas where moisture conditions are more favourable and where the soil is not saline (free of chloride). Shrubs such as *Acacia, Casuarina* and *Eremophila* can be found. Tree-like species of *Heterodendron* and *Myoporum* as well as species of *Eremophila* and *Cassia* are confined to silty soils. The most widely occurring species in Central Australia is *Acacia aneura* ("mulga"). It dominates large areas, which look like a grey sea when seen from the air. The shrub reaches a height of 4-6 m and has thin, cylindrical or somewhat flattened, resin-covered phyllodes. Its root system is well developed and penetrates the hard soil layers to a depth of about 2 m. Owing to the irregular rainfall, flowering is not connected with any particular season but rather with the occurrence of rain (see p. 212). Fruits and seeds develop after a heavy rain shower and at the same time the ground is carpeted with white, yellow and pink everlasting plants (strawflowers) belonging to the Compositae family (Fig. 133).

**Fig. 133.** Mulga vegetation in the interior of Australia near Wiluna after rain. Large bushes, *Acacia aneura*, smaller bushes, *Eremophila* species. The ground is densely covered with temporarily active everlasting plants such as *Waitzia aurea* and white *Helipterum* species (Photo: E. WALTER)

*Acacia aneura* is sensitive to salt but can survive long periods of drought. In dry habitats the bushes grow well apart, but in wet depressions they form thickets. *Rhagodia baccata* and *Acacia craspedocarpa* have recently been the subject of detailed ecophysiological studies.

The porcupine grasses (*Triodia, Plectrachne*) are another important group, collectively forming what is known as "spinifex" grassland. They are sclerophyllic species with very hard, rolled-up, pointed, perennial leaves covered with resin and they form large round cushions, or cupolas in the case of *Triodia pungens*, with a height of up to 2 m. *Triodia basedowii* dominates the sandy areas of the most arid part of western Australia. Its dense root system goes straight down for 3 m. Older cushions disintegrate and form garlands. Other characteristic genera, represented by many species, are *Eremophila, Dodonaea, Hakea* and *Grevillea* amongst others. The structure of the vegetation is determined by the kind of soil and the sheet floods which follow heavy rains, both factors leading to a complicated mosaic of vegetation. The Quaternary history derived by CROWLEY (1994) from pollen diagrams of various sea sediments shows that at the end of the last glacial period there was an increase in the amount of rain and in consequence a reduced salinity in the Australian desert areas, which again increased 5000 years ago and has been particularly noticeable since the arrival of the European settlers.

### f Namib and Karroo

Of the South African deserts the Namib and Karroo are Paleotropic. Capensic floristic elements can be found in places here. The Namib extends along the coast of southwest Africa. The fog-rich coastal Namib differs from the southern Namib in the transition region to the Karroo with the two rain periods and the actual desert between the region of the southern winter rain and the north eastern summer rain (Jürgens, pers. comm.).

The **Karroo** extends into the Orange Free State. The two rainy seasons per year favour the development of innumerable succulents: the larger species of *Euphorbia, Portulacaria* and *Cotyledon* in rocky habitats and many smaller Crassulaceae and *Mesembryanthemum* (s.l.) on and between quartz veins. The vast flat areas are covered with dwarf shrubs, mainly Compositae and Fabaceae (Fig. 134). Woody plants such as *Acacia, Rhus, Euclea, Olea, Diospyros* and even *Salix capensis* grow in the dry valleys. In the transitional region of the Upper Karroo, the grassland of the summer rain region is found growing on deep, fine-grained soils, whereas on shallow rocky ground the Karroo succulents still abound (Fig. 135).

The Namib desert on the coast of South West Africa has been selected as an example of a biome of ZB III and the subzonobiome of the fog desert since it differs vastly from all other deserts. Although it is a subtropical desert and extreme in its lack of rain, the coastal strip is remarkable for the high air humidity. There are about 200 foggy days annually and only small fluctuations in tem-

**Fig. 134.** Great Karroo near Laingsburg (South Africa) with succulent *Euphorbia, Rhigozum obovatum, Rhus burchelli* and dwarf shrubs (Photo: E. Walter)

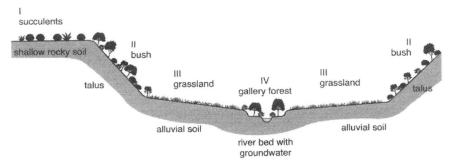

**Fig. 135.** Vegetation profile of a valley of the Upper Karroo near Fauresmith (South Africa). Distribution of the plant cover is determined by differences in soil. Bush with *Olea, Rhus* and *Euclea*

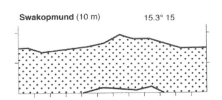

**Fig. 136.** Climate diagram of Swakopmund in the Namib. The region is almost without rain but has 200 foggy days annually

perature, similar to a maritime region. The temperature is always cool and there are only a few hot days each year. These peculiar conditions are due to the cold Benguela current (water temperature, 12–16 °C), above which a 600-m-high cold layer of air accumulates with a bank of fog, so that as a result of inversion the warm easterly air streams cannot reach the ground. Instead, a sea breeze arising daily in the south west pushes the fog and cold air into the desert (LOGAN 1960). Thunderstorms with rain can only occur if the inversion layer is penetrated. In most years this is not the case, however, and rainfall is barely measurable. Heavy rains are rare and occur only once or twice in the course of a century, as for example when 140 mm of rain was recorded in 1934–1935 or over 100 mm in 1974–1975. The annual mean for Swakopmund is 15 mm, but this figure conveys little information (Fig. 136).

The soil is slightly moistened by fog or dew, the mean daily value being 0.2 mm and the maximum 0.7 mm. The annual total of about 40 mm of fog precipitation has no effect because the moisture evaporates before it can enter the soil. The high air humidity is of benefit only to poikilohydric lichens, which brighten every available stone and rock in the fog zone with their vivid colours, and to the "window algae" which grow on the bottom of the transparent quartz pebbles where the moisture is retained longer. True fog plants such as the *Tillandsia* of the Peruvian desert (p. 242), which take up no water from the soil, are not found in the Namib. Only where wind-driven fog condenses against a rock face and the water seeps down into the crevices can plants (mostly succulents, Fig. 137) establish themselves. This happens on the isolated mountains rising from the flat platform of the Namib.

The plain rises with a gradient of 1:100 from the coast toward the east and ends at the foot of the steep escarpment of the African highlands (100 km from the coast). The fog is noticeable up to 50 km inland and still contains drops of seawater from the spray which accounts for the fact that the soils of the outer Namib are brackish.

In the Namib, perennial plants are only found in places where the soil contains water below a depth of 1 m, from the water supplies laid up in years of good rainfall. After the 140 mm of rain in 1934–1935, the desert turned green and was sprinkled with flowers, mainly ephemeral forms including a particularly large number of succulent *Mesembryanthemums*. The latter were able to store so much water in their shoots that they flowered in the following year as well, although the roots and shoot base had already dried out. They used almost the total supply of assimilates and water to form fruits and seeds (von Willert et al. 1990). In rainy years, a large number of seedlings of perennials also develop their roots rapidly, penetrating to the deeper part of the soil which stays moist for longer. These plants can only survive the ensuing decades, however, where the soil contains larger stores of water.

After heavy rains the water runs along deep, sand-filled gullies (wadis) in the direction of the sea, without ever reaching it. The water seeps down into depressions filled with alluvial soils and penetrates into the ground. The upper soil layers dry out down to a depth of about 1 m (less in sandy soils), but below this the water can remain for decades and is available for plants with roots long enough to reach it. The salt washed out of the sand in the gullies by the rainwater collects in the depressions. In this way two types of habitat are formed. One consists of non-halophilic biogeocenes in both larger and smaller erosion gullies with *Citrullus*, *Commiphora*, *Adenolobus*, and where groundwater is more plentiful, shrubs such as *Euclea*, *Parkinsonia* and *Acacia* spp. The other consists of large, shallow depressions with halophilic species including chiefly *Arthraerua* (Amaranthaceae), *Zygophyllum stapfii* (Zygoph.) and *Salsola* (Chenopodiaceae). The

**Fig. 137.** Between white marble rocks (Witport Mountains, Namibia) in the foreground flowering *Hoodia currorii*, in the background left *Aloe asperifolia* and right *Arthraerua* in fruit (Photo: W. Giess)

**Fig. 138.** *Arthraerua leubnitziae* (Amaranthaceae) in the Namib in the hinterland of Swakopmund. In the background dispersing fog clouds (Photo: K. Loris)

plants grow out of the sand that drifts onto them and this leads to the formation of small dune hillocks, which constitute the typical **nebkha** landscape (Fig. 138). Presumably, all of the plants germinate in the same year with a good rain, since they are of about equal size and will survive as long as the water supplies in the soil last. If a long period of time elapses before another good rain-year, the plants gradually die off and the dune sand is dispersed by the wind. If, however, they receive enough rain again within an appropriate time, they continue to grow.

Fog plays a substantial role in the survival of the plants since they can assimilate $CO_2$ in the water-saturated air without incurring transpiration losses and water consumption is thus low. Recently, it has been said that *Ar-*

**Fig. 139.** The dry river bed of the Kuiseb (wadi) near Gobabeb with trees of *Acacia albida, A. erioloba, Tamarix usneoides* and *Salvadora persica.* In the background dunes of the sandy Namib (S.-W. BRECKLE)

*thraerua* is able to take up water from the air (LORIS, pers. comm.).

Apart from these three biogeocene complexes with salt-free sand soils and the brackish depressions near the coast, the oases of the large dry river valleys (wadis) in the central Namib also deserve mention: the Omaruru, Swakop and Kuiseb. All arise in the highlands where there is a summer rainy season (mean annual rainfall of 300 mm) and are deeply incised into the Namib platform. The river bed is filled with sand into which the rainwater seeps, but water only reaches the sea in years with particularly good rains on the highlands. A continuous stream of groundwater is present at all times however, so that water can be obtained from wells, although some of it is brackish due to inflow from the Namib. This groundwater allows gallery forests to develop (Fig. 139), with *Acacia albida, A. erioloba, Euclea pseudebenus, Salvadora persica* or, in more brackish spots, *Tamarix* and *Lycium* species. In areas above the floodwater level, the forests can attain a great age. On the surrounding accumulations of sand, *Ricinus, Nicotiana glauca, Argemone* and *Datura*, among others, may grow while the spiny, leafless *Acanthosicyos horrida* (nara pumpkin) and *Eragrostis spinosa*, a woody thorny grass, thrive on the dunes. Where the groundwater forms ponds, *Phragmites, Diplachne, Sporobolus* and *Juncellus* can be found.

All of these plants are adequately supplied with water and are capable of considerable production. There is also a diverse and abundant fauna in oases of this kind, including birds, rodents, reptiles, arthropods and others. Elephants and rhinoceros could previously be encoun-

**Fig. 140.** *Welwitschia mirabilis* on the *Welwitschia*-flats (Vlakta) between the Khan and Swakop rivers (Photo: ERB 1987)

tered, but they have been exterminated by man and only baboons, dwelling in the rocky clefts, have survived.

The fauna of the nebkhas in the desert is poor, consisting of a few rodents, reptiles, scorpions and saprophagous beetles. More species are encountered in the isolated mountains, particularly those further inland where summer rain is more frequent, so that water accumulates between the rocks and shrubs grow in the cracks. In the sandy Namib the fauna is also more diverse.

The foregoing description applies to the outer Namib. At a distance of 50 km from the sea, the inner Namib begins. This area receives sparse summer rain and is periodically covered by grass. Desert conditions are not so extreme and the mobile game is able to find food and to take advantage of isolated water holes. This part of the desert abounds in game, including large numbers of zebra, oryx antelope, springbok, hyena, jackal, as well as ostrich and other birds. The central part of the Namib, which is uninhabited by man, has been declared a nature reserve and is being investigated from the Namib desert station in Gobabeb.

In the region between inner and outer Namib, the well-known *Welwitschia mirabilis* is found in large numbers. It grows in wide, very shallow erosion gullies in which, owing to the barely perceptible gradient (Fig. 140), the sparse summer rains accumulate and seep down into the deeper layers of the ground. The roots of *Welwitschia*, which probably go far deeper than 1.5 m, are able to exploit this water. Further down is a hard chalk layer. *Welwitschia* possesses only two ribbon-like leaves (very rare 4 leaves) that grow continuously and are extremely sensitive to running water and to being covered by sand. At present, *Welwitschia* only rejuvenates in the northern Namib.

The plant shown in Fig. 140 was photographed in 1885 by SCHENK and in 1975 by MOISEL. Mrs ERB (Swakopmund) comments: "It is striking that on all three photos only *Welwitschia* plants can be recognised and no other plants. In 1976, 1 year after Mr MOISEL took his photo, it rained a lot in this part of the Namib and many grasses grew, so that I remember this area as a swaying corn field with grazing oryx antelopes and springboks. But since then we have had no significant amounts of rain. At the beginning of the 1980s it was so dry that oryx antelopes even ate the leaves of *Welwitschia* plants. Most plants recovered during the next years. This explains why the *Welwitschia* plant at the front left, is so small, because the leaves of this plant were most probably eaten off. Also, on the larger plant it is obvious that only a small strip of the front leaf has grown again. Nearby there are also three younger plants. They could have started to grow in the rainy period of 1933–44. Only one younger plant has been found, probably from the rainy period of 1976."

The two ribbon-like leaves of *Welwitschia* grow continuously from a meristem at the leaf base on the turnip-shaped stem. They dry out at the tip to an extent dependent upon the water supply. In rainy years the surviving portion is reasonably long, but in drier years the leaves die off almost down to the meristem, thus greatly reducing the transpiration surface. The leaves are highly xeromorphic and the stomata are situated in pits. The oldest specimen to be carbon-dated was approximately 2000 years old. Transpiration and photosynthesis was researched by von WILLERT et al. (1982): *Welwitschia* is a C3 plant. A medium-sized plant requires about 1 l water per day. Calculated on the area of soil surface exploited by the roots this would correspond to 2 mm rain per year. Thus, the water supply is guaranteed even in this arid region. During long droughts, the leaf area dies back to the basal meristem.

Certain ecosystems of the Namib are unique: (1) the bare dunes south of the Kuiseb (see Fig. 139); (2) the guano islands; (3) the mating places of seals; and (4) the saltwater lagoons on the shore behind sand bars. Organic detritus blown into the dune valleys by the wind consists of grass remnants, protein-rich animal remains and dead insects (butterflies). This detritus is consumed by wingless psammophilic beetles (tenebrionids), which are eaten by small predators (spiders, solifuges) or by the larger lizards, sand-dwelling snakes and golden moles (KÜHNELT 1975).

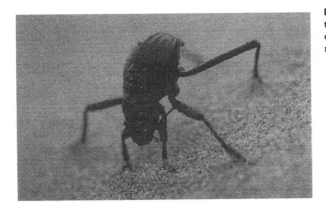

**Fig. 141.** A *Tenebrionid* beetle collecting water from fog on sand dunes in the early morning (Photo: M. Seely)

Since the sand surface warms up to a temperature of 60 °C by day, most of these animals avoid the heat by burying themselves in the cooler layers of sand during the day and emerging at night. The dew which wets the sand in the early morning provides them with water, for which they have developed unique methods of uptake (Seely & Hamilton 1976; Hamilton & Seely 1976). Some species possess comb-like extensions on their hind legs which they use to comb out fog droplets, others stand with their head down in the wind and suck fog droplets up which have condensed on their hind legs or their abdomen and thus drop down towards the head (Fig. 141). The fauna is rich in endemic species.

The guano islands are the nesting-places of the cormorants that feed on the abundant fish in the cold seawater. The excrement of these birds accumulates in the rainless climate and precludes any form of plant growth, although it is harvested by humans for guano (phosphate fertiliser). A similar situation prevails in the mating areas of the seals. The lagoons are cut off from the sea by sand bars over which the waves only break during storms. Water lost by evaporation is replaced by water seeping through the sand from the sea, so that these are aquatic habitats with a very high salt concentration. This habitat will not be considered in any further detail here.

Like the Namib, every desert has its own peculiarities and must be studied monographically. It would be beyond the scope of this volume to go into such detail (see Walter 1973; Walter & Breckle, 2002).

**Fig. 142.** Relation between the spatial distribution (**A**) and transect (**B**, next page) of the region of the Atacama Desert between the Pacific and the Andes in northern Chile (after WICKENS 1993)

## g Atacama in Northern Chile

The Peruvian-Chilean coastal desert is divided into several parts (Fig. 142 A) and is, at its most extreme, just as dry as the Namib although the fog plays a greater role here because of the steepness of the coast in places. *Tillandsia* (Bromeliaceae), the only true fog plant known to exist

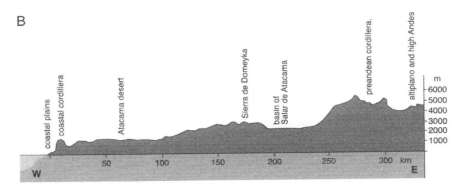

**Fig. 142 B**

among flowering plants, grows here. Although it cannot take up water from the air like the lichens, it absorbs water drops from condensed fog with the aid of special leaf scales. These plants either sit as epiphytes on columnar cacti or the rosettes sit loosely on the sandy ground.

The "garua", as the fog blanket is termed in Peru, hangs at an altitude of 600 m for months on end during the cooler season. The soil on the slopes is so wet that a carpet of herbage grows, called "loma vegetation" which is used for grazing. Although absent nowadays, woody plants formerly grew in these areas. The quantity of water from fog condensation collected under the trees in a *Eucalyptus* plantation amounted to the equivalent of 600 mm rainfall. Even in the coastal cordillera of northern Chile, the slopes exposed to fog are densely covered with 8-m-high columnar cacti (*Echinopsis atacamensis*) which are draped with lichens. Further south, near Frey Jorge, there is a true mist forest. In the neighbourhood of the large saltpeter deposits in northern Chile the desert is completely barren and only along the rivers fed by the snowfields of the High Andes is there vegetation or irrigated farmland. The inner basins are at higher altitudes. However, they are characterised by enormous salt pans right up to the higher regions of the Andes and towards southern Bolivia. The salt pans contain not only NaCl, but also other minerals which probably accumulated because of the very active volcanoes and the extremely arid climate. The extreme conditions only allow sparse growth of a few species. Only above 3500 m with the occasional summer rain does sparse growth of dwarf shrub semi-desert occur with *Baccharis*, *Fabiana*, *Parastrephia* and others followed above 4100 m by a montane grass desert (*Ichu* grass, *Fes*-

**Fig. 143.** High ranges of the Andes on the eastern edge of the Atacama: volcano Ollagüe (5900 m above sea level). The flank of the mountain is a high desert. Even at 5800 m there is hardly any remaining snow: the area is so dry that a climatic snow limit cannot be determined (Photo: S.-W. BRECKLE)

*tuca chrysophylla, F. orthophylla, Stipa venusta*) where llama and guanaco and also nandu graze. ELLENBERG (1975) describes the western slopes of the Andes as a per-arid desert extending up into the montane belt, followed by a subalpine dwarf shrub semi-desert and, above 4500 m, a tropical alpine grass semi-desert or "desert puna". However, even between 5200 and 5500 m, there are several dwarf shrubs, for example on the volcano Ollagüe (5870 m). Occasionally, bushes or trees of *Polylepis tarapacana* up to 4 m high occur in the lava rubble (WICKENS 1993). The snow limit is hardly recognisable (Fig. 143).

# Orobiome III –
# Desert Areas of the Subtropics

In extreme deserts, the air contains so little water vapour that even at high altitudes no rain can form. In the Tibesti mountains (3415 m above sea level) in the central Sahara at 2450 m, annual precipitation of only 9–190 mm was measured (over 4 years) with frequent cloud formation during the winter months. Therefore, arid conditions remained up to the higher altitudes; however the occurrence of several Mediterranean species indicate slightly more humid conditions. In gorges at altitudes of 2500–3000 m above sea level, *Erica arborea* was found and in Hoggar, at 2700 m, *Olea laperrini*, a relative of the olive tree, occurs as a relict.

In a sequence of latitude steps from the less arid Sonora Desert in southern Arizona above the *Larrea* or giant cacti desert, there is a zone with *Prosopis* grass savannas and many leafy succulents (*Agave, Dasylirion, Nolina*), then several zones with evergreen *Quercus* species and *Arctostaphylos, Arbutus* and *Juniperus* shrubs. Following these are coniferous forests: *Pinus ponderosa* spp. *scopulorum*, higher up with *Pinus strobiformis*, then *Pseudotsuga menziesii* with *Abies concolor* and only on the northern slopes of the San Francisco Peak up to almost 3700 m in northern Arizona does *Picea engelmannii* occur. In these altitudes, annual precipitation increases quickly with altitude. This does not apply to different altitudes in the Andes on the Atacama side (see above p. 244).

# Humans in Deserts

Because of the inhospitable conditions in deserts it appears remarkable that humans have lived in all deserts for a very long time. They have adapted to the ways of life in the desert, but are almost always on the move as nomads as a larger area is needed for their sustenance (Fig. 144). Settlements have always been restricted to larger oases which usually were the base for the periodically regulated movements. Animals served as a food reserve (nomads with herds of sheep and goats) and domesticated camels served for transport. In the border areas of deserts as well as in mountains, simple agriculture was possible in rainy areas (runoff, lalmi). Irrigated cultivation developed only in the regions of large incoming rivers and was the basis of the early developing cultures (Egypt: Nile; Mesopotamia: Euphrates and Tigris).

**Fig. 144.** Bedouin tents in the southern Egyptian Sahara at Wadi Allaqui, today near the eastern shore of lake Nasser on the Nile (Photo: S.-W. BRECKLE)

**Fig. 144.** Bedouin tents in the southern Egyptian Sahara at Wadi Allaqui, today near the eastern shore of lake Nasser on the Nile (Photo: S.-W. BRECKLE)

## Zonoecotone III/IV – Semi-Deserts

The boundary between true desert and semi-desert, although not always clearly defined, is to be found in the zone where the increasing winter rains lead to the "restricted" vegetation being replaced by a "diffuse" vegetation. About 25% of the total area in the semi-desert is covered by vegetation and the floristic composition of this plant cover varies just as greatly from one floristic realm to the other as is the case with the true deserts. North of the Sahara, the malakophyllous *Artemisia herba-alba* and the sclerophyllic grasses *Stipa tenacissima* (halfa grass) and *Lygeum spartum* (esparto grass) are the most abundant species. Although *Artemisia* generally grows on heavy loess or loamy soils, it has been found growing in Tunisia in places where secondary $CaCO_3$ deposits are present at a depth of 10 cm. At a depth of 5–10 cm, the soil is densely permeated by its roots, some of which even reach down to 60 cm. *Stipa* prefers high ground with a stony covering. A soil profile revealed the following: 2–5 cm of stony pavement underlain by 30 cm of loamy soil with dense root growth and below this a gravel layer with a hard upper crust, which appears to present an obstacle to the roots but probably also acts as a water reservoir. The numerous roots originating at the base of the grass tufts spread far out on all sides at a depth of 10–20 cm, so that, although the tussocks themselves are 0.5–2 m apart, their root tips are, in fact, in contact with one another. Solitary individuals of *Arthrophytum* grow between the grasses. The soils are not saline. *Lygeum spartum*, on the other hand, is characteristic of soils containing gypsum (calcium sulphate) and even tolerates a certain amount of salt. The halfa grass is cut and provides fibres for weaving, the production of coarse ropes and paper manufacture. *Stipa tenacissima* extends from south-eastern Spain to East Morocco as far as Homs in Libya; the natural habitat is open Aleppo pine forests. *Artemisia herba-alba* is found in the Middle East and has replaced the original overgrazed grassland in many places. With an increase in rainfall, isolated trees occur, such as *Pistacia atlantica* in the west and *P. mutica* in the east, or *Juniperus phoenicea*. Thin stands of trees finally lead to the sclerophyllic woodlands.

In the transitional zone in California, *Artemisia californica* and the semi-shrubs *Salvia* and *Eriogonum* species (Polygonaceae) are found. In the transitional zone in northern Chile there is a dwarf-shrub semi-desert with Compositae (*Haplopappus*), columnar cacti and *Puya* (large Bromeliaceae). This is succeeded by savanna with *Acacia caven* and the grass cover now consists of annual European grasses. The so-called renoster formation (with *Elytropappus rhinocerotis,* Asteraceae) in South Africa can be regarded as typical of a winter rain region with low rainfall. In Australia, where there are no true deserts, the transitional zone is occupied by the mallee scrub consisting of shrubby species of *Eucalyptus*, the branches of which originate from an underground tuberous stem (lignotuber; see pp. 177, 276). Open stands of *Eucalyptus* trees with an undergrowth of *Maireana sedoides* sometimes occur.

## Questions

1. How is the desert habitat defined?
2. How are aridity and humidity of a region defined?
3. What are the different types of desert?
4. Which criteria are used to classify deserts?
5. Are desert plants particularly drought- or heat-resistant?
6. Are desert plants predominantly C3, C4 or CAM plants?
7. Why are there so many halophytes in deserts?
8. Under what conditions are poikilohydric higher plants ecologically advantaged?
9. How many leaves does an adult *Welwitschia* plant form?
10. What is a contracted vegetation?

# IV Zonobiome of Sclerophyllic Woodlands (Zonobiome of the Arido-Humid Winter Rain Region)

## 1 General

This zonobiome is best divided into five floristic biome groups, according to the various floristic realms into which it falls (they form typical vegetational units, which often appear similar).

The largest of these groups is the Mediterranean, where winter rains occur from the Atlantic Ocean to Afghanistan. However, in Anatolia and further east, heavy winter frosts occur and, thus, this area must be classified as belonging to zonobiome VII. Arid zonoecotones with drought periods or stronger winter frosts usually follow from the Mediterranean climate regions of zonobiome IV (see Fig. 145) and are usually also termed Mediterranean.

The five winter rain regions (Fig. 145): (1) Mediterranean with sclerophyllic forest, macchie, garigue, asphodel stands etc., (2) Californian with scerophyllic forest, chaparral, in parts encinal etc., (3) Chilean with matorral, espinal etc., (4) Capensic with fynbos, renosterbos etc., (5) Australian with jarrah forest, sclerophyllic scrub = mallee etc.

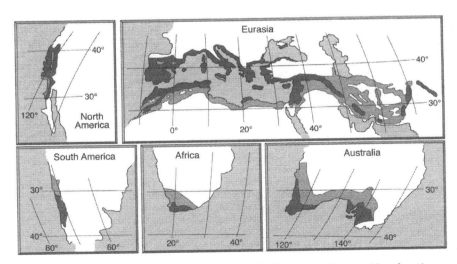

**Fig. 145.** Regions with Mediterranean climate at comparable latitudes on the west sides of continents. *Dark grey* Mediterranean climate type (zonobiome IV); *light grey* arid regions with predominant winter rains (zonoecotone III/IV; zonoecotone III/VII) (after WALTER and BRECKLE 1991)

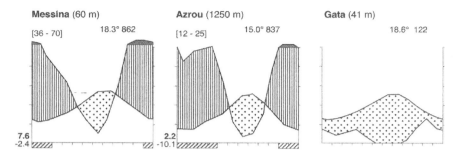

**Fig. 146.** Climate diagrams: Messina (Sicily), Azrou (montane elevation, central Atlas mountains, Morocco) and Cabo de Gata (south-eastern Spain) = driest part of Europe (desert)

Both the western and eastern parts of southernmost areas of Australia have Mediterranean characteristics but they are two distinct, separate regions (Fig. 145). Climate diagrams of the various groups are very similar except with regard to summer drought, which varies in extent. Variations in this type of climate are also very large in the western Mediterranean region (Fig. 146).

Of the 113 woody genera (with 169 species) of the sclerophyllic woodlands in Chile, only 13 genera are the same as the 109 genera (with 272 species) in California. Australia with its 66 genera (with 140 species) has only 2 genera in common with California and 3 with Chile. However, the total number of species is much higher. The usually very small regions of zonobiome IV are in some ways an exception to the rule that biodiversity increases from the poles to the equator (see Table 18).

For the corresponding but very much smaller area in the Cape region of South Africa about 8000 species are assumed, for south-west Australia also about 8000 species, but the much larger and richly subdivided Mediterranean is estimated to possess around 24,000 species. The sclerophyllic woodland vegetation of ZB IV which is typical for

Sclerophyllic woodlands dominate Mediterranean vegetation. They appear superficially very similar, but in the various regions they often belong to quite different genera.

**Table 18.** Number of genera and species in zonobiome IV of California and Chile (winter rain regions)

|  | Chile | California |
|---|---|---|
| Area in km² | 294,600 | 278,000 |
| Number of genera | 681 | 806 |
| Number of species | 3385 | 4240 |

(after ARROYO et al. 1995)

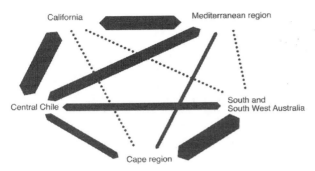

**Fig. 147.** The five Mediterranean regions. The thickness of the connecting line shows schematically the similarity of the five regions in relation to the phylogeny of the flora, phenology, morphology and type of vegetation as well as climate and land use pattern (after CASTRI 1981)

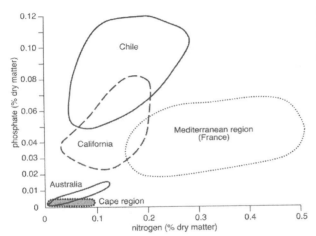

**Fig. 148.** Phosphate and nitrogen content in soil (total content in percent) in the five Mediterranean regions (after RUNDEL 1979; CASTRI 1981)

winter rain regions with sporadic frosts is able to survive cold periods for a considerable time. The growth period is in spring when the soil is moist and temperatures rise and again in autumn after the first rains. The winter period with temperatures around 10 °C or below is too cool for good growth.

Individual Mediterranean regions are geographically at great distances from each other. At first sight the vegetation and their biotopes look strikingly similar. This external similarity is particularly large between the Mediterranean, California and Chile (Fig. 147), but also between the Cape region and Australia. This duality is explicable from geological history. The climate as the primary determining factor is similar in all five areas, but their geological history is very different. Australia and the Cape region are part of the old Gondwana land and they have been eroded and leached for millions of years and consequently soils are very nutrient-deficient (Fig. 148). The other three regions are much younger and were determined by the formation of mountains; availability of nitrogen is about 10 times better and that of phosphorus about 100 times.

In the discussion of the climatic subzonobiomes of the relevant biome groups, zonoecotones must also be mentioned. There are transitions from zonobiome IV to zonobiomes V, VI and VII. The climate of ZB IV was not always the same as it is today. The widespread fossil soils, the developmental rhythm of the most common plants and other factors (fossils) suggest that the climate was still tropical with summer rain in the Tertiary (see also p. 253). The rain maximum was, in all probability, displaced to the winter months only shortly before the Pleistocene. Plants were obliged to adapt and this meant a drastic process of selection. Only Tertiary plant species of drier habitats with small xeromorphic leaves could survive. The present-day reduction of activity in summer is imposed by drought and is not seen where the plants have sufficient water at their disposal. Ephemerals and ephemeroids, which act as a vegetative buffer, confine their activity to the favourable months of spring or the wet autumn.

It is easier to comprehend the ecological behaviour of the vegetation if the historical facts are borne in mind (SPECHT 1973; AXELROD 1973). Many taxa of ZB IV are closely related to those of ZB V or ZB II. Examples are species of the genera *Olea, Eucalyptus, Quercus balout* (= *Q. ilex* s. l.) which grows in Afghanistan, with mainly summer rainfall; the chaparral of California, where rain falls only in winter corresponds to the encinal vegetation of the mountains of Arizona, with summer rain. Aspects of zonobiome IV are discussed in Castri et al. (Vol. 11 of *Ecosystems of the World*), and many references given. Although a large quantity of valuable data are presented, a synthesis of the basic relationship is lacking. This also applies to the volumes by MILLER (1981) and CASTRI & MOONEY (1973) as well as for those of ARROYO et al. (1995) and DAVIS & RICHARDSON (1995).

## Historical Development of Zonobiome IV and Its Relationship with Zonobiome V

Formation of zonobiome IV is closely connected to zonobiome V and only occurred in the late Tertiary period.

Various aspects of zonobiome IV, including its historical development, are discussed in the volume by CASTRI & MOONEY (1973). Zonobiome IV is closely related to zonobiome V since they can both be traced to a common origin extending back to the tropical vegetation of the Tertiary, which reached into the higher latitudes at that time. The further development of this vegetation to the present is summarised for California by AXELROD in the above-mentioned volume. The development in the Mediterranean region may by considered to be analogous. Fossil evidence from the Eocene at the beginning of the Tertiary, in today's temperate region of the northern hemisphere, indicates the former presence of tropical evergreen and deciduous species in a tropical climate with a pronounced summer rainy season. Studies on fossil saltwater molluscs from California indicate that the minimum temperature of the surface waters of the ocean there was approximately 25 °C, about 50 million years ago. The oceans gradually cooled during the Oligocene and Miocene, until the temperature minimum was only 15 °C by the Pliocene at the end of the Tertiary. The climate on the continents cooled correspondingly and the flora became poorer in species requiring higher temperatures. Simultaneously, the distribution of rainfall in California was altered, with a decrease in the summer maximum until it was no longer noticeable in the Miocene and had even become a small minimum by the Pliocene. During the Ice Ages of the Pleistocene, cold ocean currents developed on the west coasts of the continents, as well as climates with pronounced dry periods in summer and rain only in the winter months, typical of zonobiome IV.

During the Tertiary, the mountains in western North America and the Alpine chain in Europe rose continuously higher. The result of these changes was the development of more arid climate regions and arid localities in unfavourable locations at today's higher latitudes, which were the tropical zones of the Tertiary. Consequently, selection took place among the evergreen species, favouring species with the typical leathery leaves found in the humid tropics (often called lauriphyllic, p. 297) and drought-resistant sclerophyllic species. The development of the summer drought climate (designated as "Mediterranean") on the west coasts of the continents during the Pleistocene led to the dominance of sclerophyllic species, while the flora of woody species declined. On the east coasts of the continents, which were subject to warm ocean currents, the humid climate with summer rain and somewhat lower mean annual temperatures than zonobiome V remained unchanged. The transition from tropical humid to subtropical humid and to the temperate species-rich flora with evergreen leathery leaves is gradual on the humid east coasts of North and South America, South East Africa, South-East Asia and Australia.

Since the Tertiary species were already pre-adapted to dry locations, it was not necessary for the sclerophyllic vegetation of zonobiome IV to develop by adaptation to the summer dry season. Only a limited development of new species took place, such as in California, for example, the genera *Ceanothus* with 40 species and *Arctostaphylos* with 45 species, while others (*Adenostoma*) extended their distribution. *Arbutus* has more leathery leaves. This developmental history explains why the same genera are often found in zonobiomes IV and V, although the species differ, for example the sclerophyllic *Quercus* species of California and the evergreen *Quercus virginiana* with leathery leaves in south-eastern North America (zonobiome V). The leathery leaved sclerophyllic *Eucalyptus* species of zonobiome IV in south-western and southern Australia differ only slightly from those in the summer rain region of zonobiome V on the east coast. Just as in the west, dry calcareous soils in the east are settled by a profuse vegetation of Proteaceae, except that the species are not the same. The occurrence of the fossil "terra rossa" soils in the Mediterranean region may also be explained in a similar manner. This soil contains relicts of a tropical microscopic fauna which was able to avoid the effects of the summer drought at greater depths. The remaining fauna of zonobiome IV confirms the observations made on the vegetation (see several contributions in Castri & Mooney 1973).

As concluded by Axelrod, fossil evidence in North Africa indicates a development similar to that of the Mediterranean vegetation. The conditions in Europe, however, are somewhat more complicated since the climate of western Europe has been determined by the warm Gulf Stream since the post-glacial period. The cold Canary Current becomes effective south of the Canary Islands towards the coast of Senegal (fog coast). Zonobiome IV extends from the west, along the Mediterranean coast and further eastward because of the long coast line.

The last Ice Ages had an especially negative effect in Europe, practically exterminating the entire flora. During the post-glacial period, the flora returned from a small number of refuges, but remained poor in species. A continual series of fossils from the Tertiary to the present, such as is found in California, is missing here. It is generally concluded, however, that the development of zonobiome IV followed the same basic course wherever it now occurs and that a type of climate corresponding to zonobiome IV with zonal sclerophyllic vegetation did not exist in the Tertiary, although the sclerophyllic species certainly did exist in dry local habitats.

## Mediterranean Region

Climatic conditions prevailing in this zone can be seen in the diagrams in Fig. 146 (p. 250). Cyclonic rains occur in winter and the hot, dry summer is a result of the Azores high-pressure zone. Since some of our most ancient civilisations originated in the Mediterranean region, the zonal vegetation was long ago forced to give way to cultivation.

The slopes have been deforested and used for grazing, with resultant soil erosion, so that nowadays only varying stages of degradation remain.

There is no doubt, nevertheless, that the original zonal vegetation was evergreen sclerophyllic forest with *Quercus ilex*. Small remnants of this association have provided the following data concerning the original forests: holm oak forest (**Quercetum ilicis**).

Tree layer: 15–18 m tall, closed canopy, composed exclusively of *Quercus ilex.*

Shrub layer: 3–5 (up to 12) m tall
- *Buxus sempervirens,*
- *Viburnum tinus,*
- *Phillyrea media,*
- *Phillyrea angustifolia,*
- *Pistacia lentiscus,*
- *Pistacia terebinthus,*
- *Rhamnus alaternus,*
- *Arbutus unedo,*
- *Rosa sempervirens* and others

as **lianas:**
- *Smilax,*
- *Lonicera,*
- *Clematis*
herb layer: approximately 50 cm tall and sparse but with many species
- *Ruscus aculeatus,*
- *Rubia peregrina,*
- *Asparagus acutifolius,*
- *Asplenium adiantum-nigrum,*
- *Carex distachya* and others,
Moss layer: very sparse.

A terra-rossa profile is usually found in the chalky regions beneath these low forests, consisting of a litter layer, a dark humus horizon, and beneath this a 1–2 cm deep red terra rossa horizon containing plastic clay. On cultivated land the upper layers are missing, due to erosion, so that the red colour is visible at the surface. These are mostly fossil soils from a more tropical climatic period. Today, brown loamy soils are developing (ZINKE 1973).

**Fig. 149.** *Quercus ilex* forest
above Azrou in the central
Atlas (Morocco). *Rosa siculum, Lonicera etrusca* and
others in the undergrowth
(from WALTER 1990)

**Fig. 149.** *Quercus ilex* forest above Azrou in the central Atlas (Morocco). *Rosa siculum, Lonicera etrusca* and others in the undergrowth (from WALTER 1990)

A change in the appearance of this region takes place
in March when many of the shrubs start to bloom. The
height of their flowering season, as well as that of *Quercus
ilex,* is in May although *Rosa, Lonicera* and *Clematis* are
still blooming in June. A relatively dormant period then
follows as a result of the coincidence of the hottest and
driest seasons. Growth only recommences with the autumn rains, which may then even lead to an additional
flowering of sclerophyllic trees. *Quercus ilex* extends from
the western Mediterranean region to the Peloponnese and
Euboea. *Quercus suber* (cork oak, Fig. 153) grows in the
west (not on limestone). Growth of this species is stimulated by cultivation, particularly as competing species are
cut out of these forests. These two species are replaced by
*Quercus coccifera* in south-eastern Europe and appear in
Palestine as the tree forming race (*Q. calliprinos* shown in
Fig. 152).

   The dominant species in the tree layer of the hot lower
belts in Spain and in northern Africa are the wild olive
tree (*Olea oleaster*), carob (*Ceratonia siliqua*) and *Pistacia
lentiscus* as well as *Chamaerops humilis,* Europe's sole
palm. Of special interest in Crete are Tertiary relict habitats of a wild form of date palm, which was mentioned by
Theophrastus. A large stand grows by a small lagoon near
Vai (on Cape Sideron, north-east corner of Crete), aboveground water. In North Africa, from Morocco to Tunisia,
the distribution of *Quercus ilex* is montane (see Fig. 149),
above an intercalated coniferous belt consisting of *Tetracli-*

nis (*Callitris*) and *Pinus halepensis* (Aleppo pine). The
south-east corner of Spain, with a rainfall of only 130–
200 mm, is almost desert-like (Fig. 146, Gata).

There remain only a few places in the mountainous regions of northern Africa where typical *Quercus ilex* forest
still exists. Elsewhere, the trees are cut down every
20 years, while still young, and regenerate by means of
shoots from the old stump. This leads to the formation of
a **macchie**, consisting of bushes the height of a man. Macchie is also encountered on slopes where the soil is too
shallow to support tall forest. Sclerophyllic species, usually
shrub-like in form, may develop into big trees in a suitable habitat and can achieve a considerable age. Imposing
old specimens of *Quercus ilex* can be seen in gardens and
parks. In places where the young woody plants are cut
every 6–8 years and the areas regularly burned and
grazed, higher woody plants are lacking and open societies called **garigue** are formed, "phrygana" in Greece, "to-
millares" in Spain and "batha" in Palestine.

These areas are often dominated by a single species
such as the low cushions of *Quercus coccifera* or *Juniperus
oxycedrus* (in the east also *Sarcopoterium spinosum*
bushes) or *Cistus, Rosmarinus, Lavandula* and *Thymus*.
On limestone in the south of France the best grazing is
provided by a *Brachypodium ramosum-Phlomis lynchnitis*
community. In springtime numerous therophytes (ephemerals) and geophytes (ephemeroids) such as *Iris*, orchids
(*Serapias, Ophrys*) and species of *Asphodelus* put in an appearance on otherwise bare spots. An almost pure ***Asphodelus* vegetation** is all that finally remains in places ser-
iously degraded by continuous fire and grazing. Although
the garigue is a sea of flowers in spring, it presents a severely scorched aspect in late summer. If cultivation or
grazing is stopped then successions tending toward the
true zonal vegetation take over, as shown in the schematic
for the south of France (Fig. 150).

On sandstone or acid gravel the successions take a
course similar to that on limestone, except that the individual stages are of a different floristic composition. Characteristic species in northern Spain are, for example, *Arbutus* and *Erica arborea* with *Quercus suber* (Fig. 153).

In the east Mediterranean region the tree-like *Quercus
calliprinos* (closely related to the west Mediterranean, usually bushy *Quercus coccifera*) takes over the role of *Quercus
ilex* and becomes the second zonal type of wood (Figs. 151,
152). Progression and regression stages are similar to those
in the western Mediterranean, however, usually the species
of the most numerous genera dominate. The manifold influence of man leads to a almost unusable thorny garigue

**Fig. 150.** Schematic of the regeneration stages of degraded grazing land or cultivated land on limestone soils in the Languedoc (southern France) to holm oak forests (Quercetum ilicis) or with continuing grazing (and fire) to the *Rosmarinus-Cistus* garigue. Given is the dependence of changes on the type and intensity of usage (after WALTER & BRECKLE 1991)

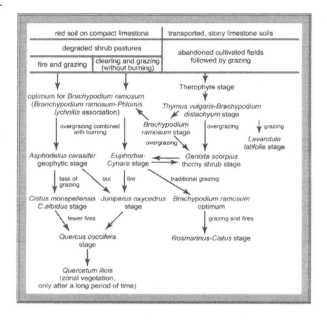

**Fig. 151.** Stages of regression and progression (regeneration) in the *Quercus calliprinos* zone on limestone, Jebel Ansariye Syria (after NAHAL 1991). *G* Continuous grazing; *GD* interrupted grazing; *F* deforestation; *U* change of vegetation

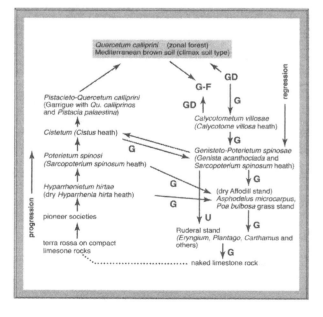

**Fig. 152.** High macchie and shrubs with *Quercus calliprinos* in Galilee (Keziv Park) rich in herbaceous species (Photo: S.-W. BRECKLE)

**Fig. 153.** Cork oak forest (*Quercus suber*) in southern Spain near Grazalema. The cork has been recently cut off the trees, the sheets of cork are collected and transported for processing. From old cork oaks the cork can be removed every 10 years (Photo: S.-W. BRECKLE)

(batha) of dwarf bushes with thorny twigs growing sideward (particularly the fire-tolerant *Sarcopoterium*) or even open heath land on rocks (Fig. 150) from which the soil has been almost completely washed off, so that in many places only the bare rock remains. A progression (regeneration) seems almost impossible without appropriate measures. In the continental Mediterranean region of southern Anatolia, *Pinus brutia* (related to *P. halepensis*) is widespread. It often constitutes the tree layer, below which sclerophyllic plants form a macchie. Since the pine is unable to regenerate in the macchie owing to lack of light, such woodlands can regenerate only after forest fires which explains why the trees are all much the same age. The natural habitat of the umbrella pine (*Pinus pinea*), which is often planted in Mediterranean regions, was probably the poor sandy areas on the coast.

## Significance of Sclerophylly in Competition

In considering the ecophysiological conditions in the Mediterranean region, the question which immediately crops up is the degree to which the plants are affected by the long summer drought. Here, a distinction must be made between sclerophyllic and malakophyllic plants, the latter being well represented by *Cistus, Rosmarinus, Lavandula, Thymus* and others. It should be borne in mind that the most favourable euclimatopes are nowadays occupied by vineyards or other cultures, the true Mediterranean species having been forced back into habitats with shallow soils, where they can be said to grow under relatively unfavourable conditions.

If the underlying rock is deeply fissured, the abundant winter rains can penetrate deeply and water is stored in the ground. This deep-lying water is available even in summer for plants capable of sending down roots through clefts in the rock to a considerable depth. In woody species, roots 5–10 m in length have been observed working their way down through the rock to the horizons which contain available water.

Observations on the cell sap concentration of sclerophyllic plants over the entire course of the growing season revealed that the osmotic potential decreases from about −2.1 MPa to about −2.5 MPa during the dry season, which means that the water balance is not disturbed to any significant degree and the hydration of the protoplasm hardly falls. When the water supply is uncertain, such a balance can only be maintained by a partial closure of the stomata and resultant limitation of gaseous exchange. Measurements of transpiration confirm that water losses in summer are three to six times greater in wet than in dry habitats. On the other hand, the cell sap osmotic potential of the stunted individuals growing in extremely dry habitats reaches −3.0 to −5.0 MPa. The euclimatopes which yield so much wine in autumn are far better provided with water and a dormant summer period due to drought was certainly not a feature of the original sclerophyllic forests.

Sclerophyllic plants are able to compete successfully in the winter rain regions with deciduous trees as well as with non-sclerophyllic, more lauriphyllic evergreen species which are sensitive to drought.

In contrast to hydrostable sclerophyllic plants, malakophyllic plants are highly labile in this respect. The cell sap osmotic potential of *Cistus, Thymus* and *Viburnum tinus* can reach −4.0 MPa in summer, with an accompanying drastic reduction of the transpiring surface, achieved by shedding the greater part of the leaves in some cases and leaving only the buds. Such species do not root deeply. Laurel (*Laurus nobilis*) is not sclerophyllic and in the

Mediterranean region its natural biotope is invariably in the shade or on northern slopes; nowadays it can only be found as forest in the altitudinal cloud belt on the Canary Islands or in a macchie in the winter rain region with no pronounced summer drought, as for example in northern Anatolia or Catalonia. *Prunus laurocerasus* shows similar preferences.

The ecological significance of sclerophylly is thus to be seen not in the ability of sclerophyllic species to conduct active gaseous exchange (400–500 stomata/mm$^2$) in the presence of an adequate water supply, but in their ability to cut transpiration down radically by shutting the stomata when water is scarce. This enables them to survive months of drought with neither alteration in plasma hydration nor reduction of leaf area. In autumn when rains recommence, plants immediately commence production again.

The situation changes at once, however, in more humid winter rain regions where the summer is not particularly dry or if the habitat is itself perpetually wet, despite a typically Mediterranean climate as is the case on northern slopes or in floodplain forests. On northern slopes sclerophyllic species are replaced first by evergreen species like laurel and then by deciduous trees. The deciduous oak, *Quercus pubescens* with its larger production of organic material, replaces *Quercus ilex*.

Deciduous trees such as *Populus* and species of *Alnus*, *Ulmus campestris* and *Platanus orientalis* are found in the floodplain forest of the Mediterranean region and in south-western Anatolia the Tertiary relict *Liquidambar orientalis* occurs. As soon as the point is reached at which the rivers dry up in the summer, however, deciduous woody species are no longer to be found. They are replaced by the evergreen sclerophyllic oleander (*Nerium oleander*).

Exact values for point 1 are not available, but it may be assumed that, in deciduous species, the contribution of leaf mass to the total phytomass is greater than in sclerophyllic species. As for point 2, the ratio is twice as large for the thin deciduous leaves as for the evergreen leaves, for point 3, measurements have shown that the rate of photosynthesis per unit leaf area varies only slightly from deciduous to evergreen leaves. Regarding point 4, evergreen leaves are of course at an advantage. This means that the deciduous species are superior in two respects and the evergreens in only one.

Exact calculations have revealed that in the humid, mild climate on Lake Garda in Italy where *Quercus ilex* and *Q. pubescens* are found, the productivity in g g$^{-1}$ dry branch weight was 22.9 for *Q. pubescens* as compared with 17.9

Productivity of plants depends largely upon their assimilation economy and is larger:

1. the larger the proportion of the assimilated material which is used for increasing the productive leaf area,
2. the larger the ratio leaf area/leaf dry weight, i.e. the smaller the amount of material required to produce a given leaf area,
3. the greater the intensity of photosynthesis,
4. the longer the time over which the leaves can assimilate $CO_2$.

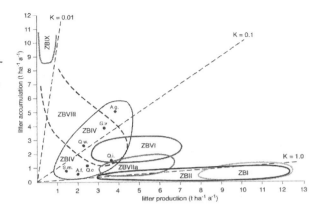

**Fig. 154.** Litter production and litter accumulation in different zonobiomes. For zonobiome IV individual species are listed (A.F.=*Adenostoma fasciculatum;* A.g.=*Arctostaphylos glauca;* G.v.=*Garrya veatchii;* Q.c.=*Quercus coccifera;* Q.i.=*Q. ilex;* Q.w.=*Q. wislizenii;* S.m.=*Salvia mellifera*). The area of some other zonobiomes is *encircled. K* is the rate of degradation if a continuous negative exponential degradation is assumed

for *Q. ilex*, thus confirming the observation that deciduous species are able to compete successfully under these conditions of climate and habitat. On steep rock faces in the same climate, where a dry summer habitat results from runoff of the larger part of the rainwater, evergreen *Q. ilex* bushes grow. In such biotopes, *Q. pubescens* is unable to compete. In addition, *Quercus ilex* is protected from cold air pockets in winter on the steep cliff slopes. Its northern limit is determined by winter cold. Of course, sclerophylly also affects the formation of soils because degradation of leaves with large amounts of wood and large amounts of raw fibres is much slower than that of malakophyllic leaves. Degradation of leaves is dependent on their mechanical strength as well as on their mineral content. Leaves that are rich in minerals are degraded faster by decomposers in the soil.

Compared to other zonobiomes the Mediterranean region with its sclerophyllic leaves is about average with respect to litter production and accumulation (because of the reduced rates of degradation by decomposers, see Fig. 154). Needles of coniferous woods in ZB VIII, of course, are mineralised more slowly because of the unfavourable climate with its long winters. In the tundra (ZB IX) plant material is mineralised even slower and therefore accumulation of raw humus occurs. Litter production and accumulation in the Mediterranean are balanced whilst in ZB I the continuous input of litter is very high but accumulation insignificant as litter is continuously degraded (k=1, Fig. 154).

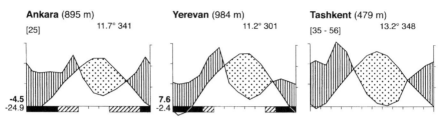

**Fig. 155.** Climate diagram of Ankara, arid Mediterranean. Homoclimates are Yerevan (High Armenia) and Tashkent (central Asia, slightly lower and warmer)

## Arid Mediterranean Subzonobiome

Small arid regions occur in the Ebro basin of north-eastern Spain (WALTER 1973) where winter cold has an influence and, in an even more extreme form, in south-eastern Spain (FREITAG 1971) the only small part of Europe which almost belongs to zonobiome III. As an example of a larger arid region, central Anatolia is described. It is a basin, 900 m above sea level and completely surrounded by mountains, which falls within the winter rain region. These mountains catch a large portion of the winter rains and in May the still wet, but already warm ascending air masses lead to thunderstorms and a rain maximum (Fig. 155). The total annual rainfall amounts to less than 350 mm. There is a pronounced summer drought and the months from December to March are cold (minimum –25 °C) with occasional intervening thaws (zonoecotones IV/VII). No forest is capable of developing under such conditions and the pine forests of the encircling mountains (Mediterranean montane belt) are succeeded, via a shrub zone with *Juniperus, Quercus pubescens, Cistus laurifolius* and *Pirus elaeagrifolia* and *Colutea, Crataegus* and *Amygdalus* (dwarf almond) species, by steppe which today is largely given over to arable land (winter wheat cultivation as "dry farming") or intensive grazing. This has resulted in degradation to an *Artemisia fragrans-Poa bulbosa* semi-desert with many spring therophytes and geophytes.

At greater altitudes, thorny cushions of *Astragalus* (Tragacantha) and *Acantholimon* (Plumbaginaceae) occur, which are especially characteristic of the cold Armenian and Iranian highlands. Originally, central Anatolia was covered by herbaceous grass steppe (*Stipa-Bromus tomentellus-Festuca vallesiaca*) reminiscent of the East European steppes (p. 376 f.), except for the Mediterranean floristic elements. The soil shows a typical chernozem profile (p. 375), although the A-horizon is not very rich in humus. The cold winter and the summer drought account for the brief growth period of only 4 months, whereby the occurrence of the rainfall maximum in May is of great significance. The most favourable season here is the spring. The first geophytes are already flowering by February and March (*Crocus, Ornithogalum, Gagea* and others), followed on overgrazed areas by numerous small therophytes which, since they only root in the upper 200 mm of the soil, have disappeared again by June. The genuine perennial steppe species are fully developed by May and do not dry out until July. Their cell sap osmotic potential is high (−1.0 to −1.5 MPa), because the soil contains sufficient water in the spring, and decreases only shortly before the plants die off. A whole series of species, including the thorny cushions, bloom during the main drought season. They are equipped with long taproots enabling them to take up water from the deeper soil horizons which are still moist in the summer: a taproot 7.65 m long has been recorded for a 30-month-old specimen of *Alhagi* (manna). Their cell sap osmotic potential, too, is above −1.5 MPa.

The periphery of the Mediterranean steppe region was settled very early by man and may be looked upon as the cradle of human civilisation. The Hittites of Anatolia were among these early settlers, as were the inhabitants of the fertile crescent formed by the mountainous slopes bordering Mesopotamia to the west, north and east. The oldest traces of grain-growing have been found in the neighbourhood of Jericho, Beidha and Jarmo. Such steppe land provided suitable conditions not only for grain-growing, but also for the support of cattle and the surrounding forests offered both game and fuel. Inhabitants of the ancient settlements have completely ruined the natural vegetation in the intervening thousands of years and in places which were once fertile country, there is now desert. Soil erosion has set in and badlands with no sign of plant life are frequently encountered. The highly varied zonoecotones in the northern part of the Mediterranean region, with its wide west-east extension cannot be discussed in detail here.

## California and Neighbouring Regions

This region in western North America, is limited by mountain ranges (Cascades and Sierra Nevada) to a narrow strip on the Pacific Coast. The winter rain region ex-

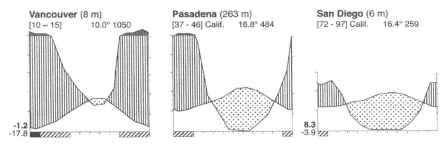

**Fig. 156.** Climate diagram of stations at the Pacific coast of North America (from N to S) in the region of coniferous forests, sclerophyllic vegetation and the transition to desert

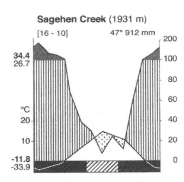

**Fig. 157.** Climate diagram of Sagehen Creek at the pass (1931 m above sea level) of the Sierra Nevada near Reno. The small peak of the rain curve in August is caused by summer thunderstorms. Absolute temperature maximum 34.4 °C, temperature minimum –33.9 °C (from WALTER 1990)

tends down the west coast from British Columbia to Lower (Baja) California, although, in the north the rainfall is so high and the summer drought so brief that the forests are hygrophilic to mesophilic coniferous forests, rich in species, and can be regarded as zonoectone IV/V (BARBOUR & MAJOR 1977).

Central and southern California together form a sclerophyllic region but Lower California is too arid for this type of vegetation (Figs. 156, 157).

The Californian zonobiome corresponds to the actual Mediterranean Californian flora, which is rich in species (Table 18, p. 250). Since the present-day flora of westernmost America is quite similar to that of the Pliocene, apparently little impoverishment took place in the Pleistocene and plant communities are therefore very rich in species. Such genera as *Quercus* and *Arbutus* are represented by a large number of species and many other genera entirely absent in Europe are found, for example, the important shrub genera *Ceanothus* (Rhamnaceae), represented by 40 species and *Arctostaphylos*, represented by 45

The north–south gradient of rainfall explains why evergreen oak forests and sometimes even mixed deciduous species occur only in the northern part of the sclerophyllic region, whereas in the south a shrub formation known as **chaparral** predominates, corresponding to the Mediterranean macchie.

shrubby species. One of the main species is the Rosaceae *Adenostoma fasciculatum* ("chamise"), with needle-like leaves; its distribution fairly exactly reflects the extent of the sclerophyllic zone.

Detailed ecological studies were carried out by Mooney and Parsons (in CASTRI & MOONY 1973) in an area of an *Adenostoma* chaparral in the mountains near San Diego (458–678 m above sea level) south of the Mojave Desert, which has been under protection for 40 years. A meteorological station at 815 m yielded the following data: mean annual temperature 14.3 °C; absolute maximum 42.5 °C; absolute minimum –7.8 °C, frost may occur between October and May; mean annual rainfall 670 mm, mainly in December to March; evaporation 1625 mm/year, mainly during the four hot summer months. In years with very little precipitation the soil may dry out to a depth of 1.2 m, below which, however, it is always moist. Lightning-induced fires are common. Temperatures may reach 1100 °C in the flames, 650 °C on the ground surface and 180–290 °C at a depth of 50 mm. Even during the dry season, *Adenostoma* accomplishes over 50% of its new growth in the first 10 days after a fire and develops 250-mm-long shoots within 30 days. All plants of the species *Quercus agrifolia* and *Rhus laurina* develop shoots. *Adenostoma* attains its densest cover 22–40 years after a fire and almost ceases growth after 60 years. The rejuvenation of such a stand occurs after the next fire. Approximately 50% of the shrub species rejuvenate by sprouting and the remainder by seed. After about 20 years the stands are again closed. In the first years after a fire steep slopes are subject to heavy erosion. The aboveground phytomass attains a value of 50 t ha$^{-1}$ while the underground phytomass is double that amount. The aboveground net production is approximately 1 t ha$^{-1}$ a$^{-1}$ in the younger stands and decreases with age. The shrubs are normally photosynthetically active all year-round. In spring an abundant ephemeral vegetation develops, some species of which only germinate after a fire. *Adenostoma* is predominant on southern slopes, while dense stands of *Quercus dumosa* grow on northern slopes. The strip of land immediately bordering the ocean north of latitude 36 °N in California does not belong to the sclerophyllic zone. Fog resulting from the cool ocean currents renders the summers cool and wet so that hygrophilic, northern tree species are found.

Unlike the macchie, the **chaparral** is the natural zonal vegetation corresponding to a relatively low winter rainfall of 500 mm. Fires are common in this region and were a natural factor even before the advent of man. Statistics of the US National Forest Service show that fire caused by lightning is very common in the chaparral region, rendering constant fire-watching necessary during thunderstorms. If a fire occurs every 12 years, the character of the chaparral remains unchanged since the shrubs can sprout afresh. However, if no fire occurs for a great length of time, then species such as *Prunus ilicifolia* and *Rhamnus crocea* infiltrate. If one fire follows another within 2 years, seedlings of those shrubs which cannot resprout after a fire are killed off and these woody species are eliminated.

The roots of sclerophyllic species reach far down into the ground because the upper soil layers are usually completely dried out in summer. By means of their roots, which may even penetrate from 4 to 8.5 m into the rock fissures, the plants are able to obtain a certain amount of water in the dry season. More details on the root system profiles may be found in KUMMEROW (1981). Availability of water can be recognised from the observation that very soon after a fire in the height of summer the shrubs begin to sprout again. After the loss of the transpiring surface a small amount of water suffices for the buds to grow. The autumn rains have no direct effect, since it takes a month for the water to sink to a depth of 1 m. In the meantime, the temperature falls so much that growth stops. In April, when the water supply is good and the temperature rising, growth is at a maximum. The old evergreen leaves continue assimilation into the spring and are shed in June, by which time the new leaves can take over their function. All chaparral species possess mycorrhizae and the *Ceanothus* species have nodules which assimilate atmospheric nitrogen. A detailed vegetation monography with much ecological data were published by BARBOUR & MAJOR (1977).

Evergreen sclerophyllic oak forests are also found as a montane belt in North America, above the cactus desert in the mountains of southern and Central Arizona, at an altitude of 1200–1900 m. This is known as the **encinal belt** and on the basis of the distribution of the different species of *Quercus* is subdivided into an upper and lower belt. The upper belt is succeeded by a *Pinus ponderosa* belt. The chaparral genera (*Arbutus, Arctostaphylos, Ceanothus*) form the shrub layer in the encinal forest. Although Arizona has two rainy seasons the vegetation is strongly reminiscent of that of California, except that the sclerophyllic forests in the mountains are much better developed and are still original. Despite the fact that summer storms supplement the low winter rainfall, summer drought is still very pronounced.

East of the Sierra Nevada, in the state of Nevada, the winter rains fall off to about 150–250 mm. At an altitude of 1300 m, the cold season lasts 6–7 months. The climate

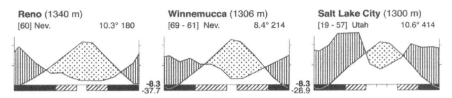

**Fig. 158.** Climate diagram of sage brush areas (*Artemisia tridentata* – semi-desert): Reno, Winnemuca and Salt Lake City (almost transitions to grass land)

**Fig. 159.** The dependency of the precipitation (*above*) on the relief (*below*), illustrated in a west to east profile through the western part of North America at approximately 38 °N latitude (from Walter 1960)

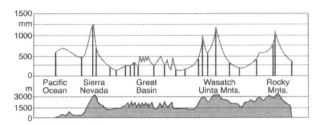

diagram of Sagehen Creek (Fig. 157) illustrates this for a mountain pass with high precipitation and forest and bog vegetation. The climate diagram for Reno (Fig. 158) reflects the conditions on the lee side of the mountains with its *Artemisia tridentata* semi-desert – also designated as "sagebrush". Figure 159 illustrates the dependency of the precipitation on the relief in this region. The *Artemisia* semi-desert occupies enormous areas in Nevada, Utah and the bordering states and is the cold-climate equivalent of the southern *Coleogyne* and *Larrea* semi-desert. *Artemisia* prefers the heavy soils of the basins and gives way to *pinyon* on elevated stony ground. Pinyon consists of low scattered tree communities with *Pinus monophylla* or *P. edulis-Juniperus* stands, including some cold-resistant chaparral species. True coniferous forests of *Pinus flexilis* and *P. albicaulis* commence at an altitude of about 2000 m in the mountain, whereas further east *Pinus ponderosa* is found. Higher up, these species are replaced by *Pseudotsuga* and *Abies concolor* and the tree line above 3000 m is made up of *Picea engelmannii* and *Abies lasiocarpa*. The dry southern slopes are often devoid of trees and *Artemisia* extends up to the alpine region, but the order of the altitudinal belts can vary greatly. The aspen (*Populus tremuloides*) is also frequently encountered: it plays an important role in wet areas with the formation of extensive clonal suckers (shoots) from the roots.

*Artemisia tridentata* (sage brush) is a half-shrub 1.5–2 m high and attains an age of 25–50 or more years. Its taproot extends about 3 m into the ground and gives off horizontal lateral roots extending far on all sides. In spring after the snow has melted and with a good water supply, the cell sap osmotic potential is very high (about –1.0 to –1.5 MPa); it soon falls to about –2.0 to –3.5 MPa and can even amount to –7.0 MPa in summer if there is an acute water shortage. At this point it sheds its older leaves like all malakophyllous plants.

The sage brush semi-desert is confined to the arid brown, semi-desert, salt-free type of soil with *Artemisia tridentata* as the dominant species, frequently accompanied by the dwarf-shrub *Chrysothamnus* (Asteraceae). The region belongs to the arid ZB VIIa. Depressions with no outflow are invariably brackish in such an arid climate. The salt pans and salt lakes in these regions are the relicts of much larger Pleistocene lakes such as Lake Bonneville in Utah, the surface of which was 310 m above the Great Salt Lake and the extensive surrounding barren salt desert. Lake Bonneville covered an area of $32,000 \text{ km}^2$, with a maximum length of 586 km and a maximum width of 233 km. The present salt desert is more than 161 km long and 80 km wide. In 1906, the Great Salt Lake measured $120 \times 56$ km when at its highest level. Its contours vary greatly, the average depth being a little above 5 m. Its salt content varies between 13.7 and 27.7%. About 80% of the salt is NaCl; the remainder consists of 20% $MgCl_2$, $Na_2SO_4$, $K_2SO_4$, $MgSO_4$ and others. Halophytes occur around the saline areas.

With the exception of the salt-excreting grass *Distichlis*, all of the halophytes are Chenopodiaceae. The entire area is one large halobiome with biogeocene complexes. The sequence corresponds broadly to the zonation of halophyte types similar to those in Central Asia. The climate in Utah is very similar to that of Ankara (Anatolia). The marked predominance of *Artemisia* in Anatolia is the result of overgrazing; such grasses as *Agropyron*, *Stipa* and *Festuca* were formerly widespread.

Vegetation around the Great Salt Lake in Utah forms very distinct zones. Hygrohalophytic *Allenrolfea* and *Salicornia* grow on the edge of the salt crust, followed by *Suaeda* and *Distichlis*. Further wide zones are occupied by *Sarcobatus*, which requires ground water and the xerohalophytic species *Atriplex confertifolia*. Species of *Kochia* and *Ceratoides* (*Eurotia*) *lanata* then lead on to a non-halophytic zone with *Artemisia tridentata*.

## Central Chilean Winter Rain Region with Zonoecotone

Chile is a long, narrow strip of land 200 km wide and 4300 km long, extending from 18–57 °S, lying at the western foot of the High Andes. It exhibits every possible transition in vegetation, from the rainless subtropical desert in the north to the very wet, temperate and subarctic

The typical vegetation of central Chile is composed of woodlands of 10–15 m high woodland species, the **matorral,** with xerophytic sclerophytes.

**Fig. 160.** Vegetation of Chile and climate diagrams (after Schmithüsen 1956). *1* northern High Andes, *2* desert region (Atacama, s.l.), *3* dwarf-shrub and xerophytic-shrub region, *4* sclerophyllic region (matorral and espinal), *5* deciduous summer green forest, *6* evergreen rain forests of the temperate zone (Valdivian rain forest), *7* tundra-like vegetation of the cold zone, *8* subantarctic deciduous summer green forest, *9* Patagonian steppe, *10* southern Andes

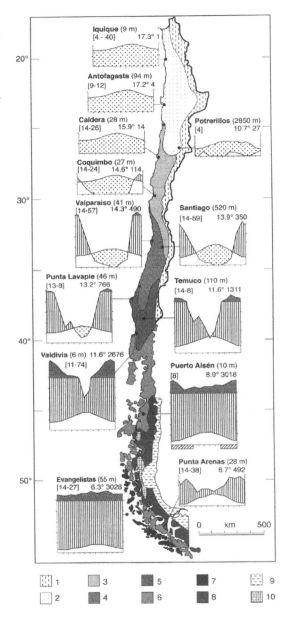

**Fig. 161.** Landscape near Santiago (Chile). In the foreground, on rocky soil, flowering *Trichocereus*. In the valley, remnants of sclerophyllic vegetation. The grasses are adventitious annual Mediterranean species (*Avena* and others) (Photo: E. WALTER)

forests in the south via a sclerophyllic region. Winter rains are prevalent throughout (Fig. 160). The cold Humboldt current flowing along the entire coast modifies the summer drought so that temperatures are lower than those of California. The mean annual temperature of Pasadena at 34 °N is 16.8 °C, whereas in Santiago at 33 °S, it is only 13.9 °C. The climates of these two regions have been compared by CASTRI (1973).

Since Chile belongs to the Neotropical floristic realm, its flora is quite different from that of the Mediterranean region or California (see p. 264 f.). Only the cultivated areas offer a similar appearance since the same species are cultivated on the farms and in the gardens in all three regions. The sclerophyllic region occupies the central part of Chile and adjoins the arid regions in the north. This is also only represented by remnants. Species occurring are xerophytes such as *Lithraea caustica* (Anacardiaceae), which causes rashes and fever if touched, the soap tree *Quillaja saponaria* (Rosaceae), *Peumus boldus* (Monimiaceae) and the Lauraceae *Cryptocarya* and *Beilschmiedia* which have a preference for wet ravines, in addition to a whole series of shrubby species. The endemic palm *Jubaea chilensis* grows in a narrowly limited area north-east of Valparaiso; columnar cacti (*Trichocereus*, Fig. 161), and the large *Puya* species (Bromeliaceae) as well as the thorny Rhamnaceae *Colletia* and *Prevoa* are found in dry, rocky habitats.

Sclerophyllic species from California, Chile and Australia look very much alike externally, but there are signifi-

**Table 19.** Types of fruit of the Mediterranean flora in central Chile, California and Australia as well as distribution in percent of the colour of fleshy fruits

| | Chile | California | Australia |
|---|---|---|---|
| Small, fleshy fruits | 34.2% | 29.1% | 12.1% |
| Small, dry fruits | 19.8% | 43.7% | 45.0% |
| Large fruits (>15 mm) | 14.4% | 6.3% | 0 |
| Anemochore (for example winged fruits) | 29.7% | 19.4% | 23.6% |
| Others (with aril hooks, thorns etc.) | 1.8% | 1.5% | 19.3% |
| Colour of fleshy fruits: | | | |
| Black/violet | 48% | 27% | |
| Red | 16% | 43% | |
| Green | 12% | 2% | |
| Others | 24% | 28% | |

(after HOFFMANN & ARMESTO 1995)

cant differences, for example in the form of their fruits (see Table 19). In Australia there are many species with fruit appendices with thorns or barbs and almost half of the species bear small dry fruits, whilst in Chile and California many species bear large fleshy fruits. Also, the colour of the fleshy fruits is very different, indicating that different animals were responsible for their dispersion.

The actual matorral area is relatively small. On the Chilean side, the Andes drop steeply: Aconcagua, at 7000 m, is only about 100 km from the coast. Talus (scree) communities predominate in the mountains and altitudinal belts are difficult to recognise. The sclerophyllic vegetation extends up to about 1500 m (Fig. 162) and shrub communities lead on to the alpine region, with the occasional appearance of the conifer *Austrocedrus* (*Libocedrus*) *chilensis*. Alpine talus plants such as *Tropaeolum* species and *Schizanthus* (a Solanaceae with zygomorphic flowers), as well as Amaryllidaceae (*Alstroemeria, Hippeastrum*) and species of *Calceolaria*, are common. Flat, cushion-like plants are characteristic of the upper alpine belt (*Azorella* and other Apiaceae).

Species occurring in the altitudinal belts of the orobiome, as well as to the south of the sclerophyllic zone, are already Antarctic elements as are the tree-form *Nothofagus* species. Immediately south of Concepción, with decreasing summer drought, *Nothofagus obliqua* forest is

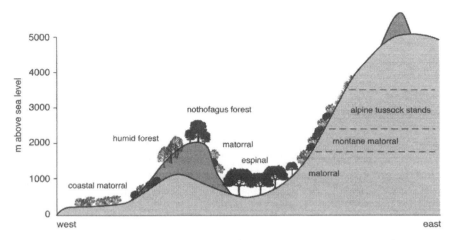

**Fig. 162.** West–east transect through central Chile with the most important vegetational forms up to the Andes (after RUNDEL 1982)

found (Zonoecotone IV/V). These trees lose their leaves in the cool winter months. Further south with a rainfall exceeding 2000–3000 mm, this forest is replaced by the evergreen Valdivian warm temperate rain forests of ZB V (QUINTANILLA 1974), which are scarcely less luxuriant than the tropical rain forests while their timber mass is probably even greater. The woody species are partly Neotropical in origin. Bamboos (*Chusquea*) are also common. Part are Antarctic elements, as for example the evergreen *Nothofagus dombeyia*. A large number of ancient coniferous forms are found, especially in montane situations. Besides *Austrocedrus* and *Podocarpus* species, *Saxegothea, Fitzroya, Araucaria araucana* (= *A. imbricata*) and *Pilgerodendron uviferum* also deserve mention. In the undergrowth are also many shrubs, for example *Weinmannia* and *Caldcluvia* (Cunoniaceae), *Raphithamnus* (Verbenaceae), *Ugni* (Myrtaceae), *Coriaria* (Coriariceae) and others. The climate is very wet and cool but frost-free, so that the evergreen forest gives way to the so-called Magellan forests which stretch almost to the tip of South America, gradually becoming poorer in species and decreasing in height until finally they are only 6–8 m tall. The westerly islands are covered by bogs with cushion-like plants (*Sphagnum* occurs, but is not important), a vegetation closely related floristically to that of the Antarctic islands. Similar Antarctic elements are found in New Zealand and in the mountains of Tasmania, indicating that these regions were formerly directly connected with each other via the Antarctic continent. The bogs can be considered to be Antarctic tundra (ZB IX).

## South African Cape

Although confined to the outermost south-western tip of Africa the South African winter rain region comprises an entire floristic realm, the Capensic. This small region is extraordinarily rich in species. In the Jonkershoek nature reserve alone, 2000 species have been recorded on 2000 ha and an equal number on the 50-km stretch from Table Mountain to the Cape of Good Hope. Included are 600 species of the genus *Erica,* 108 species of *Restio* (Restionaceae), 117 species of *Cliffortia* (Rosaceae), 115 species of *Muraltia* (Polygalaceae) and about 100 species of *Protea.* Particularly abundant among the sclerophyllic plants are the Proteaceae, a family which is otherwise well represented only in Australia, albeit by a different subfamily (a few genera also occur in South America).

The sclerophyllic vegetation of the Cape is known as **fynbos,** which is 1–4 m high and very similar to the macchie.

Many house plants of temperate latitudes originally came from the Cape (*Pelargonium, Zantedeschia* = *Calla, Amaryllis, Clivia* and others). The climate diagrams for Cape Town and Tangiers (N-Africa) are comparable, except that the annual rainfall of the former is 260 mm less, although the summer is slightly less dry (Fig. 163). The fynbos only covers a very small area, similarly to the matorral.

The only tree species, *Leucadendron argenteum* (silver tree), is of very limited distribution and is confined to the humid slopes of Table Mountain, at altitudes below 500 m. In wet ravines, forest-like stands can be found, but these

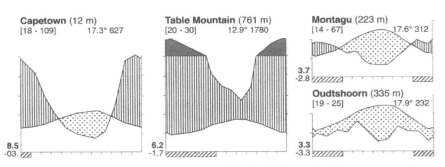

**Fig. 163.** Climate diagrams of South Africa: typical sclerophyllic region; damp montane climate (misty); transitional region; and typical karroo climate

are in fact the last outposts of the wet-temperate forests of
the south-eastern coast of Africa (ZB V). The leaves of the
*Protea* are sometimes very large and, although they have
very little mechanical tissue, are nevertheless sclerophyllic
because of their thick cuticle. As in all sclerophyllic
plants, the water balance of the Proteaceae shrubs is in a
state of equilibrium, which means that the cell sap concentration undergoes only very slight variations during
the course of the year. The deeper layers of the soil, into
which the roots penetrate, apparently contain water even
in the summer. Cape soils are acid and poor in nutrients,
which particularly suits both Proteaceae and Ericaceae
(with obligate mycorrhiza).

Fire is the most important ecological factor. In the year
immediately after a fire, innumerable geophytes (*Gladio-
lus, Watsonia* and others) of which the Cape flora possesses about 350 species, make their appearance followed
by herbaceous species and dwarf shrubs. It takes about
7 years before the Proteaceae shrubs have grown up again,
either from seedlings or as shoots from old plants.
Although these plants are capable of achieving considerable age, they lignify with time and flower less abundantly
so that periodic burning would appear to be advantageous. In this region, fire caused by lightning also seems
to constitute a natural factor, although nowadays fires are
deliberately or carelessly started by man. It is interesting
to note that bulbous plants begin to flower only after a
fire and otherwise grow vegetatively. The reason for this is
probably the sudden removal of root competition by the
bushes rather than a fertilising effect resulting from the
ash.

Rainfall increases with increasing altitude only on the
south-eastern slopes of the mountains which are affected
by the warm, humid air rising from the Indian Ocean.
The rainfall recorded at the Table Mountain station is
three times higher than that of Cape Town, 750 m lower
in altitude (Fig. 163). The Cape is mountainous, with basins scattered between the mountain ranges which often
wear a "tablecloth" or covering of cloud due to the warm,
wet winds from the Indian Ocean rising up the south-
eastern slope and dispersing again on the north-western
slope (Fig. 164). This leads to a wet mist on the high plateaux of Table Mountain and there is a tendency toward
heathland (*Restio, Erica*) and even the formation of moors
(mossy mats of *Drosera* and species of *Utricularia*). Succulents such as *Rochea coccinea* grow between dry
boulders.

Moving inland the winter rainfall decreases, particularly
in the rain shadow of the mountain ranges (Fig. 163). In the

**Fig. 164.** The "tablecloth" on Table Mountain near Cape Town. The fog not only develops on Table Mountain, but also on other higher mountain ranges on the south coast of Africa. The northern side is sunny and the southern side is covered in dense clouds (Photo: S.-W. BRECKLE)

rain-shadow area, the dry form of Cape vegetation is encountered, known as the renoster bush, dominated by the tiny bushes of *Elytropappus rhinocerotis* (Asteraceae) with characteristic bundles of roddy stems. It is the transitional region of zonoecotone IV/III which is succeeded inland by the semi-desert vegetation of the karroo (p. 234). Since the colonisation of the Cape in 1400, the sclerophyllic vegetation, or fynbos, has spread extensively. Formerly, an evergreen-temperate forest with Paleotropical elements stretched along the entire coast of south-east Africa beyond the southern tip of the continent (Cape Agulhas; ZB V).

## South West and South Australia

Zonobiome IV occurs in Australia in south-west and South Australia. Sclerophyllic vegetation is made up of *Eucalyptus* forests (jarrah) and bushes (mallee).

A **lignotuber** is an underground wooden bulb (between 5 cm to 2 m in diameter) with many resting buds from which the plant can sprout.

Perth in south west Australia lies at approximately the same latitude as Cape Town and has a very similar climate (Fig. 165). However, winter rain falls not only on the south-western corner of this continent, but also on the area around Adelaide in South Australia.

On account of the peculiar floristic situation (p. 231), the sclerophyllic vegetation differs in character from that in other winter rain regions of the earth. The tree form (*Eucalyptus* species) dominates, and Proteaceae constitute a lower shrub layer, but may achieve predominance on the sandy heaths. The leaves of *Eucalyptus* are leathery rather than hard. Many bushy or low *Eucalyptus* species which form lignotubers grow in the mallee. Lignotubers are regarded as an adaptation for surviving unfavourable events (fire, drought, coldness) and occur in all Mediterranean regions and also in other dry regions with a particularly

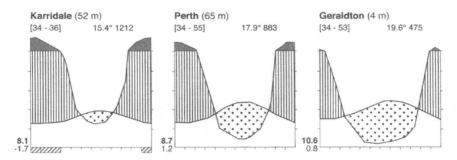

**Fig. 165.** Climate diagram from SW Australia. Stations in the karri forest, in the jarrah forest and in the bush heath (see also Fig. 182, Adelaide)

large number of species in Australia. The formation of lignotubers is a genetically fixed characteristic which may be strongly modified by environmental influences. The ecological significance of lignotubers is not always clear. In California, for example, the lignotuber forming *Arctostaphylos glandulosa* grows in the same habitat as the lignotuber-free *Arctostaphylos glauca. Eucalyptus camaldulenesis* growing in southern Australia does not have lignotubers whilst those growing further north are ecotypes with lignotubers. Several of the western Australian *Eucalyptus* species and also *Banksia* and others form lignotubers. In California *Adenostoma fasciculatum, A. sparsifolium, Ceanothus, Quercus dumosa, Rhus laurina* and others form lignotubers. In Chile, for example, *Colliguaja odorifera, Quillaja saponaria, Lithraea caustica, Cryptocarya alba* form lignotubers whilst in the Mediterranean region tuber formation is only known to occur regularly in *Quercus suber.* In all cases it is probably very important that growth can quickly occur through sprouting.

Peculiar to south-western Australia are the "grass trees" (*Xanthorrhoea, Kingia*), the Cycadaceae *Macrozamia* and species of *Casuarina.* Epacridaceae take the place of Ericaceae. Just as on the Cape in South Africa, the soils are acid and poor in nutrients, but rich in quartz containing $SiO_2$ and iron concretions which are the lateritic crusts of an earlier geological age when the climate was tropical. The parent rocks are among the oldest geological formations on earth. Indicative of the poverty of the soil is the fact that 47 species of *Drosera* (sundew) occur in the herbaceous layer of the forest around Perth. Wherever it is wet enough, bracken (*Pteridium*) is found. Rainfall increases to the south of Perth (up to 1500 mm) but decreases to the north and inland. Each change in climate

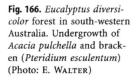

**Fig. 166.** *Eucalyptus diversicolor* forest in south-western Australia. Undergrowth of *Acacia pulchella* and bracken (*Pteridium esculentum*) (Photo: E. WALTER)

means that different species of *Eucalyptus* rise to dominance. The wetter the climate the taller the trees and therefore the greater the leaf area per hectare. Because the leaves hang vertically, plenty of light can penetrate the space around the trunks, so that the shrub layer is usually well developed insofar as it has not been reduced by frequent fires.

"Jarrah" **forest,** which is completely dominated by *Eucalyptus marginata,* is characteristic of a climate comparable to a Mediterranean type, with 650–1250 mm of rain and a summer drought. The trees can reach an age of 200 years and a height of 15–20 m (maximum, 40 m). In the wetter, southern regions *Eucalyptus diversicolor* forms the **"karri" forest,** with trees reaching a height of 60–75 m (maximum, 85 m; zonoecotone IV/V). The canopy is closed to 65% coverage and the undergrowth consists of a shrub layer and a herbaceous layer developed from the fronds of bracken which may reach 1.5 m (Fig. 166).

In the drier **"wandoo"** zone, with a rainfall of 500–625 mm, *Eucalyptus redunca* dominates. The woodland is sparser and is now almost exclusively given over to sheep grazing. In the absence of suitable indigenous grasses, *Lolium rigidum* and the annual Mediterranean clover, *Trifolium subterraneum* are sown. *T. subterraneum* buries its fruits underground and provides a source of nitrogen. Owing to the poverty of the soil, fertilisation with superphosphate is essential before sowing; because of the large areas involved, both of these processes are carried out from the air. The **mallee,** once rich in species with many

**Fig. 167.** Mallee, rich in species with several species of *Eucalyptus* and bushy Proteaceae (*Banksia*), herbs and geophytes to the west of Raventhorpe, south-west Australia (Photo: S.-W. BRECKLE)

shrubs including many Proteaceae and an enormous biodiversity of small shrubs, herbs and geophytes, has only survived in protected areas (Fig. 167).

In the zone receiving 300–500 mm of rain annually, many species of *Eucalyptus* are found scattered over the landscape (zonoecotone III/IV). Nowadays, however, this is the winter wheat zone, with farms of several hundred hectares which are completely mechanised and run by two or three men. Rust diseases render wheat cultivation in wetter zones uneconomical.

Where the mean annual rainfall drops below 300 mm, *Eucalyptus* disappears and the extensive grazing lands provided by the bush semi-desert commence (p. 231). In South Australia, there is no humid winter rain region, but otherwise the situation is similar to that in south-western Australia, only rather more complicated because mixed forest communities consisting of several species of *Eucalyptus* occur. Furthermore, the region is mountainous which again leads to a marked differentiation of the vegetation.

Apart from the forests already described, there are vast areas of heathland with 0.5–1 m tall Proteaceae. The bushes are capable of growing on the poorest of sands and even the least demanding of the *Eucalyptus* species are unable to compete with them (Peinobiome). Such areas are not cultivated and are hardly used for grazing. It is all the more remarkable that so very many species grow on such an impoverished, sandy soil. On 100 m$^2$, 90 species have been recorded, including 63 small woody species, mainly Proteaceae or Myrtaceae. *Drosera* species and a tuberous *Utricularia* also occur. Thorough ecophysiological investigations have been carried out on such a

heath in South Australia at a rainfall of 450 mm annually
and with 7 months of drought in summer (SPECHT 1958).

Soil temperatures at a depth of 15 cm varied between
4.1 and 36.0 °C and, at 30 cm, between 5.8 and 29 °C. The
root systems of 91 species were dug up. The dominating
sclerophyllic species are the shrub-like *Eucalyptus bacteri*,
nine Proteaceae, two *Casuarina* species, *Xanthorrhoea*, Leguminosae and others. The main growth period is the dry
summer since the soil remains wet to a great depth. Smaller perennial species (42% of all species) root only in the
upper 30–60 cm and grow in the spring. *Drosera* and
orchids are ephemeral species and root at a depth of only
5–7 cm. It is interesting to note that the water in sandy
soil with a wilting point of 0.7–1% is very unevenly distributed because the larger species draw the rainwater toward their stems. The composition of the heathland is determined by fire. Immediately after a fire the grass tree
*Xanthorrhoea* begins to sprout, in fact this species blooms
only after a fire. *Banksia* (Protaceae) regenerates by means
of seedlings and its share in the above ground phytomass
rises to 50% by the 15th year after a fire. The larger part
of the dry substances of 25-year-old specimens is accounted
for by the large fruits which open only after a fire.

*Banksia* is one of the many **pyrophytes** very commonly
encountered in Australia. These plants are capable of regeneration only after fire because the woody fruits do not
otherwise open, which suggests that fires caused by lightning have always played a natural role in Australia. Forest
and heath are now often burned because the woody plants
are of no commercial value and hinder grazing. In the
farmer's opinion, "One blade of grass is worth more than
two trees." – but for how long?

Many Proteaceae and Myrtaceae as well as the coniferous *Actinostrobus* and others are pyrophytes and even *Eucalyptus* species seed more prolifically after a fire. The nutrients in a heath that has not been burned for a long
time are bound up in the fruits of *Banksia*, old leaves of
*Xanthorrhoea* and the accumulating litter. A 50-year-old
community degenerates on this account and not until a
fire has caused mineralisation of the nutrients can a new
succession be initiated.

*Eucalyptus marginata* woodland is ecophysiologically
typical of a sclerophyllic vegetation. The roots penetrate
the hard lateritic layer in places to a depth of more than
2 m. There is no summer dormant period and the water
balance is maintained by partial closure of the stomata
from 10 to 15 h with a resultant decrease in transpiration.
The cell sap osmotic potential in winter was found to be
–1.6 MPa and is probably only slightly lower in summer.

Not only its flora, and thus the character of its vegetation, but also the **fauna** of Australia differs widely from that of other continents. Monotremes, the most primitive mammals, occur only in Australia. The duckbill platypus, *Ornithorynchus anatinus,* belongs to this family, it lays one to three eggs which are hatched by the female. The spiny anteater (*Tachyglossus = Echidna*), in contrast, hatches only one egg in its brood sac; this animal represents a transition to the marsupials. With only a few exceptions, the latter are also confined to the Australian continent and include both herbivores and carnivores. The best known group are the kangaroos (Macropodidae), including the large kangaroo *Macropus.* As a grazing animal, the latter has had a considerable influence on the vegetation, although exact figures are not available.

# 10 Mediterranean Orobiome

In the mountainous regions of the Mediterranean a distinction must be made between the humid altitudinal belts and the arid altitudinal belts (WALTER 1975):

a) In the **humid altitudinal belts** on the northern margins of the Mediterranean zone temperature decreases with increasing altitude and the dry season also disappears. In both cases several of the vegetational units (hypsozonal or orozonal) corresponding to the zonobiome form the altitudinal belts.

In the humid altitudinal belts, the evergreen sclerophyllic forest is succeeded by a sub-Mediterranean deciduous forest with oak (*Quercus pubescens*) and chestnut (*Castanea*). Above this beech (*Fagus*) and fir (*Abies*) form a cloud forest at the summer cloud level. In the Apennines, the tree line is formed by beech, which also occurs on Mount Etna and in northern Greece. In the maritime Alps, the beech belt is succeeded by a spruce (*Picea*) belt and in the Pyrenees by a belt of *Pinus sylvestris* and *P. uncinata.*

b) the **arid altitudinal belts** occur in the continental climate region with summer droughts noticeable up to the alpine zones. In the arid altitudinal belt there is no deciduous forest. The Mediterranean sclerophyllic forest is followed immediately by a series of various coniferous forest belts. For example, on the southern slopes of the Taurus in Anatolia, there is an upper Mediterranean belt with *Pinus brutia,* a weakly developed montane belt with *Pinus nigra* ssp. *pallasiana,* a high-montane belt with *Cedrus libanotica* and *Abies cilicica* (wetter) or species of *Juniperus* (drier), and a subalpine belt with *Juniperus excelsa* and *J.*

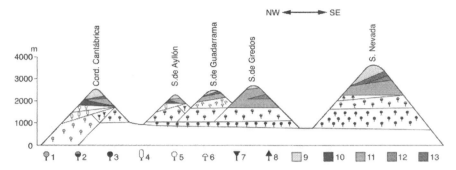

**Fig. 168.** Altitudinal belts in the high crystalline mountains of the Iberian peninsula shown in north-west-southeast profile (according to ERN 1966). *1* Deciduous oak forest (*Q. robur, Q. petraea*), *2 Q. pyrenaica* forest, *3* holm oak (*Q. ilex*) forest, *4* beech forest (*Fagus sylvatica*), *5* birch forest (*Betula pendula*), *6* pine forest (*Pinus sylvestris*), *7* mixed-deciduous forest (*Quercus, Tilia, Acer*), *8* high-altitude forest of the Sierra Nevada (*Sorbus, Prunus* etc.), *9* high-alpine grass and herbaceous vegetation, *10* dwarf-shrub heath (*Calluna, Vaccinium, Juniperus*), *11* broom heath (*Cytisus, Genista, Erica*), *12* thorn cushion belt, *13 Festuca indigesta*, dry sward

*foetidissima.* However, in the rainy north-eastern corner of the Mediterranean in the Amanos Mountains, a cloud belt with *Fagus orientalis* is found. *Cedrus libanotica* occurs also on Cyprus and a small relict is present in Lebanon at 1400–1800 m above sea level. On Cyprus and Crete, as well as in Cyrenaica, *Cupressus semper-virens* (cyprus) always occurs in the upper Mediterranean belt in its natural form with horizontal branches. The frequently planted columnar form is a mutation. In the Atlas Mountains, from the eastern High Atlas to the Tunisian border, the subalpine belt at an altitude of more than 2300 m above sea level, consists of cedars (*Cedrus atlantica*). However, the altitudinal belts vary greatly according to the course taken by the mountain ranges and the exposure of the slopes. Figure 168 shows the complicated order of the altitudinal belts in the Spanish mountain ranges.

A difference in the altitudinal belts in arid and humid regions is recognisable even above the tree line in the alpine region. Whereas the situation in the humid mountain climate is similar to that in the Alps, in the arid alpine regions the vegetation (Fig. 169) consists of thorny hemispherical cushion plants with many convergent species from different families. It is only possible to distinguish them when they are flowering or fruiting. This belt is followed by a dry, grassy belt where hygrophilic plants (mostly endemic species related to arctic-alpine plants) are found on spots kept moist in summer by melting snow.

**Fig. 169.** Thorn cushion belt with *Erinacea pungens* (Fabaceae) in the mountains of Teruel (Spain) on the Linares pass (2000 m above sea level) (Photo: S.-W. BRECKLE)

Interlinking of the Mediterranean vegetation in the mountains of South East Europe is very complicated, where transitions to ZB VI and in parts expression of ZB VII can be observed. Sub-Mediterranean deciduous forests have been almost completely degraded to a summer green bushland, the **schibljak,** because of removal of timber, deforestation by fire and grazing in woodlands. Towards the east, more and more macchie-like plants occur in addition to summer green shrubs, representatives of the East European schibljak from zonoecotone IV/VI in Bulgaria and Yugoslavia, for example *Ostrya carpinifolia, Cotinus coggygria, Fraxinus ornus, Pyrus spinosa* and others. Such a mixed vegetation of evergreen species from the macchie and the summer green species of the schibljak is called **pseudomacchie.** A survey of the Mediterranean altitudinal belts can be found in OZENDA (1975). Interesting conditions are found in the orobiomes of Macaronesia, especially on the Canary Islands, which are affected by the north-easterly trade winds.

## Climate and Vegetation of the Canary Islands

Macaronesia consists of the island groups of the Azores, Madeira, the Canary Islands and the Cape Verde Islands. The first three with winter rain and summer drought belong to zonobiome IV, while the climate of the Cape Verde Islands south of the tropic of Cancer is so dry that it must be assigned to zonoecotone II/III. The botanically most interesting and most thoroughly studied of these island

**Fig. 170.** Laurel forest on a northern slope on Tenerife at 350 m above sea level. On the *left*, the treetops of *Laurus canariensis*, deformed by ascending trade winds, can be seen (from WALTER 1968)

groups are the Canary Islands, especially the islands of Tenerife and Gran Canaria. Ever since ALEXANDER VON HUMBOLDT interrupted his journey to Venezuela, at Tenerife in 1799 and distinguished five altitudinal belts after brief observations, numerous botanists have studied the flora of this island. This has led to a bibliography of 1030 titles (SUNDING 1973). Recent plant sociological studies were carried out by OBERDORFER (1965), SUNDING (1972) and RIVAS-MARTINEZ (1987) and LUDWIG (1984) examined the flora. Ecological investigations are found in KNAPP (1973), VOGGENREITER (1974), KUNKEL (1976, 1987) as well as studies on radiative adaptation in KULL (1972) and LÖSCH (1988). The origin of these volcanic islands goes back to the Cretaceous. Gran Canaria rises to nearly 2000 m above sea level and Tenerife to a little over 3700 m. These very steep orobiomes differ from others of zonobiome IV in that they rise directly out of the ocean and lie near 28 °N and are under the influence of the trade winds. This means that the wind-exposed northern slopes have climatic conditions differing from those on the wind-protected southern slopes. Clouds from the trade winds are held back on the northern slopes, resulting in rainfall from the ascending clouds as well as fog, so that there is no summer drought. The warm, humid climate of the intermediate levels corresponds more to zonobiome V with its evergreen laurel forests (Fig. 170). In contrast, the southern slopes are especially dry at the lower altitudes and are more often swept by hot winds from the Sahara. Therefore, these islands possess local conditions such as are found in zonobiomes III–V, which are increasingly subject to frost at higher altitudes. The Pico de Teide on Tenerife is covered with an alpine rubble desert at altitudes over 3000 m above sea level, which is typical of tropical mountains.

The volcanic islands were colonised several times (mainly in the Tertiary) from the neighbouring continent of Africa which at the time was still inhabited by Tertiary evergreen forests. These tree species remained on the warm and humid northern slopes of the islands up to the present day, presenting a sort of living museum after they had become extinct on the mainland. This has resulted in floristic relationships to distant elements on the southern tip of Africa (*Ocotea foetens*), India (*Apollonias*), various other tropical regions (*Persea, Visnea* (Theaceae), *Dracaena draco*) and to the humid Mediterranean *Laurus azorica, Laurocerasus (Prunus) lusitanica, Phoenix canariensis*). On the other hand, certain elements of arid regions have found appropriate niches in suitable lower levels and rocky locations (*Launaea, Zygophyllum,* succulent *Euphorbia* and *Kleinia* species). Many species are endemic, such as the numerous succulent Crassulaceae which were previously assigned to *Sempervivum*, but are now considered endemic genera (*Aeonium* with 33 species, *Aichryson* with 10 species, *Greenovia* with 4 species, *Monanthes* with 15 species). In addition, eumediterranean elements also arrived, although probably not until the Pleistocene.

Since the colonisation of the islands 500 years ago by Spain, further Mediterranean species, including goats, were introduced. The settlements with their cultivated land expanded increasingly, thereby endangering the original vegetation, especially the unique humid evergreen laurel forest with its many Tertiary relicts. This forest is cleared for its valuable wood; the litter layer and the humus are removed for the improvement of agricultural soils, thus making regeneration of the forest in cleared areas virtually impossible. Less demanding woody plants such as *Erica arborea* or *Myrica faya* grow in such areas, or the land is reforested with *Pinus* or even *Eucalyptus*. On Gran Canaria, only 2% of the original laurel forest still exists (Fig. 171) and it is still shrinking rapidly on Tenerife. These beautiful islands have recently become even more endangered by profit-orientated mass tourism, as is the case for most impressive landscapes all over the world.

KÄMMER (1974) studied the climatic conditions on Tenerife in detail especially the significance of fog precipitation caused by trees in the cloud layer. As a result of mea-

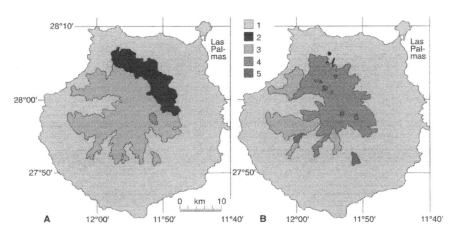

**Fig. 171.** Geographical distribution of vegetation on Gran Canaria as it was originally (**A**) and as it appears today, modified by man (**B**). *1* Succulent semi-desert (today most of the flat lowlands are under cultivation), *2* laurel forest or Myrico-Ericetum, *3* pine forest (today partly *Cistus* heath), *4* broom heath, *5* mixed stands of *Cistus* and broom (adapted from SUNDING 1972)

"Visiting the islands after an absence of 40 years, one is taken aback to find only paved fair grounds and highways everywhere. Environmental protection usually does not become effective until there is almost nothing left to protect. Today's youth has never experienced the quietness and grandeur of undisturbed nature." (WALTER)

surements carried out over a number of years, he concluded that the heavily increased ascending rains are more significant in the laurel forests than the relatively low additional precipitation provided by fog. It is probably not correct to generalise the findings of SUNDING (1972) in which a rain gauge placed in a clearing in the laurel forest registered an annual precipitation of 956 mm, while a second gauge situated beneath the trees to include water dripping from the leaves registered 3038 mm. KÄMMER estimates the annual fog precipitation as approximately 300 mm. As we know from the tropics, epiphytes rely more on the frequency with which they become wet than on the actual amount of precipitation and epiphytic mosses depend on low evaporation. The short duration of sunshine and the resulting high humidity in the cloud layer (especially in summer) are important factors for the laurel forest.

The climate diagrams in Fig. 172 provide general information on the character of the climate on Tenerife. The climate on the sea coast at Santa Cruz is that of a semi-desert, while the annual precipitation on the southern coast with only little over 100 mm, is characteristic of a desert climate. The climate of La Laguna, still below the cloud layer, is typically Mediterranean and without frosts (with the exception of the year 1869). Izaña, at 2367 m above sea level and in the upper limit of the cloud layer,

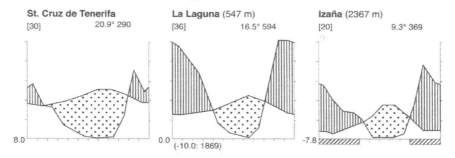

**St. Cruz de Tenerifa**
[30]       20.9° 290

**La Laguna** (547 m)
[36]       16.5° 594

**Izaña** (2367 m)
[20]       9.3° 369

8.0           0.0    (-10.0: 1869)       -7.8

**Fig. 172.** Climate diagrams: Santa Cruz at sea level, La Laguna at the lower cloud limit, Izaña at the upper forest limit (from WALTER 1968)

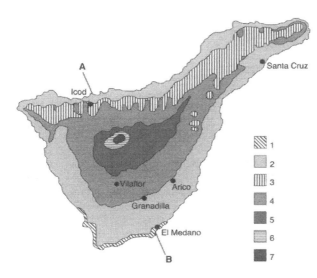

**Fig. 173.** Vegetation map of Tenerife. *1 Zygophyllum-Launea* desert, *2 Kleinia-Euphorbia* belt of the succulent semi-desert, *3* laurel forest and *Erica* belt in the north (trade wind exposed side), *4* pine forest-broom heath belt, *5 Spartocytisus*-mountain semi-desert (temperate), *6* rubble belt with *Viola* and *Silene*, *7* mountain desert with cryptogamic species (cold). **A–B** profile cross section illustrated in Fig. 174 (from WALTER 1968)

obtains somewhat less precipitation. The amount of precipitation decreases with increasing altitude so that the upper forest limit is also a moisture limit similar to the situation in Mexico. Although Izaña does not have a cold season, frosts may occur between October and April (more detailed information in KÄMMER 1982).

The climate diagrams from Gran Canaria (SUNDING 1972) indicate the same type of climate, with the most arid station on the south-eastern coast with 91 mm of rain annually, Las Palmas with 174 mm and the stations over

**Fig. 174.** NNW–SSE profile through the island of Tenerife (explanation see Fig. 173) with altitude. Z = *Zygophyllum-Launaea* desert on the sea coast at El Medano (from WALTER 1968)

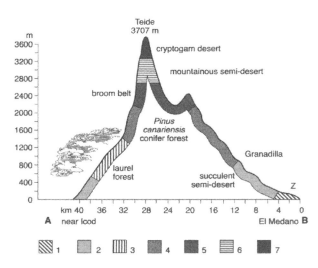

1500 m above sea level with over 900 mm. Here, the clouds often envelop the lower peaks.

The vegetation map and the profile (A–B) in Figs. 173 and 174 illustrate the distribution on Tenerife. A narrow desert-like area with Saharo-Arabian elements such as *Launaea (Zollikoferia) arborescens, Zygophyllum fontanesii* (plus *Suaeda vermiculata* on Gran Canaria) and others, is situated on the southern coast at the lee-side of the trade winds. On the steep slopes this is followed by a semi-desert with succulents and is especially well developed on the southern slopes. The montane forest belt is composed of remnants of laurel forests in the cloud zone, above which *Pinus canariensis* forests are situated (Fig. 175). On the dry southern slopes the entire forest belt is made up of the latter. This three-needled pine species is related to *Pinus longifolia* of the Himalayan Mountains.

The peak of the Teide usually rises above the clouds. Above the forest limit, it is covered with shrub-like broom species (*Adenocarpus, Cytisus* spp.). Higher up the alpine belt begins, in the lower part of which the white-blossomed broom (*Spartocytisus supranubium*) forms closed stands which become less dense with increasing altitude. Such endemic species as *Sisymbrium bourgaeanum*, the violet-blossomed *Cheiranthus scoparius* and the several-meter-high *Echium bourgaeanum* (viper's bugloss) with its reddish inflorescences, also occur.

The alpine scree begins at over 2600 m above sea level. As a result of solifluction from alternating frosts this layer

**Fig. 175.** Trade wind clouds in the pine forests (*Pinus canariensis*) of Gran Canaria on the north side of Tenerife (ca. 1400 m above sea level) (Photo: S.-W. BRECKLE)

is constantly in motion. Only individual scree creepers, such as *Nepeta teydea, Viola cheiranthifolia* and *Silene nocteolens* are able to survive in this habitat. At altitudes over 3300 m above sea level only cryptogamic species are found: several cyanobacteria (*Scytonema*), mosses (*Weissia verticillata* and *Frullania nervosa*) and lichens (*Cladonia* spp. and others).

The plant associations of Gran Canaria were studied in detail by SUNDING (1972). Altitudinal belts are identical to those on Tenerife except that they do not rise over 2000 m above sea level, therefore barely over the upper forest limit. Two colour maps in SUNDING's publication are especially interesting. One illustrates the vegetation as it appears today, and one shows the potential vegetation and probably corresponds to the original situation as far as it can be reasonably reconstructed. Mankind has altered parts of the landscape irreversibly, through severe soil erosion on deforested areas, which are no longer capable of reforestation. These maps are presented in a smaller and simplified form in Fig. 171. The map of the potential vegetation indicates a very narrow desert-like zone situated mainly on the southern and eastern sea coasts which cannot be seen. This is followed by a semi-desert of succulents which occupies over half the entire surface below 400 m above sea level on the northern slope and below 800 m on the southern slope. The remaining surface is covered by a *Pinus canariensis* coniferous forest. The evergreen laurel forest in the broadest sense (including the drier form with *Myrica faya* and *Erica arborea*), was probably only found in the lower level of the forest belt on slopes exposed to the north-east. According to SUNDING, the natural distribution of the broom belt above the forest zone was limited to the small area around the peak.

Comparing this map with the present vegetation (not including the settlements with their agricultural environs), an enormous change has taken place. The desert-like vegetation on the flat sea shore will soon be replaced by hotels or vacation homes with bathing beaches. The succulent semi-desert has expanded greatly at the cost of the forest belt and now covers 78% of the island. Broom heaths have replaced former forests in the upper ranges of the forest belt. The remaining forest is significantly smaller and now consists almost only of pine. The former extensive evergreen laurel forest remains only in a few gorges as small remnants and only appears as black spots on the small map.

Natural vegetation, therefore, is only found on the steep, nearly inaccessible rocky cliffs of the succulent semi-desert. Ecologically, this is a highly heterogeneous unit with an almost micro-mosaic structure of dry rock surfaces and shallow soils, creviced cliffs and rubble-covered slopes on which deeply rooting species are relatively well supplied with water, as well as ravines with seeping groundwater and dripping wet cliffs. This creates a situation in which the most varied ecological types find suitable ecological niches and often occur next to each other, although under completely different conditions. At the one extreme are the stem-succulent euphorbias, which are able to survive long droughts and on the other is the delicate Venus fern (*Adiantum capillus-veneris*), which occurs in shady locations on constantly wet cliffs. Beneath the ferns are mats of moss covered with a calcareous crust, which remains after the water evaporates. The small amount of NaCl in the water may also become concentrated, making possible the occurrence of the halophytic species *Samolus valerandi* next to the ferns. Even if plant sociological surveys are limited to very small areas, random lists of completely heterogeneous ecological types result, in which shallow and deep-rooting succulents and non-succulents are found, although they are dependent on entirely different niches. The presence of annual therophytes is of no significance since they develop in clearings where they are free of competition during the short rainy season when all soils are moist.

The occurrence of certain ecological types can only be explained with the aid of a detailed ecological analysis including information on the root and water distribution in the soil during different seasons. An analysis of this type is quite tedious and requires very careful observations with well-directed field experiments during all seasons and over a period of several years. It is in this altitudinal belt of the succulent semi-desert that the palm *Phoenix canariensis* grew. Wild specimens of this palm no longer exist, although it may be found in the parks in the range of Zonobiome IV or V. It is more ornamental than the closely related date palm (*Phoenix dactylifera*), but its fruit is not edible. It was certainly originally situated in sunny locations with easily accessible groundwater such as in the water-rich ravines. The well-known dragon tree of the Canary Islands (*Dracaena draco*) probably also grew in similar habitats, yet today it is only found planted in gardens.

## Man in Mediterranean Regions

For thousands of years man has had a very large influence in the European-African Mediterranean region, starting from the early high cultures in the Middle East (see Table 20). Deforestation some thousand years ago (for example by the Phoenicians in Dalmatia) caused the loss of the original sclerophyllic forests and regeneration is no longer possible because of the eroded soils. Formation of soil on bare rocks takes thousands of years. Grazing and early agriculture in the orient caused a strong selection of species and thorny and poisonous plants have extended.

At the beginning man probably did not influence biodiversity too much, although some species were introduced and promoted. For example, the olive tree probably came some thousands of years ago from northern East Africa and/or southern Arabia, but is today regarded as the characteristic tree for the Mediterranean region. From the

**Table 20.** Timescale of influence of man in Mediterranean ecosystems: numbers given are years b.p.

|  | Mediterranean | Australia | South Africa | Chile | California |
|---|---|---|---|---|---|
| First occurrence of man: hunter/gatherer, use of fire | 400,000 | 40–70,000 | 500,000 | 11,000 | 14,000 |
| First occurrence of domestic animals | 10–6000 | 150 | 20,000 | 400 | 400 |
| First occurrence of agriculture | 10–6000 | 150 | 300 | 1000? | 150 |
| Intensive agriculture | 2000–1000 | 50 | 300–200 | 400 | 50 |

(after GROVES et al. 1983)

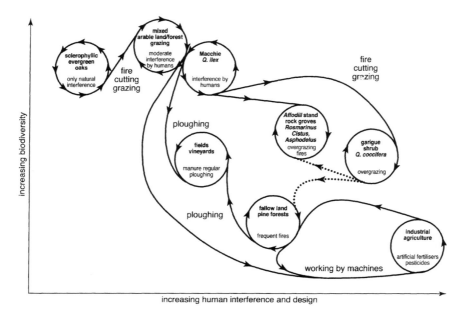

**Fig. 176.** Biodiversity, dynamics of the ecosystem and human influence on Mediterranean formations in the western Mediterranean region (after BLONDEL & ARONSON 1995)

New World came agaves and opuntias, from South Africa aloes and crassulas, from Australia acacias and eucalypts. In Portugal the 'eucalyptisation' has increased the frequency of forest fires almost to a dangerous extent. In many parts of the Mediterranean region forms of use have evolved which were well adapted to the ecological conditions of the region. Cork and holm oak stands, widely found in the western Mediterranean, were used for firewood, for grazing animals and the cork was stripped off; their stands were very fire resistant. They were interspersed with other crops in small areas resulting in additional fire protection. Today, many of those cultured areas have been left and bushland grows there, in other areas the fast growing *Pinus pinea* or *Pinus maritima* have been planted, thus increasing the danger of forest fires.

With the increasing human influence, biodiversity and the dynamics of ecosystems (the number of functional groups, interspecies interactions etc.) have decreased significantly, as shown in the schematic in Fig. 176. However, the rich mosaic of various types of usage, the mixture of small agricultural uses, animal husbandry, forestry, transhumance etc. of the late Middle Ages to the beginning of the last century probably displayed the largest biodiversity (BLONDELL & ARONSON 1995). Even the macchie, garigue and asphodel rock heaths often show high biodiversity. Increased degradation and intensive use and industrialisation of agriculture have caused considerable depletion in many landscapes in recent times.

## Questions

1. *What are the differences of macchie, garigue, batha, phrygana, pseudomacchie, schibljak, matorral, chaparral, espinal, fynbos and mallee?*
2. *What are Mediterranean homoclimates? Examples?*
3. *Despite similar annual precipitation and similar annual mean temperatures summer rain regions (ZB II, savannas) and winter rain regions (ZB IV, sclerophyllic forests) differ considerably. Why?*
4. *Is a lignotuber hereditary?*
5. *What is the difference between sclerophylly and lauriphylly?*
6. *Which winter rain regions do not belong to ZB IV? Why?*
7. *How long does it take for a zonal sclerophyllic forest to establish on the Dalmatian karst (bare limestone)? What are the stages in which this could happen?*
8. *Where do hedgehog-like hemispheric and thorny cushions occur as a component of the vegetation?*
9. *Why is terra rossa not a zonal soil?*
10. *Under what conditions do summer green (for example Quercus pubescens) and evergreen oaks (for example Q. ilex, Q. suber) have competitive advantages?*

# V Zonobiome of Laurel Forests (Zonobiome of the Warm-Temperate Humid Climate)

## General

Zonobiome V cannot be sharply delineated since it is a transitional zone between the tropical–subtropical and the typical temperate regions, although too large to be considered an ecotone. Two subzonobiomes can be distinguished:

1. A very humid subzonobiome with rainfall at all times of the year with a minimum in the cool season. The principal vegetational period is invariably wet and very humid due to high temperatures. These regions lie on the eastern sides of the continents between latitudes 30 and 35 in both the northern and southern hemisphere and are influenced by trade and monsoon winds. Temperatures drop quite severely in the cool season and there may even be frost, but there is no cold season with temperatures below 0 °C. Nevertheless, the vegetation spends the winter in a resting state.
2. The other subzonobiome lies along the western seaboard of the continents, nearer to the poles than the first, adjoining the wet subzonobiome of ZB IV. In this subzonobiome winter rains are also predominant, but the summer drought is usually missing.

Both subzonobiomes are characterised by lauriphyllic tree species and/or large pine woods, both rich in relict forms from the Tertiary.

In North America the subzonobiome with winter rain stretches along the coastal regions from northern California to southern Canada. It is the zone of the *Sequoia sempervirens* forests. Further north this vegetation is succeeded by forests with *Tsuga heterophylla*, *Thuja plicata* and *Pseudotsuga menziesii* (Fig. 177). *Prunus laurocerasus*, *Rhododendron ponticum* and *Araucaria excelsa* flourish in gardens, indicating mild winters. Still further north the temperature gradually drops, the climate becomes wetter and diurnal and annual fluctuations in temperature are

**Zonobiome V** is a transitional biome. sZB V (sr): on the eastern sides of the continents, transition from ZB I and ZB II with summer rains to the temperate regions with slight frost; sZB V (w): on the west side of continents, transition from ZB IV with winter rain to ZB VIII with very maritime climates.

**Fig. 177.** Damp, oceanic coniferous forest with *Pseudotsuga menziesii, Tsuga heterophylla* and *Thuja plicata*
on the Hoh River (Olympic
National Park) (Photo: H.
WALTER) (cf. climate dia-
gram of Vancouver,
Fig. 156)

small. The maritime, frost-sensitive Sitka spruce becomes
predominant. The zone extends along a meridian up to
the subarctic in Alaska, but regions corresponding to ZB
VI or ZB VIII are barely recognisable. It is an extremely
humid, maritime ecotone in which land cultivation is im-
practicable and the population is therefore sparse.
    Studies within the International Biological Programme
(IBP) have examined these coniferous forests which are
probably the most productive in the world, especially the
Douglas fir ecosystems (*Pseudotsuga*). A volume containing 11 contributions on the preliminary results obtained
from studies in the years 1971–1978 has been published
(EDMONDS 1982), although summaries of these results
have not yet appeared. An overview of the expansion of
evergreen forests has been given by KLÖTZLI (1987). An
analogous situation is found in southern Chile. The subzonobiome with winter rainfall, but no summer drought,
corresponds to the luxuriant Valdivian evergreen rain forests previously mentioned (see p. 273). The Magellanic
forest, which continues to the south with both evergreen
and deciduous *Nothofagus* species and well-developed
bogs, constitutes the perhumid transitional zone to the
subant-arctic Tierra del Fuego and the islands.
    The tall, frost-sensitive coniferous species of the Pacific
coast of North America are wholly lacking in western Europe, where they died out during the glacial periods of
the Pleistocene (fossils in the lignite of the Rhine area).
The nearest counterparts to this subzonobiome in Europe
are the heath formations on the coasts of northern Spain

and south-western France (Les Landes). The perhumid transitional zone is as split up as the coasts of western Europe. It comprises Wales, western Scotland and the island groups, including subarctic Iceland and the wettest parts of the Norwegian west coast with the Lofoten Islands, and extends to the Arctic. Heather moors with birch and willow are the predominant form of vegetation at the present time (p. 314).

The south-western tip of Australia with winter rains and no summer drought (karri forest, p. 278) also belongs to this subzonobiome. An extremely perhumid zone, however, consisting solely of Tasmania (with small *Eucalyptus* species and bogs) and the south-west portion of the South Island of New Zealand with Stewart Island, constitutes the transition to the subantarctic islands.

A completely isolated region belonging to this subzonobiome is found in northern Anatolia with Colchic forests in which *Rhododendron ponticum* and *Prunus laurocerasus* are native. These are offshoots of the luxuriant forests of the Colchic Triangle between the Caucasian Mountains and the Black Sea with evenly distributed rainfall which may amount to as much as 4000 mm annually. Although the evergreen undergrowth has persisted, the tree layer with the **Tertiary relict forest** species *Zelkowa, Pterocaria* and *Dolichos* and the lianas (*Vitis, Periploca*) is deciduous. Despite isolated outbreaks of cold, citrus fruits are grown. A similar situation is encountered in the Hyrcanian relict forests on the south coast of the Caspian Sea, with *Parrotia* (Hamamelidaceae) and *Albizia julibrissin* (Mimosaceae) and others as relict species.

## Tertiary Forests, Lauriphylly and Sclerophylly

If the sclerophyllic vegetation of ZB IV developed in recent geological times from laurel leaves, it should be possible to find relict species with laurel-type leaves in the Mediterranean region. *Laurus nobilis* occurs in wet areas and protected habitats particularly in the western Mediterranean. Even *Arbutus* is more of a laurel type than a sclerophyllic type. Lauriphyllic species occur azonally in canyon forests or orozonally as mist forests. Sclerophyllic plants with woody elements in their leaves (sclereids) generally replaced the lauriphyllic plants, but were unable to do so in very temperate habitats and in permanently humid regions. Laurel woods only occur in large areas as zonal vegetation in East Asia and south eastern United States. Today, however, many of the evergreen temperate forests only occur in very small residual areas, their biodiversity (for example in China, Korea or Japan, but also in the south-eastern United States) is remarkably high, as well as in southern Brazil.

The areas degraded to heathlands in northern Portugal are regarded as impoverished residual stands of ZB V. Euxinic and hyrcanian relict forests are characterised by relict species of the Tertiary period. This also applies to a far greater extent to the other ZB V regions, where many genera are related to the Tertiary species known through fossils. Today, laurel forests only occur as part of sZB V (s) on the east coast of continents and are classified by KLÖTZLI (1987) as thermophilic (20–25 °C mean average monthly temperature during the vegetative period) and frost-sensitive (minima just under –10 °C) as well as drought-sensitive (almost no arid months during the course of the year). Zonobiome V is delimited from the subtropical/tropical rain forests which have more or less evenly distributed precipitation and temperatures, from sclerophyllic forests which have lower and sporadic precipitation (winter) and regular fires, and from summer green forests which have colder winters with late frosts and often drier summers.

## Humid Subzonobiomes on the East Coasts of the Continents

Because of the trade or monsoon winds on the eastern sides of the continents, there is an almost continuous sequence of wet subzonobiomes of ZB II to ZB V via a wet subtropical zonoecotone II/V and to ZB VI via zonoecotone V/VI. However, since the landmasses of the southern hemisphere do not extend as far towards the pole as those of the northern hemisphere, and owing to the interruption of North and South America by the Caribbean Sea, the various zones are not easily distinguishable. The tropics are taken to end where frost occurs or where, even in the absence of frost, the mean annual temperature is below 18.3 °C and cultivation of tropical plants such as coconut, pineapple, coffee and others is no longer economically worthwhile and only tea, citrus fruits and a few palms remain.

Frosts already occur within ZB V, but the mean daily minima of the coldest month are still above 0 °C, i.e. no cold season appears on the climate diagram. The annual means are slightly above or below 15 °C, and the forest trees are at least in part evergreen, whereas only a few shrub species remain evergreen in zonoecotone V/VI. In

**Fig. 178.** Relicts of lauriphyllic forests on the island Ullung-do (South Korea) in some valleys, on the upper slopes followed by beech forest with *Fagus multinervis* as relict forest (Photo: S.-W. Breckle)

**Fig. 179.** Open young beech forest on the upper slopes of the island Ullung-do (South Korea) with many stems of *Fagus multinervis*. In the undergrowth there are still some individual evergreen lauriphyllic species (for example *Euonymus japonica*) and several herbaceous species (see also Albert 1997) of the genera *Helleborus, Hepatica, Maianthemum* and also *Sasa* (Photo: S.-W. Breckle)

contrast, in ZB VI, a cold season of 2–5 months is characteristic and the woody species shed their leaves in autumn.

In eastern Asia, which is exposed to the East Asian monsoon and therefore has a ZB II, the humid subzonobiome of ZB V covers an unusually large region. Its northernmost limit at 35 °N just reaches the southern tip of the Korean peninsula with the many islands, turns in the Japanese Sea towards the north and island Ullung-do (see Figs. 178, 179) and runs through the southern part of the main Japanese island of Honshu. The forests of this region consist of evergreen Fagaceae (*Cyclobalanopsis, Quercus* and *Castanopsis*), Myrsinaceae (*Ardisia*) and Lauraceae (*Machilus*) and other forest-forming trees. Shrubs commonly used as decorative plants in northern Italy (Insubria) *Aucuba japonica, Euonymus japonica, Ligustrum japonicum*, as well as the frost-sensitive *Camellia* originate from here. Further north, deciduous tree species become predominant (NUMATA et al. 1972; Fig. 179).

In China, the northern limit of the subzonobiome is slightly displaced inland toward the south, owing to the influence of outbreaks of cold winds in winter resulting from the Siberian high pressure zone. The southern limit which borders the evergreen tropical–subtropical forests of southern China, is much less sharply defined. Canton itself is still within ZB II. Figure 180 shows the classification according to AHTI & KONEN (1974). They also discuss the orobiome in Japan.

The southern tip of Florida, in the south-east of North America is still subtropical, but even in Miami and Palm Beach slight frosts may occur. The forests of evergreen oak (*Quercus virginiana*) stretch northward along the coast into North Carolina. The total area covered by ZB V is not large, since inland outbreaks of cold extend as far south as the Gulf of Mexico. Extensive sandy areas are occupied by psammobiomes with pine forests containing *Pinus clausa, P. taeda, P. australis* and others, with an undergrowth of evergreens in places. In addition, there are the extensive *Taxodium-Nyssa* swamp forests (hydrobiomes) and two types of helobiome, the evergreen *Persea-Magnolia* moor forests as well as heathland moors with the Venus fly trap (*Dionaea muscipula*). Large expanses of the coast itself are taken up by salt marshes (halobiomes).

In South America, the evergreen forests of eastern Brazil extend far to the south, from tropical to subtropical and even warm-temperate. The tropics end on the coast between Porto Alegre and Rio Grande. Even in Misiones and Corrientes in northern Argentina, the forests are subtropical, and along those great rivers the Paraná and Uruguay, the forests penetrate the pampas region as gallery forests. On the coast, ZB V ends near La Plata and ZB VI is altogether lacking.

On the high plateau more than 500 m above sea level in southern Brazil, the coniferous forest is composed of *Araucaria angustifolia* and has to be allocated to ZB V. Generally speaking, the forests in this region have been greatly reduced in area by clear-cutting.

The south-eastern coast of Africa is also exposed to the south-east trade wind. The obstacle presented by the Drakensberg Mountains causes large amounts of precipitation and tropical-subtropical forests are found in the coastal regions as far as East London. The region along the southern coast can be described as warm-temperate. In

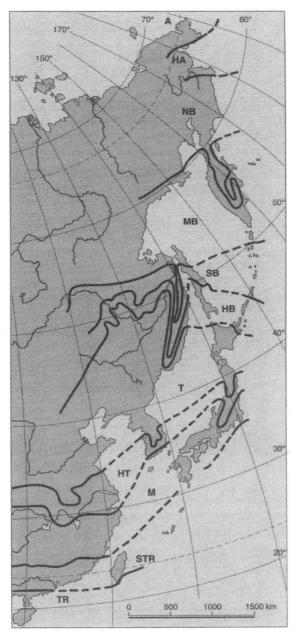

**Fig. 180.** Bioclimates of eastern Asia (from AHTI & KONEN 1974). *TR* Humid tropics; *STR* humid subtropics, *M* maritime warm-temperate ZB V, *HT* ZE V/VI and *T* temperate ZB VI. *HB* Hemi-boreal mixed forest zone. *SB, MB* and *NB* southern, middle and northern boreal zones (= ZB VIII), *HA* and *A* hemi-Arctic and Arctic zones (= ZB IX)

earlier times, the forests stretched without interruption up to the eastern slopes of Table Mountain near Cape Town. However, the larger part has been cleared or is secondarily occupied by the fynbos of ZB IV. The only large forest reserve, near Knyshna, contains tall, ancient specimens of *Podocarpus* trees and a large number of evergreen deciduous trees, including the "stinkboom" (*Ocotea foetans*), which yields valuable timber.

The situation in Australia and New Zealand will be dealt with in the next sections.

Scarcely any ecological studies have been made on ZB V and therefore no details concerning its ecosystems can be given. Investigations would be particularly difficult since the majority of the forests contain a large diversity of species and conditions are very favourable for growth. The decisive factor is undoubtedly competition and this is not easy to estimate.

 ## Subzonobiome on the West Side of Continents

In Oregon and Washington in the western USA slightly frost-resistant coniferous forests growing in humid conditions reach a height of 100 m. The relict species *Sequoia sempervirens*, in parts mixed with *Abies grandis, Pseudotsuga menziesii*, replaced further north by *Tsuga heterophylla* and *Thuja plicata*, creates the upper tree layer. In the lower layer many deciduous trees are represented (*Acer macrophyllum, Alnus rubra* and others). Many trees are abundantly covered by epiphytic ferns, mosses and lichens. These coniferous forests are photosynthetically active during the whole year and should be seen as Tertiary relict forests. Obviously, they were hardly affected by the Ice Age because of the north-south running mountains. Further south larger refuge areas for vegetation remained during the ice age, so that in contrast to the west–east mountain barriers in Europe, extension to the north could proceed rapidly. Therefore, western Europe lacks a vegetation corresponding to zonobiome V, even though the climate would nowadays allow such vegetation. Remnants of some relict species can be found in the mountains near Algeciras (Campo de Gibraltar) where the evergreen *Rhododendron ponticum* ssp. *baeticum, Quercus lusitanica* and *Prunus lusitanica* as well as the epiphytic fern *Dalvallia canariensis* and the ancient fern *Psilotum nudum* occur.

In the euxinic forests of northern Anatolia only deciduous trees are found. However, several evergreen species (*Prunus laurocerasus, Ilex, Buxus, Daphne pontica, Vacci-*

**Fig. 181.** *Nothofagus* forest with *N. dombeyi* and other *Nothofagus* species in the national park Nahuelbuta with large *Araucaria* (*Araucaria araucana* (Photo: J. RENZ)

*nium arctostaphylos, Ruscus* and others) occur in the undergrowth. A similar situation exists in the Colchis on the east bank of the Black Sea and the hyrcanian forests on the southern banks of the Caspian Sea.

In southern Chile the Valdivian rain forest corresponds to zonobiome V (w). This forest is rich in species and its luxuriance is reminiscent of tropical rain forests and permanent humidity. Several relict coniferous trees occur (amongst others *Fitzroya cupressoides, Austrocedrus chilensis, Podocarpus nubigenus, Dacrydium foncki, Araucaria araucana*), however, these species are never dominant. Forests are formed by species of *Nothofagus* (Fig. 181); the deciduous *N. obliqua* can grow up to 40 m and the evergreen *N. dombeyi*, as well as *Eucryphia cordifolia* and others reach heights of 35–40 m (see p. 273).

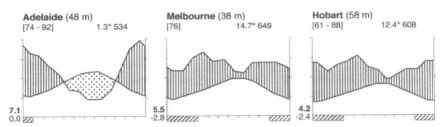

**Fig. 182.** Climate diagrams from the sclerophyllic region of South Australia and the warm-temperate regions of Victoria and Tasmania

**Fig. 183.** *Eucalyptus regnans* high forest near Healesville, north of Melbourne (Victoria), in the undergrowth large tree ferns (Photo: S.-W. BRECKLE)

## 5 Biome of *Eucalyptus-Nothofagus* Forests of South-Eastern Australia and Tasmania

Wet tropical–subtropical evergreen rain forests on Australia's east coast have some Indo-Malayan elements foreign to the Australian realm. They extend as far as southern New South Wales on rich, usually volcanic soils. Only in southern Victoria and in Tasmania does the Australian element dominate with the genus *Eucalyptus*, combined with some Antarctic elements. Here, in this humid climate which lacks a cold season (Fig. 182), *Eucalyptus regnans* may reach a height of 110 m (earlier reports of 145 m can no longer be confirmed). Today, trees of 75 and 95 m height are found (Fig. 183). *Eucalyptus gigantea* and *Eucalyptus obliqua* attain similar heights. The most important of the Antarctic elements are the evergreen *Nothofagus cunninghamii* and the tree fern *Dicksonia antarctica* and several other species in Tasmania. The composition of these forests depends on the frequency of forest fires:

1. In the wet parts of western Tasmania where no fires occur, a tree stratum of *Nothofagus* and *Atherosperma moschata* (Monimiaceae) develops to a height of 40 m and below it is a 3-m-high stratum of the fern *Dicksonia* which is able to grow even when receiving only 1% of the total light. Hymenophyllaceae and mosses abound as epiphytes in the wet forests.

2. If forest fires occur every 200–350 years, mixed forest develops, consisting of three strata. Apart from the two strata mentioned above there is a further loftier stratum consisting of the three largest species of *Eucalyptus* (75 up to 90 m). That the trees in this stratum are all of one age is an indication that germination of their seedlings took place simultaneously over an extensive area after a fire. Although the *Eucalyptus* and *Nothofagus* strata are destroyed by fire the fruits can still open, and the undamaged seeds are dispersed and germinate. The more rapidly growing *Eucalyptus* overtakes *Nothofagus*, with the result that two tree strata are formed. Although the tree ferns lose their leaves in a fire, they are able to develop new ones at the tip of the stem. Regeneration of *Eucalyptus* below *Nothofagus* is impossible owing to lack of light and can take place only after another fire.

3. If forest fires occur once or twice in a century, then *Nothofagus* is replaced by other more rapidly growing, but shorter tree species such as *Pomaderris, Olearia, Acacia*.

4. A pure, low *Eucalyptus* vegetation results where fire occurs every 10–20 years.

5. Still more frequent fires lead to a degradation of the forests. An open moor results, with "button grass" *Mesomelaena sphaerocephala*, (Cyperaceae) and scattered Myrtaceae shrubs, together with *Drosera, Utricularia* and Restionaceae.

## Warm-Temperate Biome of New Zealand

New Zealand's forests warrant special mention. Although both islands are relatively near to the Australian continent and were probably directly connected with it in the geological past, this connection must have been interrupted before the flora of the Australian realm was fully developed. There is not a single native species of *Eucalyptus* or *Acacia* in New Zealand and the Proteaceae are represented by only two species.

In the north of North Island, there are still subtropical forests consisting of coniferous *Agathis australis* and palms and along the coast there are even mangroves with low *Avicennia* bushes. The forest species are melanesic elements of the Paleotropic realm. Forests of this type occur even on South Island, although its climate is definitely temperate, despite the absence of a cold winter season in the lower-lying country. The coniferous genera *Podocarpus* and *Dacrydium*, which are distributed throughout the entire southern hemisphere, are very common here. The Antarctic element, represented by five evergreen *Nothofagus* species, plays an important role in the forests of both North and South Islands. These mutually exclusive forest species form a mosaic for which there is no satisfactory ecological or climatological explanation. The plant cover gives the impression of not being in a state of equilibrium with its present-day environment. Rather than being differentiated by soils or other ecological factors, it seems that the vegetation reflects mostly historical factors. Seventeen hundred years ago, North Island was partly covered by a thick layer of volcanic ash. The first pioneers were Podocarpaceae disseminated by birds; they are gradually being replaced by forests containing tropical elements as well as by *Nothofagus* forest in some of the mountainous regions. In the Pleistocene, South Island was covered by large glaciers, so that since *Nothofagus* spreads very slowly, the process of re-colonisation is still going on.

In the extremely humid fjord country of south-western New Zealand, where the rainfall exceeds 6000 mm, the *Nothofagus* forests are similar in nature to those of southern Chile. A peculiarity is, however, represented by the **bare strips**, 2–6 m wide, suggestive of avalanches, found in the middle of the forest on steep slopes. When the weight of the tree layer on the rocky slopes becomes too great, then the entire vegetation, inclusive of roots and soil, slides down owing to the force of gravity. The bare rock which remains is then re-colonised by lichens, mosses and ferns, followed by shrubs and finally trees, until the process repeats itself.

The imported European red deer presents a great danger to the forests of New Zealand, where originally, the only mammals were bats. It is impossible to control the multiplication of the deer and they hinder the regeneration of the *Nothofagus* forests, so that the danger of erosion and flooding is greatly increased. The Australian opossum (kuzu), also imported for its fur, is equally dangerous. It has confined itself to a tree species growing at the tree line in the high mountains and completely strips the trees of their leaves, thus bringing about their death and increasing the danger of soil erosion on the steep slopes. New Zealand provides an example of the extreme danger of disturbing the natural equilibrium by introducing new plants or animals. The damage done is often irreparable.

## Questions

1. *What is the difference between malakophyllic, lauriphyllic and sclerophyllic leaves?*
2. *Which vegetational types does one distinguish in ZB V?*
3. *In which zonobiomes do Tertiary relicts appear and why?*
4. *Can it be explained why the highest deciduous and coniferous trees occur in zonobiome V?*
5. *What is the maximum height of a tree and which factors limit the height?*
6. *Why are there no laurel forests in Europe?*

# VI Zonobiome of Deciduous Forests (Zonobiome of the Temperate Nemoral Climate)

## Leaf Shedding as Adaptation to Cold Winters

A temperate climatic zone with a marked but not too prolonged cold season occurs only in the northern hemisphere (Fig. 184); apart from certain mountainous districts in the southern Andes and in New Zealand it is absent from the southern hemisphere. The phenomenon of facultative leaf-shedding has already been discussed in the context of the tropics. There, the leaves are shed only when the water balance is disturbed by a lengthy period of drought and this takes place in order to decrease water loss (see p. 144).

The triggering factor causing the yellowing of leaves in autumn just before the first frosts is not completely known. It could partly be due to the shortening of day length. It is striking that the change in leaf colours of various types of trees happens in a relatively short time span, according to the phenological calendar, in central Europe between the 10th and 20th of October. There is no sharp distinction between west or east or between lower or higher altitudes in mountains. Trees near street lamps remain green longer.

In **zonobiome VI**, the zonobiome of temperate deciduous trees, shedding of leaves is an adaptation to the cold season. However, shedding of leaves is not facultative, but obligatory, i.e. it also occurs if the trees are sheltered from the winter cold in a greenhouse.

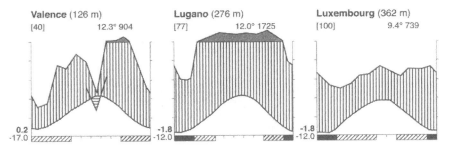

**Fig. 184.** Climate diagrams from the sub-Mediterranean zone (with no cold season), the warm and damp deciduous forest zones and the central European beech forest zone (*right*)

An evergreen broad leaf is neither resistant to cold nor to winter drought, that is to say to sustained temperatures below freezing. In a central European climate below 0 °C, *Prunus laurocerasus* (cherry laurel) invariably freezes during severe winters and even a light frost causes the leaves to emit $CO_2$ in the light, which means that although respiration continues, photosynthesis is blocked. *Ilex aquifolium* (holly) is Atlantic in its distribution and *Hedera helix* (ivy) sub-Atlantic; both species thus avoid the eastern continental regions with their cold winters. The same holds true of the broom species *Ulex* and *Sarothamnus*. The evergreen Arctic rhododendron (*Rhododendron lapponicum*) and cowberry (*Vaccinium vitis-idaea*) in central Europe can survive the cold only beneath a covering of snow.

The loss of thin deciduous leaves in winter and protection of buds from water loss represents a saving of material compared to the freezing of thick evergreen leaves. It is essential, however, that leaves newly formed in spring have a sufficiently long, warm summer period of at least 4 months to produce enough organic material for growth and maturation of the lignified axial organs and formation of reserves for the fruits and buds of the following year. Even in their bare winter condition twigs lose water, the extent varying from species to species. For this reason, the central European beech is not found in the zone with a cold East European winter, although oak extends as far as the Urals. In extreme continental Siberia the only deciduous trees are small-leaved birch (*Betula*), aspen (*Populus tremula*) and mountain ash (*Sorbus aucuparia*) which has small pinnate leaves.

Where the summers are too cool and too short, evergreen conifers replace the deciduous species. The xeromorphic needles of the conifers are more resistant to cold in winter and when warmer weather returns in spring they are capable of starting photosynthesis immediately. In this way the short vegetational season can better be exploited. Whereas deciduous trees require a vegetational season lasting at least 120 days with a mean daily temperature above 10 °C, conifers manage with 30 days. However, the resistance of conifers varies from species to species. Yew (*Taxus*) does not extend further east in Europe than ivy (*Hedera*). Pine (*Pinus sylvestris*) and spruce (*Picea abies*) are very resistant and *Abies sibirica* and *Pinus sibirica* (*Pinus cembra*) survive in Siberia. However, the deciduous, needle-leaved larch (*Larix dahunica*) extends furthest of all into the continental Arctic region (up to 72°40′N) where it exploits the short summers and is very productive. Whether species with deciduous or those with evergreen assimilatory organs are more successful in competition and rise to dominance, thus depends upon external conditions and the ecophysiological characteristics of the species themselves (see also ZB II, p. 167).

## Effect of Cold Winter Periods on Species of the Nemoral Zone

In zonobiome VI winter cold periods, even if very short, thus play an important role in the adaptation of plants. Plants are damaged in cold winters due to one of two causes:

1. Direct damage due to freezing of tissue water. This is termed **frost damage.**
2. Desiccation of aerial organs, which even at low temperatures, transpire to a certain degree. When the conducting vessels are blocked by ice, insufficient water reaches the aerial organs to meet these transpiration losses and the organs dry out. This is called **frost (or winter) drought** or desiccation damage.

Plants are not equipped with any means of protection against the effects of low temperature, their own temperature being close to that of the surrounding air. Their only possible adaptation is to become hardened. If the resistance of plant organs to cold is tested in summer by placing them in a refrigerator at various temperatures, e.g. below 0 °C for 2 h, it can be shown that only a few degrees suffice to cause irreversible damage. The same plant organs tested in winter, however, tolerate much lower temperatures without undergoing damage because they have developed a so-called hardiness. **Hardening** is a physiological process taking place in autumn with the onset of shorter days and the first cool nights. In spring, when the weather becomes warmer, the opposite process, **dehardening**, is initiated.

Hardening is connected with certain physicochemical changes in the protoplasm. The stability of membranes increases [for example because of additional sulphur bridges (-S-S-)] as well as the viscosity of the plasma and plasmolysis becomes concave rather than convex. This change is accompanied by a sudden decrease in cell-sap osmotic potential of $-1$ MPa due to an increase in sugar concentration. Protoplasm in its hardened state is more or less inactive. The resistance of buds of European deciduous trees to cold in their winter condition can increase from $-5$ °C in autumn to over $-25$ °C or even $-35$ °C in January or February. A larger increase in resistance to cold is developed in a cold winter than in a mild one. Among related species of a genus, the resistance is greater the further the species advances into the continental region.

This hardening is a very complex process, involving several stages of development. The first stage is a result of the shorter days in autumn and leads to a period of rest. Further hardening is achieved when the temperature decreases to just above 0 °C. It is more intensive in species subject to very low temperatures when the first severe frosts occur. If parts of hardened plants are suddenly exposed to extremely low temperatures, resulting in a "vitrification" of the protoplasm (without the formation of ice crystals), it is even possible to freeze them in liquid nitrogen (at −190 °C), or even at −238 °C. They must, however, be warmed in several stages until they thaw, so that no plasma-damaging formation of ice crystals is able to occur. After such an experiment, plants of the cold climate zones, which are capable of hardening, remain alive. Near the coldness pole in eastern Siberia, the forest vegetation is normally subject to temperatures of −60 °C or below in the winter. Tropical species, or even those from zonobiomes IV and V, cannot be hardened.

As a rule, "hardiness" suffices to prevent frost damage to native trees in Europe in a cold winter, although exotic species with no ability of "hardening", imported from warmer climes, often suffer. If an early frost occurs before hardening has set in or if there is late frost after dehardening has commenced, frost damage is widespread. Late frost damage is most commonly found in young, newly opened leaves and it can also cause damage to the cambium if the trees are already "in sap" and the protoplasm in an active state.

The eastern limit of the distribution area of the beech may possibly be conditioned by the frequent late frost damage of young leaves which reduces the competitive ability of the trees. An increase in resistance to cold by hardening can also be demonstrated in herbaceous forest plants even though they are not exposed to extremely low temperatures because they are covered by litter and snow. Resistance to cold in the evergreen leaves of, for example, *Anemone hepatica* reaches only to −15 °C, that of the better protected flower buds to −10 °C, and that of the rhizomes only to −7.5 °C.

Damage due to frost drought is more difficult to detect. Shedding of the intensely transpiring leaves, bud protection by hard bud scales and protection of the twigs by a layer of cork prevent the loss of large quantities of water by deciduous trees in winter. Nevertheless, a certain amount of transpiration can be measured in the bare twigs in winter. In experiments under identical frost conditions, transpiration is higher in deciduous trees than in evergreen conifers and higher in deciduous species of southern origin than in those of northern. These transpiration losses become dangerous in spring, when the intensity of the incoming radiation increases and the air temperature rises while the ground is still frozen hard. Buds and twigs sometimes dry out as a result. Evergreen species such as holly (*Ilex*) and broom-like shrubs such as *Sarothamnus* or *Ulex* are especially sensitive in this respect. Frost damage usually occurs at the coldest time of the year and frost drought is more common as spring approaches and on warm southern slopes. The latter should not be confused with late frost damage.

## Distribution of Zonobiome VI

The climate of ZB VI, with a warm vegetational season of 4–6 months with adequate rainfall and a mild winter lasting 3–4 months, is especially suitable for the deciduous tree species of the temperate climatic zone. Such trees avoid the extreme maritime as well as the extreme continental regions and favour what is termed the nemoral zone. In the northern hemisphere a climate of this kind, with the rainfall maximum occurring in summer, is found in eastern North America and East Asia, between the warm-temperate and the cold- or arid-temperate zones. It is also found in western and central Europe north of the Mediterranean zone where, as a result of the influence of the Gulf Stream, winter rains are replaced by evenly distributed rainfall or by rainfall with a summer maximum and the cold season is relatively short.

The Mediterranean winter rain region with sclerophyllic vegetation covers a very considerable distance from west to east and is replaced by different vegetational zones to the north. In the maritime region on the Atlantic coast to the north-west of Gibraltar, *Rhododendron ponticum* with the ferns *Woodwardia* and *Drosophyllum* in the undergrowth, and elements of evergreen warm-temperate laurel forests occur. This type of vegetation, however, is rapidly succeeded in the Iberian peninsula by Atlantic heaths which extend in the coastal region as far as Scandinavia (Fig. 185) and are replaced in the north by birch forests. True laurel forest is only found on the humid windward side of the Canary Islands (Tenerife) as mist forest or in the very similar zonoecotone IV/V of northern Anatolia. Further to the east, a sub-Mediterranean zone is interca-

**Fig. 185.** Map of western Europe indicating heath areas (from Hüppe 1993)

lated between the Mediterranean and nemoral zones. Although there are still winter rains, the summer drought is no longer pronounced and frost occurs regularly in all months of the winter (Fig. 184, Valence).

In the sub-Mediterranean zone all of the tree species are deciduous, e.g. *Quercus pubescens, Fraxinus ornus, Acer monspessulanum, Os-trya carpinifolia* and the frequently cultivated edible chestnut (*Castanea sativa*). Woody evergreen species are absent, apart from *Buxus*. Consequently, this region cannot be termed a Mediterranean zone but should preferably be considered as belonging to the deciduous-forest zone or as zonoecotone IV/VI.

To the north-east of this sub-Mediterranean zone is the steppe zone, which is only replaced by various kinds of forest further north. In the Middle East, the Mediterranean sclerophyllic zone is succeeded by Mediterranean steppes and semi-deserts.

## Atlantic Heath Regions

Atlantic heaths (Fig. 185) represent stages in the degradation of deciduous forests, a process which has been going on since prehistoric times. Destruction is by now so complete that the heaths are often considered to be the true zonal vegetation. The historical development can be seen in the pollen diagram (Fig. 186). Moors developed as a result of deforestation, shown by the increase in charcoal from burning forests, with rapid formation of large areas of *Calluna*.

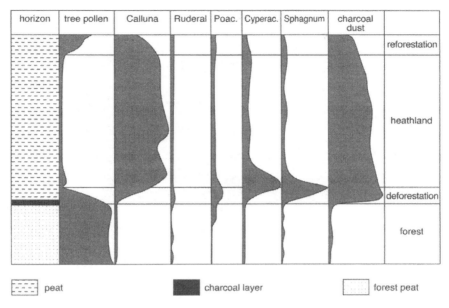

| horizon | tree pollen | Calluna | Ruderal | Poac. | Cyperac. | Sphagnum | charcoal dust | |
|---|---|---|---|---|---|---|---|---|
| | | | | | | | | reforestation |
| | | | | | | | | heathland |
| | | | | | | | | deforestation |
| | | | | | | | | forest |

peat          charcoal layer          forest peat

**Fig. 186.** Development of heaths in post-glacial times from a pollen diagram for a peat bog (from HÜPPE 1993)

The fact that the soils are extremely poor and acid and capable of supporting only a weak kind of heath vegetation was formerly attributed to leaching, as a natural consequence of the humid climate. What has already been said with regard to tropical rain forests (p. 118 f.) can also be applied here. As long as the natural forest vegetation remains untouched, leaching of nutrients from the biogeocenes does not occur, and the reserves of nutrients are mainly stored in the aboveground phytomass. However, as soon as the forest is cleared and burned, most mineralised nutrients are lost and an impoverished soil results. If the ensuing heath vegetation is exploited or repeatedly burned, reforestation, problematic in any case, is rendered impossible. In uncolonised extreme oceanic regions on the Pacific coasts of north-western North America, in the south-west of South America, and in Tasmania and New Zealand, there are areas which have similar temperatures and two to four times the rainfall of the Atlantic heath regions. However, in those areas forests grow in undisturbed luxuriance, with no sign whatever of degradation due to leaching of nutrients. It is not easy to reconstruct the original composition of the West European forests. In all probability, oak was the most abundant spe-

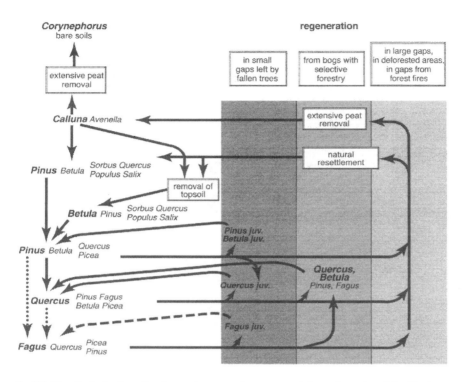

**Fig. 187.** Hypothetical schematic of present regeneration stages and forest dynamics of the Lüneburger heathland (northern Germany) on nutrient-deficient sands with differing human interference (after LEUSCHNER 1993). Processes inhibited by high game populations are indicated by *interrupted lines*, those with fragmented forest cover with a *dotted line*

cies (*Quercus petraea* and *Quercus robur*) together with birch (*Betula*) in the north, as well as the evergreen species *Ilex aquifolium*. Heath (*Calluna*), as an independent community, occurred only on shallow or peaty soils in the clearings and otherwise formed the forest undergrowth. Only after destruction of the forests did it take possession of the entire area. After grazing and the removal of peat are stopped, the regeneration of forests occurs differently according to conditions in particular areas. In removing peat, the upper 10 cm of the raw humus deposits was cut out in squares (turfs or sods) and used in stables as litter and then applied as natural fertiliser on fields. This was pointed out by LEUSCHNER (1993; Fig. 187) who reconstructed stages of rejuvenation and regeneration depending on the original conditions before the formation of heathland (Fig. 188). He

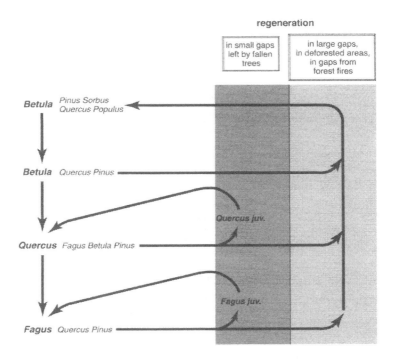

**Fig. 188.** Hypothetical schematic of regeneration stages and dynamics of forest of the Lüneburger heathland under natural conditions before the destruction of the forest (ca. 800 b.c.; after LEUSCH-NER 1993)

determined the factors leading to regeneration of forests and heaths and, conversely, necessary for the maintenance of heathlands.

In the southern part of the coastal zone, many species of broom dominate (*Ulex, Sarothamnus* and *Genista*), accompanied by various species of *Erica*. In central districts broom species become scarce, *Ulex europaeus, Sarothamnus scoparius* and *Genista anglica* being left as the most important representatives. At the same time, the dominance of Ericaceae greatly increases, especially *Calluna vulgaris, Erica cinerea* and *E. tetralix. Empetrum, Vaccinium, Phyllodoce* and *Cassiope* dominate in the north.

One quarter to half of Scotland is covered by *Calluna* heath, which is regularly burned. The iron podsol soils often have a cemented B-horizon forming a hardpan. *Calluna vulgaris* is absolutely dominant. It is a dwarf shrub, achieving a height of about 50 cm, which develops a dense web of roots in the upper 10 cm of soil with a few roots going down 75–80 cm to the hardpan. Its very small leaves, which are sessile on short twigs, are mainly shed in autumn thus reducing the danger from winter drought in the cold season. Annual litter production of a dense community can amount to 421 kg ha$^{-1}$. If the heath is burned every 30 years, development of the vegetation falls into the following three phases, each lasting 10 years:

1. The reconstructive phase of the dwarf shrub layer after the fire. Part of the nutrients is bound in the litter.
2. The phase of maturity, during which litter production increases and the rate of increase in phytomass drops.
3. The phase of degeneration, during which litter production remains constant, but decomposition increases until a state of equilibrium is attained. After 35 years, the standing phytomass amounts to 24,000 kg ha$^{-1}$ and the litter to 17,000 kg ha$^{-1}$.

As a rule, the heath is burned again after 8–15 years, without the phase of degeneration having been reached. In climates as humid as that of Scotland, fires are caused only by man. Natural fires caused by lightning very rarely occurred in the original forests so that degradation has only taken place where man has intervened. All of the transitional forms from heath to moor are found. Four stages associated with increasing moisture are listed, the species being arranged in order of decreasing abundance:

1. *Erica cinerea – Calluna vulgaris – Deschampsia flexuosa – Vaccinium myrtillus.*
2. *Calluna vulgaris – Erica tetralix – Juncus squarrosus.*
3. *Erica tetralix – Molinia coerulea – Nardus stricta – Calluna vulgaris – Narthecium ossifragum.*
4. *Erica tetralix – Trichophorum caespitosum – Eriophorum vaginatum – Myrica gale – Carex echinata.*

In Scotland the heath is utilised for hunting and also for extensive sheep grazing, with 1.2–2.8 ha/animal. The supply of nitrogen from grazing and from the anthropogenic nitrogen in the atmosphere strongly stimulates growth of grasses even though in earlier times, too, there was a cycle in heathlands between a *Calluna* and an *Avenella* phase (see Fig. 189).

In Germany, farming was formerly common (buckwheat cultivation) on the "Lüneburger Heide," which is also purely anthropogenic in origin. The heath was cut for peat and this process hindered reforestation. Nowadays the heath is no longer cultivated and forests have grown up from birch and pine seeds blown in by the wind. Scrub is growing on the heathland or the heath is being systematically reforested.

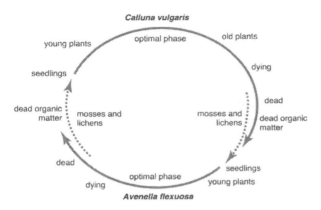

*Calluna vulgaris*
young plants — optimal phase — old plants
seedlings — dying
dead organic matter — mosses and lichens — dead
mosses and lichens — dead organic matter
dead — seedlings
dying — optimal phase — young plants
*Avenella flexuosa*

**Fig. 189.** Possible cyclic change in dominance of *Calluna* and *Avenella* on heathland in the Netherlands (after KAKAGMANN & FANTA 1993)

Apart from heaths, **blanket bogs** occur frequently in extreme maritime regions, where the climate is oceanic, with small fluctuations in temperature. In Ireland, for example, January temperatures are 3.5–3.7 °C and those for July are 14–16 °C. Frosts may occur, but snow covers the ground only 3–10 days of the year. Rainfall amounts to 350–1000 mm annually and is evenly distributed, varying from year to year by 25% at the most. Owing to the prevalence of cloud the amount of sunshine is only 31% of the possible maximum. Under such conditions, the danger of bog formation after deforestation is very great. Since a low, herbaceous vegetation loses less water owing to transpiration than does the tree stratum of the forest, a rise in groundwater level is noticeable after deforestation in humid regions. This, in turn, favours the growth of peat mosses, principally *Sphagnum* species, although *Rhacomitrium lanuginosum*, too, is widespread. In regions where it rains on more than 235 days of the year, bogs may cover the entire area even in this rolling landscape. Such blanket bogs are found in West Ireland, Wales and Scotland, where the largest encompasses 2500 km$^2$.

In regions further removed from the Atlantic coast, heath formation presents no danger, since despite the fact that *Calluna* has very small leaves with a thick cuticle and stomata lying in hairy grooves, heather species are very sensitive to frost. The leaves of *Calluna* differ from true xeromorphic leaves in that the mesophyll has a very loose structure. Transpiration is relatively active in summer if the water supply is adequate and in shady habitats values can equal those of wood-sorrel (*Oxalis acetosella*), calculated on a fresh-weight basis. When water is scarce, however, transpiration is sharply decreased. These properties, nevertheless, do not suffice to prevent water losses during the long periods of frost. Even in the mild winters in Heidelberg in southern Germany, *Calluna* dries out because it lacks a protective covering of snow. In the north it is only found where there is a covering of snow each year.

Inland, in western Europe, heath is to be found in patches on the western slopes of the low mountains with an oceanic climate (Ardennes, High Venn, Eifel, Vosges and even on the Feldberg in the Black Forest). It also extends as a narrow strip along the southern coast of the Baltic.

## Deciduous Forests as Ecosystems (Biogeocenes)

### a General

A deciduous forest is a multi-layered plant community, often consisting of one or two tree strata, a shrub stratum and an herbaceous stratum. Numerous hemicryptophytes grow in the latter, as well as many geophytes, which develop only in the spring. Illumination on the forest floor is too weak for the development of therophytes, i.e. annuals. A mossy ground cover is lacking since it would be covered by the falling leaves and mosses are only found on rocks or tree stumps projecting above the ground. These groups of plants constitute synusiae (see p. 86 f.).

There are no virgin forests on euclimatopes in the western European deciduous forest zone, perhaps with the exception of the area around Bialowiez in eastern Poland. However, beech no longer occurs in these woods (Figs. 190–192).

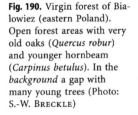

**Fig. 190.** Virgin forest of Bialowiez (eastern Poland). Open forest areas with very old oaks (*Quercus robur*) and younger hornbeam (*Carpinus betulus*). In the *background* a gap with many young trees (Photo: S.-W. BRECKLE)

**Fig. 191.** Virgin forest of Bialowiez (eastern Poland). In a gap of the mixed forest area many ash trees, hornbeam and oaks grow close together (Photo: S.-W. BRECKLE)

**Fig. 192.** Virgin forest of Bialowiez (eastern Poland). In winter large herds of bison are found there, where they are kept "semi-wild". On their gathering places in the old oak forest with very high trunks, the shrub and herbaceous layer is considerably degraded (Photo: S.-W. BRECKLE)

In central Europe the structure of forests is determined by the type of management practised. From the forestry point of view it is the woody species that are of importance and the herbaceous layer is influenced only indirectly. If forest grazing is practised, on the other hand, it is the herbaceous layer which is changed by the selective grazing of cattle which are also a danger to the tree saplings. High forests run on a rational basis approach virgin forest, although they differ basically in the small number of species in the tree stratum, in having trees all of the same age, in the lack of rotting wood on the forest floor and in their homogeneous structure; virgin forest is usually of a mosaic-like structure (see pp. 31, 127).

Managed beech forests are pure stands with only a herbaceous stratum in addition to the trees. Oak forests,

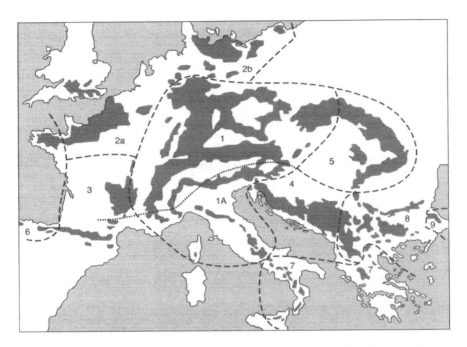

**Fig. 193.** Areas of beech forests in Europe. *1* Typical central European beech forest (*Fagion medioeur-opaeum*); *1A* meridional expression; *2a* northern French and southern English beech forests; *3* central French/southern Atlantic beech forests (*Scillo-Fagion*); *4* Dinarian beech forests (*Fagion dinaricum*); *5* beech forests of the Carpathian mountains (*Fagion dacicum*); *6* hyper-Atlantic beech forests (*Illici-Fagion*); *7* Mediterranean beech forests (*Geranio-Fagion* in southern Italy and *Fagion hellenicum* in Greece); *8* beech forests in the Balkans with *Fagus moesiaca; 9* Pontian beech forests with *Fagus orientalis* (after OZENDA 1994)

on the other hand, are usually mixed stands of various deciduous tree species and possess a shrub stratum (Figs. 190, 191). Among the various types of deciduous-forest biogeocenes, a western mixed forest in Belgium, beech and spruce forests in the Solling area in Germany and eastern oak forests on the forest-steppe margin have been investigated in detail.

The active layer in deciduous forests is the tree canopy, where both direct radiation from the sun and diffuse radiation are to a large extent transformed into heat. Only a minute portion of the daylight penetrates the forest. Beech is the dominating forest tree in ZB VI of central Europe; in ZB VI of East Asia and North America beeches occur, but the number of tree species is very much higher because of the glacial refuge history (see p. 254) and the

**Fig. 194.** Biosphere reservation Schorfheide (Germany). Tall oak-hornbeam forest on unsettled glacial sand and moraine deposits (Photo: S.-W. BRECKLE)

**Fig. 195.** Open pine forest with invading birch and oak wood, in the Senne, south of Bielefeld (Germany) in the Augustdorfer dune area with late glacial fossil dunes. Seedlings have initial difficulties with the competition from the dense carpet of *Avenella flexuosa* (Photo: S.-W. BRECKLE)

number of types of forest is much higher (PETERS 1997). *Fagus* species only occur as relicts in very small areas (see Fig. 179). *Fagus sylvatica* is the most widely spread of all 12 types of beech and forms several different types of beech forests (see Fig. 193) reaching from Scandinavia to northern Spain and from England to Turkey.

Besides beech forests there are other types of forests in central Europe, which are frequently much more influenced by man. Examples of these forests are the oak-birch forests on very acid soils (Fig. 194). During the last century coniferous forests were promoted strongly by forestry departments so that spruce and forests (Fig. 195) are the norm into which beeches and oaks slowly migrate.

After 1980 the death of spruce forests was observed and since 1990 significant damage has also been observed

**Fig. 196.** Montane, almost dead spruce forest in the western Harz mountains (Germany), the highest classification of forest damage. In the *background* the Brocken mountain (Photo: S.-W. BRECKLE)

in deciduous trees. Damage to whole forests on slopes was first seen in the Erzgebirge and the Sudeten mountains (Czech Republic) and then in the Harz mountains (Germany; Fig. 196). The cause of the damage in forests is not precisely known. It is probably not possible to blame air pollution and the changed input of toxic substances or too much nitrogen exclusively. Additional acceleration of nutrient leaching from leaves and the upper soil layers (cryptopodsolisation) increases the effects as much as exhaustion of soil through monocultures.

### b  Beech Forest in Solling (Germany)

Within the framework of the IBP (International Biological Programme) and successors, three beech and three spruce forests and differently fertilised meadows and one arable field were researched and compared during the period 1966–1986. In other countries characteristic vegetational types were also examined over many years. Here, as an example, the results concerning the structures and processes in a nemoral forest of zonobiome VI, in Solling, are presented. Comparative information from research on a continental oak forest on the Vorskla, the left tributary of the middle Dniepr (IBP project of the former Leningrad University) is given. This mixed oak forest extends over 1000 ha for forestry research and 160 ha is an almost primeval forest consisting of 300-year-old trees.

The beech forest in Solling (ELLENBERG et al. 1986) is a beech forest growing on acid soils (Luzulo-Fagetum) about 500 m above sea level. Soils are predominantly weak podsolic, brown earths, from loess layers above variegated (new red) sandstone. The Solling beech forest has a char-

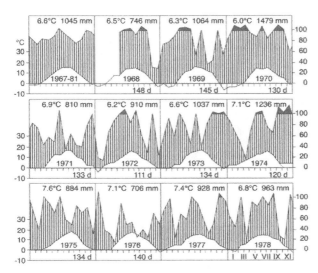

**Fig. 197.** Climate diagram (average of the years 1967–1981) and climatogram of 1968–1978 from Solling. *Below right*: The number of days in the year with mean temperatures above 10 °C (from ELLENBERG et al. 1986)

acteristic maritime climate, however, with significant differences between individual years. Mean precipitation from 1967–1981 was 1045 mm/year, corresponding to the typical high rain fall in the German low mountain ranges (Fig. 197). The driest year 1976 only had 706 mm, the wettest more than double, 1479 (1970). Drier periods occur now and then, however, they are not part of a typical seasonal event. Thus, at any time in the year there are days with a relatively dense cloud cover and resulting low radiation. Some typical days of global radiation are shown in Fig. 198. With cloud-covered skies (Fig. 198, 3 July 1972) global radiation is often less than one tenth of that on a clear day (Fig. 198, 13 July 1972). This is thus the limiting factor for photosynthesis of beech trees.

Weather is predominantly determined by west and south-west winds (Fig. 199). East winds occasionally occur with high pressures in winter, north winds hardly ever occur.

### c Ecophysiology of the Tree Stratum

The size of a tree renders it an unsuitable object for experimentation. Its form is, to a large extent, dependent upon its surroundings. The crown of a solitary tree is usually dome-shaped or spherical, whereas if the tree is part of a dense stand, the crown is usually very small. Since the leaves are arranged in several layers, the outer

**Fig. 198.** Daily radiation (global radiation) in early summer and summer for 1972 for three clear and three cloudy days (from ELLENBERG et al. 1986)

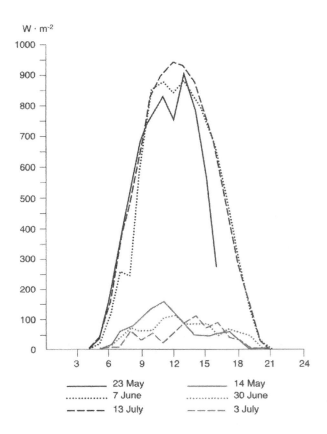

|  |  |
| --- | --- |
| ———— 23 May | ———— 14 May |
| ·············· 7 June | ·············· 30 June |
| — — — 13 July | — — — 3 July |

**Fig. 199.** Daily and hourly average values (in percent) of predominant wind directions in the Solling area (from ELLENBERG et al. 1986)

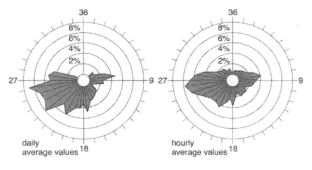

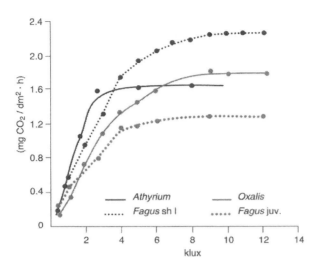

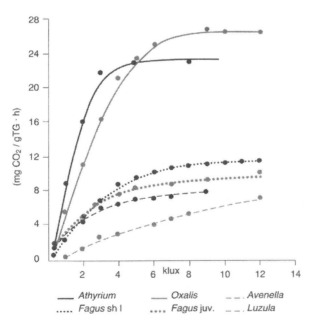

**Fig. 200.** Light saturation curves of photosynthesis (*below* related to mass; *above* related to leaf area) of some species of the beech forest in the Solling forest. *sh l* Shade leaves, *juv* juvenile growth (from ELLENBERG et al. 1986)

ones are exposed to full daylight and the inner ones grow in the shade. A distinction is therefore made between **sun leaves** and **shade leaves** which differ in anatomical-morphological and ecophysiological properties and are linked by intermediate forms.

Sun leaves are smaller and thicker, have a denser venation and possess more stomata per square millimetre on the undersurface. In other words, sun leaves are more xeromorphic than large, thin shade leaves. Structural differences between sun and shade leaves result from differences in the water balance at the time when the buds for the next spring are being formed. Twigs exposed to the sun transpire more actively, as indicated by an increase in cell-sap concentration. Cell-sap osmotic potential of sun leaves of a beech tree is $-1.6$ MPa and that of shade leaves is $-1.2$ MPa (see also p. 49). The $CO_2$ assimilation also differs. In laboratory experiments, it has been established that shade leaves respire less actively than sun leaves in darkness: in beech, shade leaves give off only 0.2 mg $CO_2 \, dm^{-2} \, h^{-1}$ compared to 1.0 mg for the sun leaves. This explains the finding that, in spring, the light compensation point (where respiration = gross photosynthesis) of shade leaves is 350 lx and that of sun leaves 1000 lx. Photosynthesis increases in proportion to light intensity until a maximum is reached (Fig. 200), which for shade leaves is 20% of maximum daylight and for sun leaves about 40%. Thus shade leaves are better able to utilise lower light intensities and sun leaves higher intensities, although even the sun leaves do not appear to utilise available daylight to the full.

Sun leaves are oriented vertically, parallel to the incoming rays, whilst shade leaves are horizontal.

The above figures apply to leaves oriented at right angles to the incoming light. Sun leaves near the apex of the tree, however, are nearly always rather steeply inclined. This protects the leaves from overheating and helps to reduce water losses and also means that more light can penetrate the outer canopy, to the advantage of the lower leaves. It also means that the radiation in the morning (from the east) and in the evening (from the west) is used better. Leaves in the deep shade are always at right angles to the incoming light, thus even with a LAI of five or more they are able to achieve positive production on average.

An exact production analysis was carried out by SCHULZE (1970), who directly measured the $CO_2$ assimilation of beech (*Fagus*) in its habitat. Measurements during a single day are shown in Fig. 201. It can be concluded that the annual production of sun leaves and shade leaves per unit dry weight during the vegetative period is the same because shade leaves remain active longer in autumn (Fig. 202).

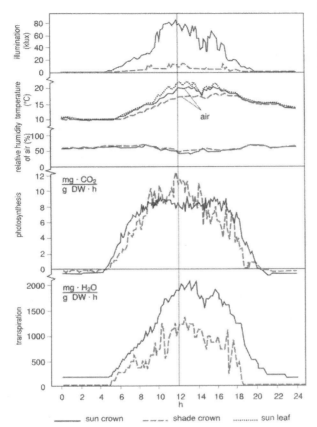

**Fig. 201.** Course of microclimatically and ecophysiologically important parameters for sun and shade leaves of beech in Solling on a clear day (from ELLENBERG et al. 1986)

——— sun crown ——— shade crown ·········· sun leaf

If illumination is continuously below a certain minimum, respiration is no longer compensated for by photosynthesis, losses of material occur and the leaves turn yellow and are shed. This minimum, expressed in percentage of full daylight, varies from one tree species to the next. A "shade" tree with a dense crown, such as beech, has a low light minimum (1.2%) and "light" trees, such as birch and aspen, with a thinner crown, have a higher light minimum (11%). Figures for species such as maple and oak lie somewhere between those mentioned above. This light minimum is valid for the crown of the tree and does not necessarily coincide with that light minimum which must be exceeded if tree seedlings are to develop on the forest floor, although the values are correlated. Beech seedlings

**Fig. 202.** Daily values of $CO_2$ gas exchange of beech (*Fagus sylvatica*) and spruce (*Picea abies*) during the year (after ELLENBERG et al. 1986)

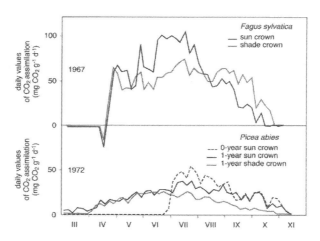

require little light, whereas birch seedlings need at least 12–15% of total daylight. An oak forest in full leaf reflects 17%, but without leaves only 11% of the incoming radiation is reflected which is significantly less than for meadows or cultivated land (25%). Half way up the stand or on the floor of young stands in full leaf only 1.2% or 0.6% of the daylight is measured, in contrast, values are about 20% or 2% for old stands.

Light conditions are of vital importance to trees in competition with one another. Light-demanding trees grow up within a few decades in a clearing and under their canopy shade-tolerating trees germinate and gradually grow higher, producing in turn a canopy so dense that the "light" trees are incapable of regeneration. In time it is the species tolerating the most shade which achieves dominance, providing that the other local conditions are appropriate.

Zonal forests in central Europe consist of beech (*Fagus sylvatica*). Only on very poor soils, places where groundwater is high, or in the drier biotopes, is beech unable to compete successfully. In the western parts of eastern Europe where the climate is too continental for beech it is replaced by another shade-loving species, the hornbeam (*Carpinus betulus*). Still further east hornbeam is replaced by oak (*Quercus robur*).

The mean daily temperature of the canopy in summer is 2 °C higher than that on the forest floor; the mean daily maximum, 11 °C higher and the mean daily minimum about 2–3 °C lower. Mean air humidity is 98% on the ground and drops as low as 77% with increased height.

Wind velocity in the forest is low and, since the forest floor is protected from direct radiation, the air in the forest remains cooler during the day than that in open stands.

Productivity of a forest depends to a large extent upon its leaf area index (LAI), that is to say, upon the ratio of total leaf area of the tree stand to the ground area covered by it. This ratio is limited to a maximum value above which it may not rise since otherwise the lower, overshadowed leaves would not be able to maintain a positive material balance. This maximum, however, not only depends upon light intensity, but also decreases in the face of inadequate supplies of water or nutrients. The LAI of a pure oak stand is 5 to 6 (higher in wet years) and in mixed stands with a good water supply, inclusive of all tree species and shrubs, it can exceed 8.

Production of dry matter for a 40-year-old beech stand in Denmark is ($t\ ha^{-1}\ a^{-1}$):

- gross production of the assimilating leaves = 23.5
- respiratory losses (leaves 4.6; branches 4.5 and roots 0.9) = 10.0
- annual production (leaves 2.7; branches 1.0; litter and roots 0.2) = 3.9
- wood production (aboveground 8.0; belowground 1.6) = 9.6

On average, $6\ t\ ha^{-1}$ of the maximum of $8\ t\ ha^{-1}$ of trunk wood is utilisable, which is equivalent to $11\ m^3$ wood. The same weight of wood is produced by spruce, but it occupies a mean volume of $17\ m^3$. Figure 203 shows how production changes with age of the beech stand.

The beech forests in Solling also achieve an annual production of $10\ t\ ha^{-1}$, of these about 3 t are leaves. Flowers and fruits yield varying amounts in different years (mast years and lean years). Production of twigs is about 10% of the annual production, the production of roots about 10% of the aboveground production. The mass of dead wood is hardly smaller than the increase in wood production during the same period, i.e. the net phytomass increase is practically zero as expected for a virgin forest in its optimal phase. Primary production per year is $8.9\ t\ ha^{-1}$ and if the herbaceous layer is included $9.6\ t\ ha^{-1}$. Subterranean production was not measured. Primary production is a little less than in western deciduous forests because of the semi-arid climate. Annual production of leaf mass and area increases rapidly in the first 20 years, but as soon as a dense canopy is achieved, leaf mass and LAI remain almost constant. The canopy is raised above the soil surface by the increased height of stems whilst leaves with fallen

According to GORYSCHINA (1974), the phytomass for mixed oak forests in Russia is as follows: aboveground $306.7\ t\ ha^{-1}$ (leaves 3.7; twigs and branches 71.2; trunks 230.8); belowground 124.9; total $431.6\ t\ ha^{-1}$; plus 0.7 for the herbaceous layer.

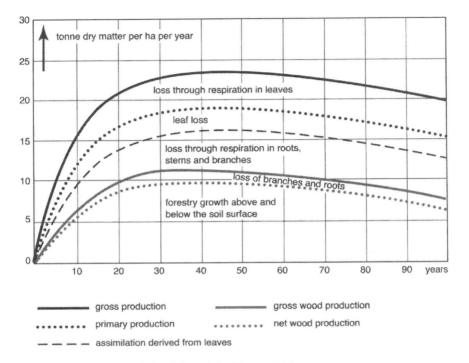

**Fig. 203.** Production curve of a beech forest (after WALTER 1990)

branches form the litter which, together with the dying roots, make up the total waste. Only the mass of wood produced is stored, so that the standing phytomass of the forest increases steadily, but progressively more slowly until the forest is very old. The standing phytomass of the forest may exceed 200 t ha$^{-1}$ for 50-year-old stands and 400 t ha$^{-1}$ for 200-year-old stands.

For oak forests the following average increase in wood production was found in relation to the age of the stand (in brackets): 3.8 t ha$^{-1}$ (13), 3.6 t ha$^{-1}$ (22), 4.3 t ha$^{-1}$ (42), 4.7 t ha$^{-1}$ (56), 0.4 t ha$^{-1}$ (135), 0.0 t ha$^{-1}$ (220). Increasing diameter of the stem (BHD: breast height diameter) is related to mass of trunks. The relation is shown in Fig. 204.

Litter accumulates in the forest until an equilibrium is reached, i.e. until annually as much litter is mineralised as new litter is produced. Some of the most important nutrients (N, P, K, Ca) are bound in litter. Enormous litter deposits are therefore unsuitable and the use of litter is particularly damaging, as in this process nutrients are completely

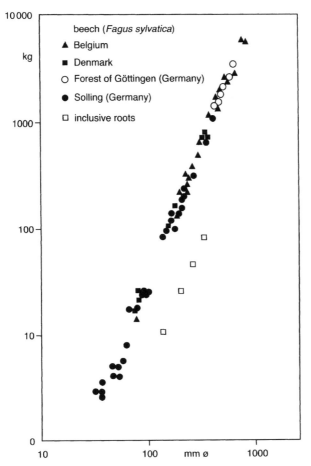

**Fig. 204.** Dependence of aboveground biomass (below ground shown with *open squares*) on stem diameter (BHD) in beech of different stands (after EL-LENBERG et al. 1986)

removed, calciumcarbonate particularly. This causes forest soils to become nutrient-depleted and to acidify quickly with a subsequent loss of wood production. Nitrogen compounds are mineralised during litter breakdown. Most of the nutrients from the lower, decomposing humus layer are available to tree roots. Thus the life in the soil is of particular importance to the habitat of forests, as indeed is the water supply. On the other hand, animals above ground are not very important, even insect excrement (frass) is only a few percent of turnover (see p. 350 f.).

For deciduous trees litter accumulates periodically and remains more or less the same from year to year. More than 90% of beech leaves drop in October (Fig. 205, top).

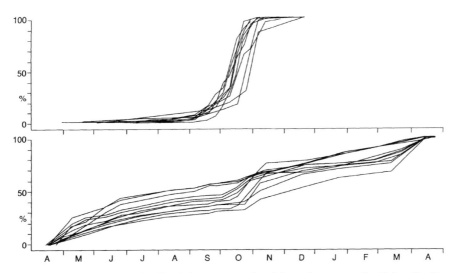

**Fig. 205.** Annual leaf fall (in %) of beech (*top* 1967–1975) and drop of spruce needles (*below*; fine litter of three spruce stands 1968–1971) during a year in the Solling forest shown as a family of curves (after ELLENBERG et al. 1986)

For spruce, on the other hand, dry or yellow needles drizzle down all year long, but even here there is also a small maximum of needle fall, a little later than the drop of beech leaves (Fig. 205, bottom).

## d Ecophysiology of the Herbaceous Layer (Synusiae)

The microclimate of the forest floor is vastly different from that of an open habitat. When the forest is in leaf, the light intensity on the forest floor is weaker, the temperature is more moderate and the humidity of the air and upper soil layers is greater than that outside the forest. For these reasons, the herbaceous plants of the forest are shade-tolerating and hygrophytic and their cell-sap concentration is very low. Thus, there is a greater hydrature of the plasma and therefore they wilt faster in full sunlight.

On a clear day, light conditions on the forest floor can be very heterogeneous. Single rays of sunshine falling through the tree canopy cause sun flecks and, as the sun moves across the sky or the branches are moved to and fro by the wind, these flecks change their position and intensity. If a leaf of a herbaceous plant is hit by a sun fleck, illumination can rise 10- or even 30-fold, a factor of great significance for the plant's photosynthesis. It is therefore preferable, when determining the amount of light received by herbaceous plants as a percentage of total daylight, to carry out the comparative measurements on a bright day with a more or less even cloud cover. Such measurements, however, provide us with no more than preliminary information. It would be better to measure the sum total of the daylight falling on a certain place on the forest floor with the aid of an automatic light meter. Before the trees come into leaf the herbaceous stratum is very adequately illuminated, but as the trees come into full leaf the situation deteriorates quickly.

The favourable light conditions prevailing before the trees come into leaf are exploited by the **spring geophytes** (*Galanthus, Leucojum, Scilla, Ficaria, Corydalis, Anemone* and others). They profit from the fact that, even in April, the litter layer in which they root is warmed up to 25–30 °C because of its uninterrupted exposure to the sun's rays. Geophytes benefit from the fact that the heat capacity of the air-containing litter layer is small and, as a result, the temperature conductivity is very good. Trees come into leaf later because the deeper soil layers in which they root are slow to warm up. In the short, early-spring season, geophytes flower and fruit and store up reserves in their underground storage organs for the coming year. When the trees are in leaf, the leaves of the geophytes turn yellow and a dormant period begins for them. Leaf death is not, however, due to the deeper shade, but is the expression of an endogenous rhythm; the leaves die even sooner in light. Apparently, geophytes are "just the right" plants to fill the vacant ecological niche in deciduous forests.

Spring geophytes of this kind are also termed **ephemeroids**. Although their vegetational period is just as brief as that of the annual ephemerals, they are perennial species with underground storage organs. The ecological behaviour of spring geophytes is very similar and their developmental cycles are almost identical, so they can be considered to form a functional unit termed a **synusia** by ecologists. Synusiae have no independent material cycle and are thus merely constituent parts of the various ecosystems (see p. 86).

The synusiae of deciduous forests have been investigated in the Russian mixed oak forest on the Vorskla (GORYSCHINA 1974; WALTER 1976). The following five examples of synusiae in deciduous forests are distinguished:

1. Ephemeroids: *Scilla sibirica, Ficaria verna, Corydalis solida, Anemone ranunculoides.*
2. Hemiephemeroids: *Dentaria bulbifera.*

3. Early-summer species: *Aegopodium podagraria, Pulmonaria obscura, Asperula (Galium) odorata, Stellaria holostea.*
4. Late-summer species: *Scrophularia nodosa, Stachys sylvatica, Campanula trachelium, Dactylis glomerata, Festuca gigantea.*
5. Evergreen species: *Asarum europaeum, Carex pilosa.*

The various synusiae grow in different light phases on the forest floor by morphological/physiological adaptations. *Aegopodium*, for example, first develops small light leaves, but by summer has developed large shade leaves which are replaced in autumn by very small, xeromorphic, cold-resistant leaves persisting throughout the winter (*Aegopodium* has no winter resting stage). Exactly the same situation can be seen in *Stellaria* and *Asperula*, except that the different types of leaves develop successively on the same vertical shoot axis.

The assimilate economy, or rather the way in which assimilates are used, varies considerably from one synusia to the next. *Scilla* uses all of the assimilates stored in the bulb for formation of the flowering shoot and leaves and only towards the end of the short vegetational period are the newly formed assimilates directed into the young bulb for use in the following year. *Dentaria*, on the other hand, quite soon begins to lay up reserves in the rhizome and thus requires more time for flowering and fruit formation. *Aegopodium* exhausts its small reserves in order to produce the light leaves which then assimilate $CO_2$ intensively so that, by the beginning of May, they are able to lay up new reserves while at the same time delivering sufficient assimilates for formation of the shade leaves, after which they die off. In its large tuber, the late-summer plant *Scrophularia* stores much water, but very few organic reserves for the formation of the first leaves. Assimilation of these leaves in the shade is so low that it takes until autumn before the shoot is fully grown and in flower and the fruits ripen. The assimilate economy of *Asarum* is characterised by the evergreen leaves from the previous year assimilating after the winter and then dying off during spring, long after photosynthetic activity of the young leaves is fully established.

The total phytomass of the herbaceous layer is not large, but its significance for the ecosystem lies in the fact that it is rapidly mineralised and thus contributes to a quick cycling. Leaf litter of trees, however, disintegrates slowly and the nutrients contained in it are not available until the following year.

The majority of species of the herbaceous layer are hemicryptophytes, which means that their regenerative

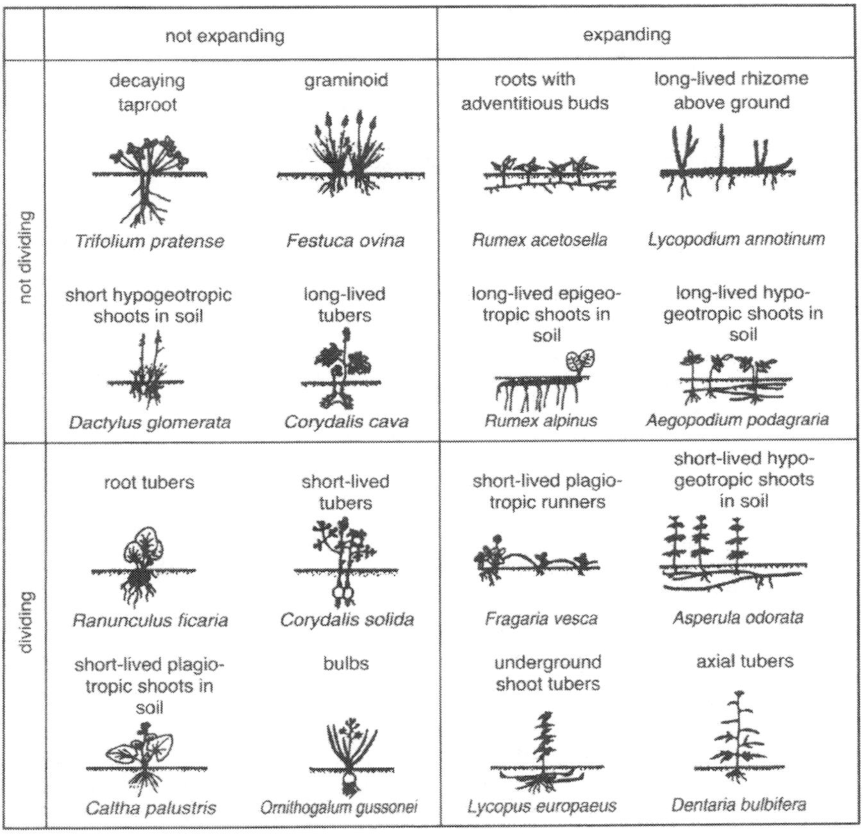

| | not expanding | | expanding | |
|---|---|---|---|---|
| **not dividing** | decaying taproot | graminoid | roots with adventitious buds | long-lived rhizome above ground |
| | *Trifolium pratense* | *Festuca ovina* | *Rumex acetosella* | *Lycopodium annotinum* |
| | short hypogeotropic shoots in soil | long-lived tubers | long-lived epigeo-tropic shoots in soil | long-lived hypo-geotropic shoots in soil |
| | *Dactylus glomerata* | *Corydalis cava* | *Rumex alpinus* | *Aegopodium podagraria* |
| **dividing** | root tubers | short-lived tubers | short-lived plagio-tropic runners | short-lived hypo-geotropic shoots in soil |
| | *Ranunculus ficaria* | *Corydalis solida* | *Fragaria vesca* | *Asperula odorata* |
| | short-lived plagio-tropic shoots in soil | bulbs | underground shoot tubers | axial tubers |
| | *Caltha palustris* | *Ornithogalum gussonei* | *Lycopus europaeus* | *Dentaria bulbifera* |

**Fig. 206.** Different clonal structures with indication of their distribution and life span (after VAN GROENENDAEL et al. 1996)

buds form at the base of the shoot and spend the winter just below the ground surface, protected by a covering of autumn leaves and sometimes snow as well. However, many of the representatives are clonal, which means that they multiply in different ways by division, i.e. vegeta-tively. Division of the mother plant, formation of runners, nodules on shoots or roots etc. serve to maintain the spe-cies and their distribution. Many of the clonal types were listed comparatively by GROENENDAEL et al.; examples from central European deciduous forests are given in Fig. 206.

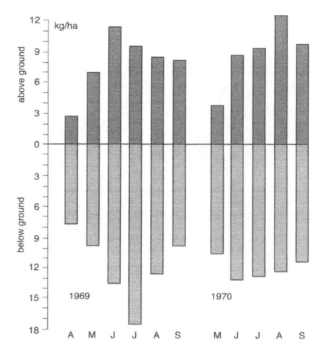

**Fig. 207.** Changes of aboveground and belowground biomass of the herbaceous layer in the beech forest in Solling in 1969 and in 1970, when the vegetation period started later and there was more rainfall during the year (after ELLENBERG et al. 1986)

The total biomass of herbaceous plants is usually very small (Fig. 207), however, it is rapidly turned over. The reaction of shoot and root is not the same in every year and is dependent on the supply of water and nutrients as shown in Fig. 207 and even the number of shoots deviates significantly from year to year (Fig. 208). The herbaceous layer is thus able to adapt quickly to changing conditions.

In mast years, as in 1971, the number of beech seedlings is very large, however during the next years the number decreases. The remaining few (from a million beechnuts perhaps 0.1–1 young plants grow) are sufficient to guarantee a new generation of trees.

Illumination values have been determined for many species of the herbaceous layer in forests. These plants possess an **illumination maximum** ($L_{max}$), since they are not found in full light, and an **illumination minimum** ($L_{min}$), since they avoid the deepest forest shade. The following limiting values, in percentage of total daylight, are given as examples: *Lamium maculatum* 67–12%; *Lathyrus vernus* 33–20%; *Geranium robertianum* 74–4% and *Prenanthes purpurea*, 10–5% (sterile to 3%). The $L_{max}$ value

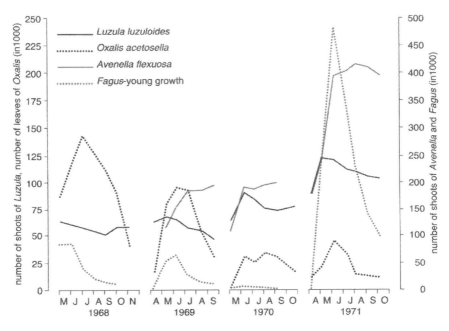

**Fig. 208.** Number of shoots of some species of the beech forest in four consecutive summers from 1968. *Avenella* was not counted in 1968 (from ELLENBERG et al. 1986)

is dependent upon the water supply. Hygrophilic species require damp soil and cannot tolerate a high saturation deficit of the air such as would occur under conditions of full illumination. $L_{min}$ is the starvation limit for plants. Light intensity is just enough to make it possible to produce the necessary materials for development. Generally the dead zone in forests is below 1% of daylight. There only the fruit bodies of heterotrophic fungi are found or holosaprophytes amongst the flowering plants, for example, the bird's nest orchid (*Neottia nidus-avis*).

A further very important factor is the competition of tree roots. Supply of water is a very important factor in dry forest regions at the borders to the forest steppes. Trees with their higher cell-sap concentration than that of herbaceous plants are able to develop lower water potentials in their absorbing roots and thus are better equipped to draw water from the soil than herbaceous plants. As a result, the floor of beech forests is bare (Fagetum nudum). If roots of trees are severed, thus excluding them from competition, herbaceous plants develop, which proves that water and not light is the limiting factor. Trees also ex-

**Fig. 209.** Compartment model of water relations in a level ecosystem in the Solling forest (*As* subterranean runoff; *E* evaporation; *I* interception; *N* precipitation; *Nd* drips of the crown; *Ns* runoff on stem; *dR* size of storage; *T* transpiration). The width of *arrows* indicates size (after ELLENBERG et al. 1986)

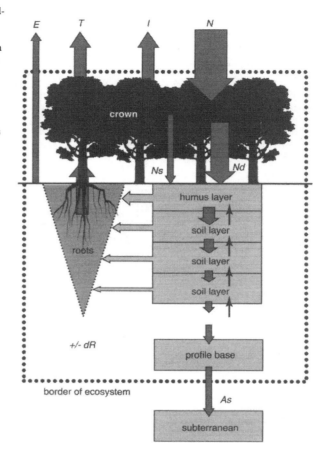

tract nutrients, especially nitrogen, from poor and very shallow soils and herbaceous plants are obliged to make do with what is left. Only plants with low nutrient requirements, like *Luzula luzuloides*, *Avenella flexuosa*, *Potentilla sterilis* and *Vaccinium myrtillus* are found in such forests.

### e Water Relations

In the Solling area, precipitation is the only input of water. Output consists of evaporation and run-off, which have different components (Fig. 209). In the water relations balance (see p. 55) internal water flows have to be consider-

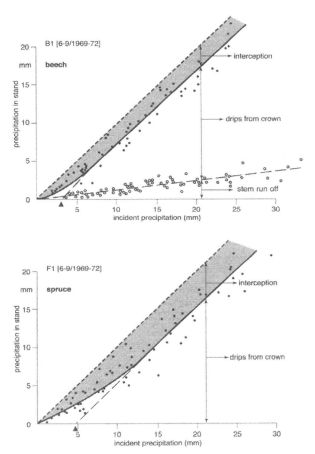

**Fig. 210.** Interception, drip from crown and stem runoff on stem during the months of full foliage for beech and spruce depending on precipitation. For spruce, stem runoff is negligible (after ELLENBERG et al. 1986)

ed. In forests, for example, water drips from the canopy and down the stems. The latter is important for beech, but negligible for spruce (Fig. 210).

For beech about 17% of precipitation falling on the forest is retained on average by the canopy during the summer period (green in summer), for spruce about 27% during the whole year (evergreen). The rest drips down or runs down the stems. In dry years the interception is significantly higher than in wet years.

Snow which has accumulated on the forest floor melts slowly in spring and the snow water seeps into the ground almost completely. Transpiration of the tree layer is so strong that under the forest no water enters the ground-

In a wet year (1970), spruce forest allows up to 880 mm water to seep into the water table, in a dry year (1971) only 232 mm. The corresponding values for beech forests are: 1970: 973 mm; 1971: 304 mm. These amounts of water contribute to replenishing the groundwater.

water. Water run-off from the herbaceous layer is about five to six times lower. A well-developed deciduous forest in the forest steppe area uses practically all the water supplied by precipitation, on the other hand, a beech forest in central Europe used only 50–60% of the precipitation. During the summer months, there is no surplus, even here in humid central Europe.

Corresponding to the water turnover model in Fig. 209, in the Solling area, water flowing through the layers of the soil was measured using tensiometers (soil matrix potential), lysimeters (with suction devices) and tritiated water. The infiltration of water into the soil, in as far as it is not taken up by plant roots, seeps to depths of 1.65 m under beeches and to 1.20 m under spruce, during the course of a year.

If the chemical energy bound during production is held in relation to the incident energy on 1 ha of forest, values of about 2% for the gross production and 1% for primary production are achieved. One third of the energy is used for transpiration, a total of 80% for evaporation and interception of water, the rest is changed into heat.

Coupling $CO_2$ assimilation with transpiration, as both processes are regulated by the same factors, can be expressed quantitatively by the **transpiration coefficient**. Herbaceous plants (for example wheat) consume 540 kg water to produce 1 kg plant material. Beech in the Solling area has an average transpiration coefficient of only 180, spruce 220. Gas exchange measurements help to explain the greater efficiency of beech (see Fig. 202). Corresponding to the strength of illumination, humidity and $CO_2$ concentration, leaves regulate their stomata so quickly that they adapt immediately to any change. Furthermore, photosynthesis of the thin shade leaves of beech is just as effective as that of the sun leaves although the latter leaves have less dry matter. This enables beech leaves to utilise sunlight much better than oak, which only produces three layers of sun leaves, whilst beech produces additionally three to four layers of shade leaves. Biomass production of the herbaceous and moss layer on acid soils is therefore insignificant, whilst it can be considerable in mixed oak forests. In Solling there is almost never a lack of water for beech.

## f Long Cycle (Consumers)

The role of animals in an ecosystem is determined predominantly by their food relation. Food chains are of such diversity and their connections of such complexity that they have not yet been completely elucidated for any single ecosystem. Plants are attacked by a variety of para-

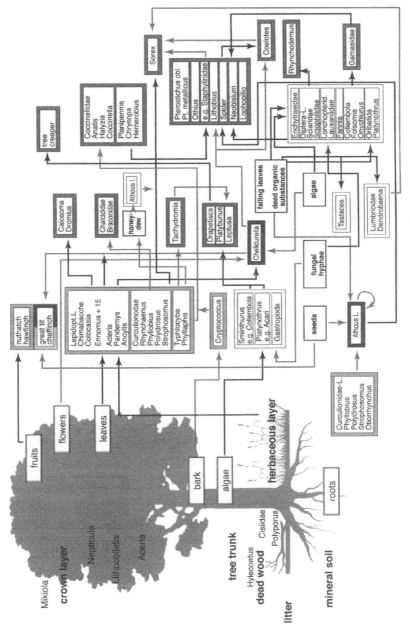

**Fig. 211.** Significant nutritional interrelations in a beech forest in Solling (after ELLENBERG et al. 1986)

**Fig. 212.** Density of animal groups of the beech forest in Solling with indication of main food sources (logarithmic $y$-axis of individuals m$^{-2}$; with additional data for nematodes and aphids (after ELLENBERG et al. 1986)

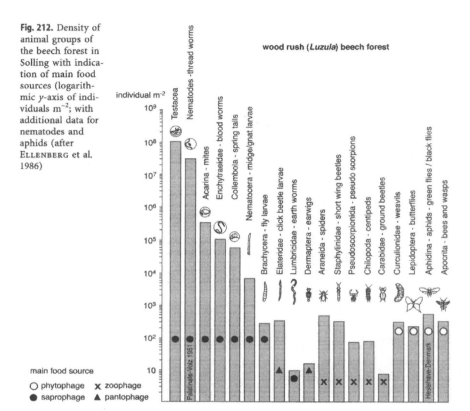

sites, chiefly fungi, and a large number of insect pests. Different plant organs serve as food for various herbivores that constitute the food source of predators of the first order, both large (birds and mammals) and small (invertebrates). These predators, in turn, are consumed by predators of the second order, e.g. birds or shrews that catch predatory insects. Some quantitatively significant relations of food in the beech forest of Solling are shown in Fig. 211. However, it must be kept in mind that the relations given are not clear-cut, often only facultative or even episodic. In total two main paths can be seen: a phytophage nutrient chain, depending on living plant materials, mainly leaves of beech, and a saprophage nutrient chain depending mainly on the dead organic material on the forest floor.

Only a very small part of the chemical energy in the food of animal organisms is transformed into secondary

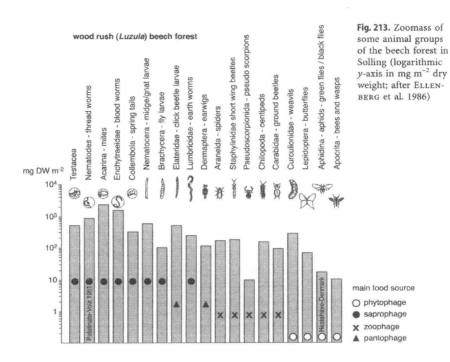

**Fig. 213.** Zoomass of some animal groups of the beech forest in Solling (logarithmic $y$-axis in mg m$^{-2}$ dry weight; after ELLEN-BERG et al. 1986)

wood rush (*Luzula*) beech forest

main food source
○ phytophage
● saprophage
X zoophage
▲ pantophage

production, i.e. into animal body substance. The larger part is lost in excrement or used up in respiration.

Upon closer inspection, leaves or other plant organs are often seen to be damaged. Twenty different insect species are known to live off the leaves, buds, bark and wood of the oak alone, and the number of gall-forming insects specialising on oak or beech trees is extremely large. For the Solling area, a listing of animal groups and their food materials was worked out over many years of detailed research. The most important groups and the number of individuals are listed in Fig. 212. It is obvious that microscopically small types will occur in enormous numbers. In Fig. 213 the biomass of individual groups of organisms in relation to area is given for comparison. Many of these species live in the soil or old wood.

Research on **old trees** in Bavaria showed that about 2000 species of beetle live in old wood. There are about 8000 native species of beetles. Beetle communities in an oak tree change considerably during its life as shown in Fig. 214. Many of the inhabitants of old wood are relict species of original forests. These species were fairly common about 2000 years ago (six or eight tree generations,

Keeping old trees and historically old stands of forest saves much time and requires a turnabout from unnecessary "measures of cultivation" and "excessive revitalisation of trees". An understanding of natural processes, also rational working with nature and no sticking to "the duty of safe travelling" is required (LEICHT 1996).

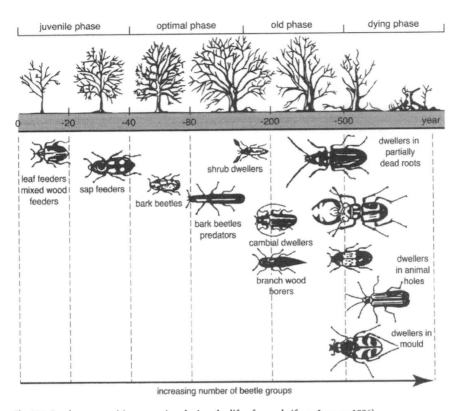

**Fig. 214.** Beetle communities occurring during the life of an oak (from LEICHT 1996)

in the time span of evolution a few moments) and have now been pushed back to a very few habitats. At that time, dead wood and old trees probably were the most common substrates and for this reason many species of small animals invaded this habitat (LEICHT 1996). The diversity of the habitat structure of an old tree is shown in Fig. 215. However, there are hardly any old trees left.

### g Decomposers in Litter and Soil

Most of the annual waste of a deciduous forest forms a layer of dying and yellow leaves (litter) on the ground which is immediately attacked and broken down by micro-organisms, fungi and bacteria. Saprophages which include small animals such as insect larvae and other ar-

**Fig. 215.** Overview of the habitat structures in an old tree and dead wood (from LEICHT 1996)

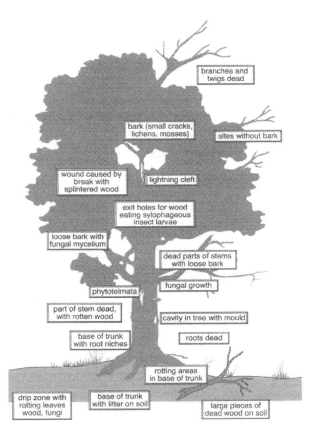

thropods but are mainly earthworms, in whose casts bacteria are particularly active, feed on the litter and break it down, rendering it more accessible to micro-organisms. Exact quantitative studies on the activity of such animals in three deciduous forest stands have been done by ED-WARDS et al. (1970).

Leaf fall begins gradually in autumn at the end of October when the days begin to shorten and assimilation of the shade leaves is no longer adequate. Sugars, organic acids and pigments are washed out of the leaf litter by rain and the dead leaves turn brown.

The lower the C:N ratio in the litter, the more rapidly mineralisation proceeds. By next June, birch litter has lost about four fifths of its dry weight; lime tree litter, about

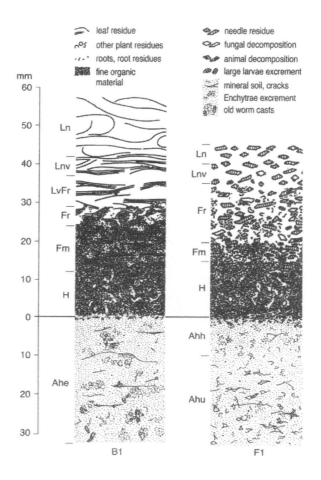

one half, and oak litter, which is not so readily broken down, only about one quarter.

Mineralisation of litter does not proceed to completion: humus is produced which, when saturated with calcium, forms the layer of mull containing large numbers of lumbricids (earthworms) or, under acid conditions, forms the layer of mould containing oribatids (mites) and collembola (springtails). In extreme cases, when the reaction is very strongly acid, a raw humus layer accumulates in which large quantities of fungal hyphae, but no animals, are present. The upper organic humus layer (H) and litter (L) shows the difference between a beech forest and a

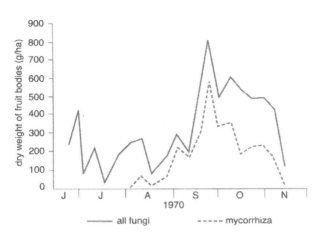

**Fig. 217.** Oscillations of dry weight (g/ha) of the fruit bodies of mycorrhiza and all higher fungi between June and November in the wet year 1970 in the beech forest of Solling (after ELLENBERG et al. 1986)

spruce forest (Fig. 216). In spruce humus the composition is lighter, there are fewer fine roots, fungus mycelia and animals. The decomposition of litter up to Fm (the mould layer of medium decomposition) takes about 4.5 years in a beech forest, about 2.5 years in a spruce forest.

SATCHELL (from DUVIGNEAUD 1974) reported the activity (i.e. respiration) of the individual groups of soil organisms in an English oak forest on chalky soil as follows:

- Invertebrates (diptera, collembola (springtails), oribatids (mites), molluscs, enchytraeds, lumbricids (earthworms), nematodes (thread worms), and protozoa): a total of 361 kcal m$^{-2}$ a$^{-1}$.
- Bacteria and actinomycetes: 77 kcal m$^{-2}$ a$^{-1}$.
- Most important is the activity of fungi: in the litter layer 543, in humus 220 and in the A and B horizon 380, a total of 1143 kcal m$^{-2}$ a$^{-1}$. The mass of the micro-organisms is relatively low in comparison to the mass of invertebrates.

Ninety percent of the dead wood dropping to the forest floor is destroyed by micro-organisms, chiefly fungi. Fungi have another important function as mycorrhizal partners for trees. In the Solling forest, almost half of the higher fungi are mycorrhizal fungi. The maximum development of fruit bodies is in September (Fig. 217); for other fungi development of fruit bodies can also take place in summer. This fact is explained by the different development balance between the root of the fungus and the tree in the mycorrhizal symbiosis.

**Fig. 218.** Annual net primary production in 1000 MJ ha$^{-1}$ (*open columns* estimated) of the experimental areas in Solling (beech forest, spruce forest, meadow with yellow oats, sown grazing land). *NPK* Full compliment of fertiliser: *PK* without nitrogen; *O* no fertilisation (after Ellenberg et al. 1986)

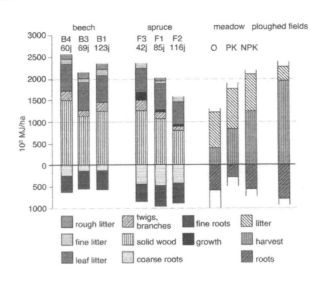

## h Ecosystem in the Solling Forest

Finally, in addition to the detailed discussion of some phenomena of the nemoral beech forest, a comparison of the productivity of different areas and the energy throughput is given. Annual net production of the experimental area in the Solling area is shown in Fig. 218. It is surprising that the productivity of fertilised meadows and fields is the same as that of the forest. Under the climatic and soil conditions in the Solling area, different stands of plants have almost the same productivity, which is expressed in Fig. 219 as energy content. This flow of energy through the important compartments shows that amongst heterotrophs decomposers have the highest energy turnover. This is characteristic for the short cycle of organic substances, whilst for the long cycle (via herbivores and carnivores) there is relatively little turnover. This is also known from other terrestrial ecosystems. It is interesting that the number of zoophages is even larger than that of herbivores; the nutritional pyramid is thus "falsified" by the high percentage of saprophages.

Solling, as a low medium-sized mountainous region, has to some extent montane features; precipitation is significantly higher than in the lower lying area, there is frequent cloud formation and the temperature is a little lower. In other medium-sized mountainous areas, and even more so in the northern Alps, the effect of altitude on the orozonal sequence becomes more significant.

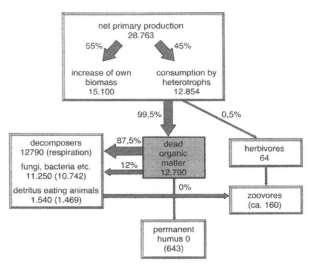

**Fig. 219.** Energy flow in the Solling beech forest calculated in parts from the respiration values (number in kJ m$^{-2}$ a$^{-1}$; *numbers in brackets* assume that 5% of the organic substances is changed to permanent humus; for carnivores it is assumed that they eat about 10% of the saprophages and herbivores (after ELLENBERG et al. 1986)

# Orobiome VI – Northern Alps and Alpine Forests and Tree Line

The Alps are the major barrier separating central Europe (ZB VI) from southern Europe (ZB IV). The geological formations of the Alps are the "crystalline" central Alps and the bordering limestone alpine chains of the northern and southern Alps (Fig. 220). These formations, of course, also affect the flora and vegetation.

## a Altitudinal Belts

Altitudinal belts of orobiome VI are well formed at the northern edge of the Alps. The mean annual temperature drops with increasing altitude in mountainous regions and the vegetational period becomes shorter. Whereas direct insolation increases with altitude, diffuse insolation decreases with the result that the temperature differences between northern and southern slopes become more pronounced. Owing to the ascending air masses, precipitation increases rapidly with altitude on the northern margins of the Alps, e.g. Munich at 569 m above sea level receives 866 mm, whereas Wendelstein at 1727 m receives 2869 mm.

Corresponding changes can be observed in the vegetation of the individual vegetational belts in the northern Alps:

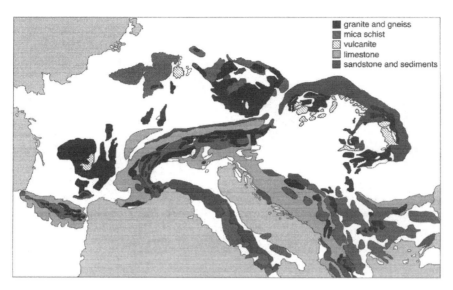

**Fig. 220.** Schematic of the geological composition of mountains in central Europe (from Ozenda 1994)

| Belt | Vegetation |
| --- | --- |
| Nival | Cushion plants, mosses and lichens |
| → **Climatic snow line at 2400 m above sea level** | |
| Alpine | Alpine meadows |
| Subalpine | Dwarf trees and shrubs |
| → **Tree line at 1700 m above sea level** | |
| High-montane (oreal) | Spruce forests |
| Montane | Beech and fir forest |
| Submontane | Beech forest |
| Colline | Mixed oak forest |

Since the Alps are interzonal, the succession of altitudinal belts on the southern margins is typical of a humid orobiome IV and the tree line is formed by beech (*Fagus*). However, in the sheltered valleys with their continental type of climate, the sequence is different in that there is no deciduous belt and a pine (*Pinus sylvestris*) belt precedes the spruce (*Picea abies*) belt, above which a larch (*Larix*) – stone pine (*Pinus cembra*) belt extends up to the tree line. The tree line and the snow line are 400–600 m higher here as a result of less cloud cover and stronger radiation (Fig. 221). Various parts of the Alps are distinguished as Helvetic (northern Alps), Pennine (central

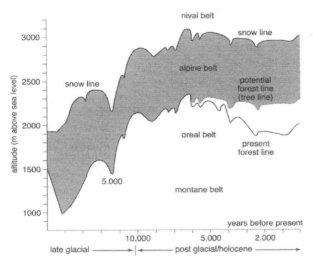

**Fig. 221.** Fluctuations of the tree line and snow line in late and post-glacial times in the Swiss central Alps (after OZENDA 1994). Because of human activities the current tree line is lower than expected for various trees

Alps), and Insubrian (southern Alps). The corresponding altitudinal belts, in order of decreasing altitude, are as follows:

| Helvetic | Pennine | Insubrian |
|---|---|---|
| **Altitudinal belt** | (Continental) | **Altitudinal belt** |
| (Central European) | Alpine belt | (Sub-Mediterrnean) |
| Alpine belt | Larch – stone pine | Alpine belt |
| Spruce forest | Spruce forest | Beech forest |
| Beech forest | Pine forest | *Quercus robur* forest) |
| Oak forest | – | Sclerophyllic (scattered) |

## b  Forest Belt

In the central Alps the uppermost forest belt consists of European larch (*Larix decidua*) and the stone pine (*Pinus cembra*), which is related to the Siberian subspecies. The larch plays the part of the light-demanding pioneer species and is eventually replaced by the five-needled *Pinus cembra* which is able to tolerate shade better. On paths left by avalanches larch may continue down to lower altitudes. Above the tree line the two-needled dwarf mountain pine (*Pinus montana*) occurs, but is replaced by the shrub-like alder *Alnus viridis* on wet habitats. The ecology of the spruce biogeocene will be dealt with later (pp. 355, 422).

**Fig. 222.** Forest and tree line in Vorarlberg near Bludenz. View from the top of Saladina (Freiburger hut) towards the south onto the opposite northern slopes (Photo: S.-W. BRECKLE)

**Fig. 223.** Spruce forest line with wind swept and stunted spruce on the Tegelberg near Füssen (1600 m above sea level) (Photo: S.-W. BRECKLE)

The northern margins of the Alps have been studied with regard to the factors responsible for setting the **upper limit of the spruce forests** (Fig. 222). With increasing altitude, the period of vegetation is shortened, the summers become cooler and the winters are both colder and longer. Although these climatic changes take place gradually, in contrast, the tree line in high mountains is very sharply drawn. The powers of growth of the trees seem to decrease quite suddenly and only a very narrow zone consisting of stunted, low forms provides the transition from forest to treeless alpine belt (Fig. 223).

The question arises as to whether the short summer or the long winter is responsible for the cessation of tree growth: it appears that both factors are important. With a period of vegetation of less than 3 months, young needles are incapable of maturing properly and their cuticle cannot attain the required final thickness. As a result, during the long winter with temperature inversion and much sunshine and particularly in the strong sunlight of the spring when the ground is still frozen, large water losses occur, as indicated by increased cell-sap concentration which decreases the osmotic potential by $-6.5$ MPa and more. Damage typical of frost drought can be observed and the needles drop off. Beneath a covering of snow, such events cannot take place; this explains the ability of the stunted forms to survive some distance above the tree line. It is apparently the combined effect of the two factors – the shortened vegetation period and the increased danger of frost drought – that is responsible for the abruptness of the tree line at a certain altitude. *Pinus montana*, which grows above the limits of the spruce forests, manages with a shorter period of vegetation. However, about 100 m further up, the phenomenon repeats itself. The needles are unable to mature, suffering damage from frost drought, and the upper *Pinus montana* limit is just as sharply defined as that of the forest.

The factors responsible for setting the limits of **polar spruce forests** (see p. 445 f.) have not been investigated, but they are probably similar, apart from the fact that sunshine plays no part in the damage caused by frost drought during the polar night. This factor is probably replaced by the drying effect of the strong, cold winds, as is borne out by the observation that in sheltered valleys, the tree line pushes further north than on the watersheds. The situation is the opposite in the Alps, since cold air is trapped in valleys and temperatures are lower than on the mountain ridges from which the cold air flows downward.

At its highest, the tree line in the central Alps lies at 2000–2150 m and is, as already mentioned, formed not by spruce, but by deciduous larch and evergreen *Pinus cembra* (stone pine), which has relatively delicate needles. Continuous measurements of climatic factors and photosynthesis have been made here throughout the entire year, so that the productivity of larch and stone pine can be compared. During the cold winter, photosynthesis is at a standstill even in the evergreen stone pine, but in spring the evergreen needles rapidly become active. However, at these altitudes, the larch is not green until the middle of June and begins to turn yellow by the end of September. Not more than 107 days are at its disposal for production compared with the 181 days available to stone pine. However, young larch has three to six times the mass of needles possessed by young stone pine and, despite the brief

**Aperzeit**, from Latin *aper-tus* = open and German *Zeit* = time, means the time without snow cover.

vegetation period, it assimilates 47% more $CO_2$ per gram of needles. Therefore, the total production of a 4-year-old larch is 4.5 times, and that of a 12-year-old, 8.5-times that of *P. cembra* of the same age. Only from the 25th year onward is the quantity of needles produced by the larch smaller than that of pine and larch begins to lag behind in its growth, particularly on raw humus soils. In time, the *Pinus cembra* succeeds in establishing itself as a shade-enduring species. The relationship of larch to *P. cembra* is similar to that of pine to spruce (see p. 421).

Wood found in the subfossil humus deposits in the subalpine belt indicates that during the warm post-glacial period all belts were 400 m higher than they are today. Dwarf shrubs that are covered by snow in winter are thus partly relicts of the former forests. The forest and snow limits during the post-glacial times are given in Fig. 221. Because of the thick covering of snow in the alpine belt, the duration of the snow-free season, the **aperzeit**, is more important for the low alpine vegetation than air temperature.

Snow cover depends to a large extent upon relief, wind direction and exposure. Snow collects in the hollows and forms cornices on the lee side of ridges, but is blown off the windward side. If the windward side is sunny as well, then the snow melts and the habitat is snow-free ("aper") all year-round. The plants here (Loiseleurietum) are exposed to the same extremes of frost desiccation as those in mountainous tundras and are accompanied by exactly the same lichens. A shady, windward slope, however, is not warmed by solar radiation. In the presence of large drifts of snow at the foot of a slope facing north, the snow-free season is reduced to a minimum on the so-called snow patches ("schneetälchen") or is completely lacking, the snow remaining throughout the summer. The snow-free season can, however, vary in length from year to year on the same habitat according to the snowfall. Its average length decreases with increasing altitude and is theoretically zero at the climatic snow line. In individual cases, however, such as on steep rock faces, the snow-free season can be very long, even above the snow line. This is why flowering plants are found in the Alps in the nival belt above the climatic snow line.

In any case, the microclimate, particularly on sunny days, is propitious with regard to temperature. If leaves are exposed to direct sunlight, their temperature can be as much as 22 °C higher than that of the air. Every mountain climber is familiar with the warm niches which can be exploited by low, ground-hugging plants. In cloudy weather temperature differences tend to be equalised.

From what has already been said it is clear that as far as the vegetation is concerned there is no such thing as a standard climate in the steep alpine belt. Instead, this region is split up into very small climatic units which can differ vastly from one another within a very short distance, as for example on the sunny and shady sides of a boulder. The way in which the snow is distributed in winter is of primary importance and must be known in order to be able to estimate the length of the snow-free season and understand the pattern of vegetation.

Temperature inversions and cold air pools play an important role and can cause a reversal of the order of attitudinal belts (beech above spruce). Even at the height of summer, outgoing radiation in dolines can be accompanied by night frost and trees are then unable to grow. Another source of disruption of altitudinal belts is to be found in avalanches, in whose wake the alpine vegetation continues to grow far down into the forest belt since here it is not subject to competition with the forest vegetation. Even on the shallow, impoverished dolomite soil which undergoes little erosion, alpine exclaves can be found in the middle of the forest belt and surviving habitats of alpine species in the bogs of the alpine foreland are familiar to the botanist. In habitats such as these the unpretentious but slowly growing alpine species are less inhibited by competition from other plants.

### c Alpine and Nival Belt

The ecology of the Alps has been studied in detail. The process of developing hardiness to frost in evergreen species takes an annual course similar to that observed in deciduous forest species. Hardiness is developed in late autumn and the process is reversed in spring. Although spruce needles are killed by temperatures of −7 °C in summer, they survive −40 °C in winter. Despite the fact that alpine species grow much higher up in the Alps than the spruce, the maximum frost hardiness achieved by alpine species is usually lower (less than −30 °C) since they are protected from the lowest winter temperatures by a blanket of snow. Only *Loiseleuria*, growing on windy habitats free of snow in winter, develops a greater degree of hardiness. Although the wind prevents the lowest temperatures from being reached, it enhances the danger of winter drought. If *Loiseleuria* is suspended in the open in winter it dries out within 15 days, despite its xeromorphic structure. In its snow-free natural habitat, however, it normally grows tightly pressed to the ground and the sun thaws any snow held between its shoots, so that occasional water uptake is possible. Dwarf shrubs are not exposed to the dangers of winter drought beneath their blanket of snow.

The water budget is fairly well balanced in summer on account of frequent rainfall and plants are only subjected to substantial evaporation for a few hours when radiation is intense or a strong wind blows. The effects of wind are reduced near the ground. Even in habitats that appear dry on the surface, such as talus or scree slopes and rock faces, the soil carries abundant water. In this kind of habitat plants develop extensive root systems or taproots capable of penetrating into the rocky crevices, even though their root systems are normally very shallow and confined to the upper soil layers. The propitious water balance is reflected in the high cell sap osmotic potential of -0.8 to -1.2 MPa and even in xeromorphic species such as *Dryas*, *Carex firma* and *Androsace helvetica*, it never drops below -2.0 MPa. Here again, it would probably be more correct to speak of a peinomorphosis (see p. 442) resulting from nitrogen deficiency than of xeromorphosis, since the uptake of nitrogen is, in fact, more difficult at low soil temperatures. Luxuriant, hygromorphic herbaceous plants are found only on N-rich habitats such as the areas frequented by livestock.

If the total water lost by the plant cover of alpine meadow communities is calculated, the figure of 200 mm a$^{-1}$ is obtained. Evaporation depends above all, upon the wind and is for this reason influenced by the relief, although in the inverse direction to snow deposition.

The short growing season in the alpine belt gives rise to the question of adequate production just as in the Arctic. The days are shorter than in the Arctic, but the light intensity is stronger and nocturnal temperatures are lower. Under favourable conditions of illumination, 100–300 mg $CO_2$ dm$^{-2}$ can be assimilated in 1 day. One month of good weather suffices for the accumulation of sufficient reserves for the coming year and the ripening of seeds, however, since the growing season lasts 3 months adequate production is in any case ensured. Primary production in plant communities depends largely upon the density of vegetation.

The following values were found:

- closed mats                 50–276 g m$^{-2}$
- Dryadeto-Firmetum      91 g m$^{-2}$
- Salicetum herbaceae    85 g m$^{-2}$
- Oxyrietum                    15 g m$^{-2}$
- on limestone scree       1 g m$^{-2}$

Photosynthesis is less intense in dwarf shrubs than in herbaceous species. However, since the total leaf area of the former is larger and the period of vegetation longer in the lower alpine belt, dwarf shrubs achieve a higher primary production.

The most unfavourable conditions of all occur on the snow patches ("schneetälchen") on the north-facing slopes in the siliceous rock regions. Snow melts very slowly from the edges of such patches and the ground gradually becomes exposed. This means that, within a very small area, zonation can be recognised in which the ground is free of snow ("aper") for ever-shorter periods. The soil of such habitats is rich in humus, slightly acid and invariably well-provided with water from the melting snow, but for this reason also relatively cool. If the snow-free period lasts for 3 months, mats of *Carex curvula* develop. If the growth season is reduced to 2 months, the willow *Salix herbacea* predominates, a woody species of which only the tips of the shoots are above ground so that its leaves form a compact sward. *Salix herbacea* only bears fruit after a winter with particularly little snow, when the snow-free season amounts to 3 months. There is also a scattering of very small plants such as *Gnaphalium supinum, Alchemilla pentaphylla, Arenaria biflora, Soldanella pusilla, Sibbaldia procumbens* and others. Given an even shorter growth period, only mosses manage to grow because they do not produce flowers and fruits. The most common of these is *Polytrichum sexangulare* (*P. norvegicum*). If the snow-free season is too short for such green mosses, then only a liverwort, *Anthelia juratzkana*, which looks like a mouldering crust, is found. This liverwort grows in symbiotic relationship with a fungus and is saprophytic to a certain extent. This latter zone does not become snow-free at all if there has been an unusually large amount of snow in the preceding winter. On firn areas in the nival belt the sole living organism is the alga *Chlamydomonas nivalis* which gives the snow its rosy hue.

Since bare rock predominates in the alpine belt of the Alps, its chemical composition is of great significance for the vegetation because it governs the characteristics of the soil. The floristic differences between the calcareous Alps and the crystalline siliceous central Alps are very pronounced. Accordingly, a distinction is made between limestone-demanding basophil species and limestone-avoiding acidophil species. **Vicarious species** like the alpine rhododendrons are often encountered: *Rhododendron hirsutum* on limestone and *R. ferrugineum* on siliceous rock or acid humus soil.

The ecosystems of the dwarf shrub heaths were studied in the years 1969–1976 as part of the International Biological Programme (IBP) on the Patscherkofel near Innsbruck, Austria. Three experimental areas above the tree line were analysed in detail (LARCHER 1979):

A *Vaccinium* heath (1980 m above sea level) in a basin protected from wind and sheltered from winter snow: *Vaccinium myrtillus* 3, *V. uliginosum* 2, *V. vitis-idaea* 1, *Loiseleuria procumbens* 1, *Calluna vulgaris* 1, *Melampyrum alpestre* 1, mosses 1, lichens 1.

B *Loiseleuria* heath (2000 m above sea level); dense stands, often snow-free in wind-exposed locations: *Loiseleuria* 5, *Vaccinium uliginosum* 1, *V. vitis-idaea* 1, others only +, lichens (*Cetraria islandica* 1, *Alectoria ochroleuca* 1, others only +).

C Open trellis-like and lichen-rich *Loiseleuria* stands (2175 m above sea level) in extreme wind-exposed locations: *Loiseleuria* 3, stunted *Vaccinium uliginosum* 2, *V. vitis-idaea* 1, *Calluna* 1, others +, mosses +, lichens (*Cetraria islandica* 2, *C. cuculata* 1, *Alectoria ochroleuca* 1, *Cladonia rangiferina* 1, *C. pyxidata* 1, *Thamnolia vermicularis* and others +).

The climate is cold, with an annual temperature of little over 0 °C. Frosts may occur in any month (the absolute

**Table 21.** Production parameters of alpine vegetational units; living standing phytomass, dead portion and litter are given in g m$^{-2}$ dry matter of dwarf shrub heath (A), the dense *Loiseleuria* heath (B) and the open *Loiseleuria* stand (C)

| Experimental area | A | B | C |
|---|---|---|---|
| Living aboveground phytomass (max.) | 983 | 1105 | 748 |
| Adhering dead material | 263 | 123 | 72 |
| Living underground phytomass | 2443 | 2200 | 803 |
| Dead underground material | 1549 | 608 | 56 |
| Total living phytomass | 3426 | 3305 | 1551 |
| Together with dead material | 5238 | 4036 | 1679 |
| Surface litter | 819 | 1080 | 931 |
| Shoot/root ratio | 1:2.5 | 1:2.0 | 1:1.1 |
| Percentage of assimilating parts in living phytomass | 55% | 68% | ? |
| Aboveground net primary production (t ha$^{-1}$ a$^{-1}$) | 4.8 | 3.2 | 1.1 |

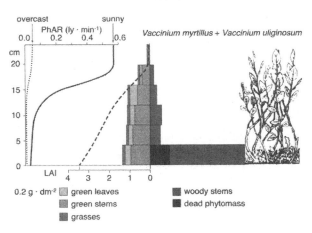

**Fig. 224.** *Centre* Phytomass stratification of the *Vaccinium* and *Loiseleuria* heath (*left* assimilating portion; *right* non-assimilating and dead portions). *Left* Cumulative leaf area index (*dotted line* LAI) and light decrease (PhAR) in the stand (according to CERNUSCA 1976; from LARCHER 1977)

minimum is around −20 °C) although the daily maximum temperature may reach 20 °C in the summer months. In experimental area A, snow lies approximately 6 months of the year; in experimental area B snow lies for 4–6 months and in area C it only lies in certain locations and only for short periods. The microclimate in A and B is somewhat warmer, while in C extreme temperature differences occur. The duration of $CO_2$ assimilation is approximately 100 days for the deciduous species and 140 days for the evergreen species. The composition of the stands as well as the photosynthetically active radiation (PhAR) and the cumulative leaf area index (LAI) are illustrated in Fig. 224. For further data of the ecology of production see Table 21. The wind is significantly decelerated in the dwarf shrub heath even during strong storms, so that the humidity remains high. The annual precipitation in this area is approximately 900 mm although it is 100 mm on average in each summer month.

On shale-like biotite gneiss, soils are sandy, acidic iron podsols with thick raw humus layers and are only poorly developed in stand C. They were formed from earlier stone pine forest stands. The humus is mineralised only slowly (nitrogen supply: approximately 3–4 kg ha$^{-1}$; in stand C only one third as much).

The phytomass is probably constant, except for certain fluctuations, meaning that the stands are in an ecological equilibrium with their surroundings; any increase in phytomass is inhibited by consumption (game animals, ptarmigan, arthropods) and by certain losses of material in winter (freezing and desiccation of parts above the snow).

Photosynthetic capacity per unit surface area is similar for the leaves of deciduous and evergreen dwarf shrubs. In relation to the dry weight of the leaves, it is similar for deciduous dwarf shrubs and soft-leaved deciduous woody species as well as for evergreen dwarf shrubs and conifers. The optimal temperature range for photosynthesis is between 10 and 30 °C for Ericaceae and thereby corresponds to normal temperatures on overcast and clear days in such stands; temperature minimum for assimilation of $CO_2$ in super-cooled leaves is –5 to –6 °C. Leaves are almost never subject to overheating or to limitation of photosynthesis due to water shortage. Although water supply is sufficient during the growth period, and the total amount of transpired water is in the range of 100–200 mm, restricted transpiration has been observed during periods of föhn (warm, dry south winds). Cuticular transpiration is very low during the winter.

Heat damage in the summer only affects individual shoots located above the loose stones or over raw humus soils devoid of vegetation. Damage from freezing in winter is possible only in snow-free conditions. Hardening protects plants from frost damage, although late frosts, after the process of dehardening has begun, can be dangerous. Damage due to frost drought is difficult to recognise and usually the damage is the result of the combined effects of several factors. The Arctic-alpine species *Loiseleuria procumbens* and *Vaccinium uliginosum* are completely frosthardy. During the main period of growth, respiration is significantly increased. At this time, the fat-storing *Loiseleuria* reduces its respiration coefficient to 0.8–0.9 until the phase of intensive growth is completed and then it returns to 1.

During the growth period, the efficiency of net primary production, in percent of the photosynthetically active radiation, is 0.9% for the dwarf shrub heath, 0.7% for the dense *Loiseleuria* heath and 0.3% for open stands.

Although species of Ericaceae store much fat as well as starch these substances are only partially mobilised. Most of this material remains in dead parts of the plant. On the first days of recurring frosts dwarf shrubs react immediately by transforming a large portion of their stored starch into sugar and *Loiseleuria* turns red due to the presence of anthocyanin.

Further investigations were carried out in the nival belt, i.e. above the climatic snow limit, on the Hohe Nebelkogel in the Stubaier Alps under very difficult conditions (Moser et al. 1977). One cabin had to be transported by helicopter, carefully isolated and well-grounded, since it was often in the thunder clouds. This belt has no closed vege-

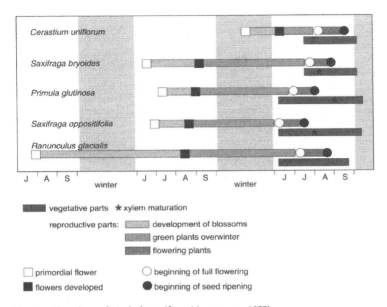

vegetative parts  ✳ xylem maturation

reproductive parts:
  development of blossoms
  green plants overwinter
  flowering plants

☐ primordial flower              ◯ beginning of full flowering
■ flowers developed              ● beginning of seed ripening

**Fig. 225.** Phenology of nival plants (from MOSER et at. 1977)

tational cover. In the 0.5-ha research area 3184 m above sea level, a flat ridge with seven species of flowering plants and several species of cryptogams, a northern slope with very sparse vegetation and a southern slope with eleven phanerogamic species on a level surface were chosen.

The climatic conditions do not correspond to those of the high Arctic and in the summer they are more similar to those in the páramos in the tropics. On clear days the leaf temperature is often over 15 °C and sinks to below 0 °C at night without impairing photosynthetic activity. The 24-h summer days of the Arctic with the sun low on the horizon, however, have relatively constant temperatures.

Of the three chosen locations the southern slope had the most advantageous light and temperature conditions. The phenology of the most important species is presented in Fig. 225. While *Saxifraga oppositifolia* flowers early (the floral organs are frost-resistant), *Cerastium uniflorum* blossoms late.

*Primula* spp. and *Ranunculus glacialis* store their assimilates in the form of starch which is transformed to sugar in winter, while *Saxifraga* spp. store fats. The location of the stored substances is illustrated in Fig. 226. It is in-

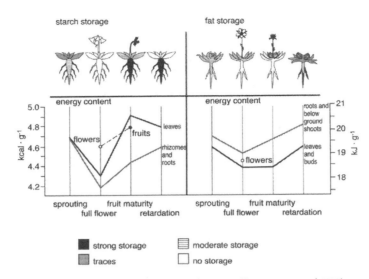

**Fig. 226.** Energy content of two nival species and the storage of reserves (from MOSER et al 1977)

teresting to note that transport of assimilates from leaves to underground storage organs in *Ranunculus* occurs even during intermittent periods of bad weather, which is reversed after the weather improves. This is important since each snow cover in the summer could potentially last until the following summer. In general, the growth period on the southern slope lasts approximately 3 months, although due to the often poor weather conditions only 60–70 (15–100) days are suitable for production. At some locations plants may not become snow-free at all during some years.

*Ranunculus glacialis* achieves half of its production during the few light and warm days and half on the many cool days with sparse light due to a low snow cover or fog. This species is capable of photosynthesis in the temperature range between –7–38 °C.

Assimilation is greatest during full flowering and fruit formation and may attain up to 0.056 g dry matter/dm² leaf surface/day in *Ranunculus glacialis* and 0.063 g in *Primula glutinosa* under optimum conditions; under unfavourable conditions these plants assimilate between 0.015 and 0.020 g. In the course of the growth period *Androsace alpina* increased its surface cover by 13.5%. The average rate of net assimilation was 0.058 g dry matter/dm² mat surface per day during this period. Because of the poor plant cover, primary production is extremely low at the ni-

No other mountain range on earth has been so thoroughly investigated as the complicated mountain system in the centre of western Europe known as the Alps.

val level. Under optimum conditions and a cover of 10%, production is estimated at 0.66 g dry matter $m^{-2}$ $day^{-1}$.

The following fundamental classical works deal with this system: C. SCHROETER, *Das Pflanzenleben der Alpen*, 2nd ed., 1288 pp., Zürich 1926; J. BRAUN-BLANQUET and H. JENNY, *Vegetationsgliederung und Bodenbildung in der Alpinen Stufe der Zentralalpen* (Denkschr. Schweiz. Naturf. Ges. 63:183–349, 1926); H. GAMS, *Von den Follatères zur Dent de Morcles* (Beitr. Geobot. Landesaufn. Schweiz 15, 760 pp., 1927); E. AICHINGER, *Vegetationskunde der Karawanken* (Pflanzensoz. vol. 2 329 pp., Jena 1933; and R. SCHARFETTER, *Das Pflanzenleben der Ostalpen* (Vienna 1938, 419 pp.).

SCHMID (1961) distinguishes the following belts corresponding to altitude:
1. *Quercus pubescens* (on limestone) and *Quercus robur-Calluna* with *Castanea* (on acidic formations) in the warm altitudinal belts,
2. *Quercus–Tilia–Acer* mixed deciduous forest in the warm and mild temperature belts,
3. *Fagus–Abies* in the cool temperature belt,
4. *Picea* conifer forest in the harsh lower cold temperature belt,
5. *Vaccinium uliginosum–Loiseleuria*, including the entire alpine or cold belt which extends to the much elevated tree line.

Added to these in the continental central Alpine valleys is a *Pulsatilla* steppe belt with *Pinus sylvestris* in the lower regions under the *Picea* belt followed by a *Larix–Pinus cembra* belt above this up to the high elevated forest limit; the dry föhn valleys are characterised by *Pinus sylvestris* with *Erica carnea*.

The network of weather stations in the Alps is more complete than in any other mountain range and includes a number of high altitude stations. H. REHDER (1966) took advantage of these for the development of a climate map of the Alps and its adjacent areas. Also of great ecological importance is the work of K. F. SCHREIBER (four maps, 1:200,000, 1977), which distinguishes 18 altitudinal temperature belts (three hot belts on the southern limits of the Alps in the Tessin and three warm, three mild, three cool, three harsh and three cold belts, followed by the alpine and the nival belts which are not further subdivided). A 1:500,000 map identifies the föhn areas of Switzerland in which the development of the vegetation may be up to 3 weeks ahead of other regions.

There are also a large number of vegetational maps available. Besides the many specialised maps there are also general maps with the most important altitudinal belts. The following concern the eastern Alps: H. MAYER, *Karte der natürlichen Wälder des Ostalpenraums* (1977) and his publication, *Wälder des Ostalpenraums* (1974); H. WAGNER, *Karte (1:1,000,000) der natürlichen Vegetation* in the Österreich Atlas (1971); P. Seibert, *Übersichtskarte der natürlichen Vegetationsgebiete von Bayern 1:500,000 mit Erläuterungen* (1968), which includes the northern limits of the Alps and thereby leads to the next section of that mountain range.

The central Alps are presented and discussed in the vegetational maps of E. SCHMID (in four parts, 1:200,000, 1961).

An especially large number of ecological vegetation maps with detailed information at a scale of 1:100,000 (to 1:10,000) for the western Alps, and to some degree for other regions, are published regularly by P. Ozenda in the *Document de Cartographie Ecologique* (Grenoble). This series also includes the altitudinal belt ascending from the Mediterranean coast in the south (therefore, within orobiome IV). Such a large and unique cartographic work is especially worthy of mention. The detailed colour vegetational maps provide exact information on the altitudinal belts and their dependence on exposure and geological formations.

In *Vegetation Südosteuropas* by I. HORVAT, V. GLAVAC and H. ELLENBERG (1974), the vegetation of the Dinarian Alps and the adjacent mountains of the Balkan Peninsula is discussed.

## 7 Zonoecotone IV/VII – Forest-Steppe

Deciduous forests of the temperate zone are confined to climatic regions of an oceanic nature where there are no sharp extremes of temperature and rainfall is more or less evenly distributed throughout the year, usually with a summer maximum. Steppe and desert occupy the continental regions which are much more extensive in the northern hemisphere. In a continental climate the temperature amplitude is greater and the summers are hotter but the winters are much colder, so that the annual mean temperature is lower than that in oceanic regions of the same latitude. This is accompanied by a decrease in annual rainfall and more arid summers.

The zonoecotone between deciduous forest and grassy steppe in eastern Europe is the forest steppe. It is not a

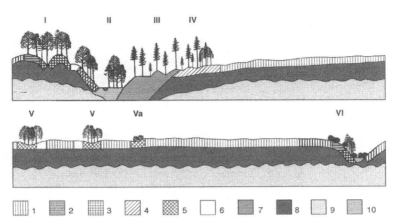

**Fig. 227.** Relation between vegetation, soil and relief in the forest-steppe (after WALTER 1990). *1* Deep, poorly drained chernozem with meadow-steppe; *2* degraded chernozem; *3* dark-grey forest soil (both well drained); *4* porous sandy-loamy forest soil; *5* light-grey forest soil; *6* solonez on flat terraces or around depressions with no outflow and with soda accumulation; *7* fluvio-glacial sands; *8* moraine deposits or loess-like loam; *9* pre-glacial strata; *10* alluvium in river valleys. **I** Oak forest on well-drained elevations or on slopes; **II** floodplain forests (oak and others); **III** pine forests on poor sands with *Sphagnum* bog in wet hollows; **IV** pine–oak forests on loamy soils; **V** aspen groves in small hollows (pods), in spring containing water that seeps away only slowly (soil in central portion is leached); **Va** the same, but with willows; **VI** ravine–oak forest, with steppe shrubs at upper margins

homogeneous vegetational formation like the tropical savanna, but rather a macromosaic of deciduous-forest
stands and meadow steppe. At first, the former predominate, the steppes forming scattered islands. However, the
more arid the climate becomes, the more the situation
tends to be reversed, until finally, small islands of forest
are left in a sea of steppe. Relief and soil texture (Fig. 227)
determine the predominating vegetation. Forests are found
on well-drained habitats, slightly raised ground, the sides
of the river valleys and porous soils, whereas meadow-
steppes occupy badly drained, flat sites with relatively
heavy soil. Grasses and tree seedlings compete with one
another and if (as happens in the course of reforestation
experiments) the young tree plants are protected for the
first couple of years from competition with roots of
grasses, they are able to grow on the steppe although they
are incapable of regeneration. In previous times, fires
caused by lightning and grazing by big-game herds encouraged the growth of the steppe. However, we can only
speculate on the real role of big game herds. The grazing
density of domestic animals (sheep, goats, cows) is prob-

**Fig. 228.** Mosaic of steppe, shrub and forest in Dobrudscha (Romania). Grazing with goats and sheep keeps larger areas as steppe and dry meadows open which are rich in species. Shrubs grow only slowly towards the outside, rejuvenation would hardly occur without grazing (Photo: S.-W. Breckle)

ably much higher than that of the original big game herds. Nevertheless, it is assumed that in some regions the flora and mosaic structure is very similar to the original (Fig. 228). Nowadays, however, the steppe is entirely given over to farming.

The forest zone, forest-steppe zone and steppe zone of eastern Europe are readily distinguishable from one another on a climatic basis. Climate diagrams for the forest zone reveal the absence of a period of drought, whereas diagrams for the steppe zone always indicate the presence of such a period. Although no drought is to be detected on the diagrams for the forest-steppe zone, a dry period is always recognisable which is not the case for the forest zone (Fig. 229, p. 372).

During the post-glacial period, the boundary between forest and steppe shifted. In the soil profile beneath the present-day forest stands it is possible to find "krotovinas" (Fig. 232, p. 375), the deserted burrows of steppe rodents (ground squirrels). Since these animals never inhabit forests it must be assumed that before the forest-steppe was inhabited by man the forests were in the process of advancing, because the climate became more humid after a warm optimum had prevailed. Later shifts in the boundaries cannot be detected because of the large degree of human interference.

Replacement of the forest zone by steppe in continental regions is governed by the supply of water. Water turnover is almost exclusively confined to the upper 2 m of the soil and no water percolates to the deep-lying groundwater. An oak forest uses up all of its water and the deeper soil is always dry. Such is the situation in the euclimatopes, but on the southern slopes with drainage and high evaporation, steppe develops because the soil water is insufficient to support forest. In August and September even the grass steppe dries out, water supplies being insufficient to cover losses due to transpiration. This is not harmful to the grasses, although the trees suffer damage if they lose their leaves too early or if entire branches die off.

Precipitation decreases and temperature increases towards the southeast in the forest-steppe, patches of forest becoming smaller and smaller and increasingly confined to the northern slopes, until finally, on the southern-most limits of the forest-steppe, only oak and sloe (*Quercus* and *Prunus spinosa*) bushes are left in the gorges. In the forest-steppe, grasses and tree seedlings compete with one another. CLEMENTS & WEAVER (1920) demonstrated this on what was, in the 1920s, still original tall grass prairie of Nebraska (Fig. 235, p. 379) corresponding to the forest steppe and showed that planted tree seedlings could only hold their own if all the grass roots were removed from the vicinity.

The amount of water required by a forest increases with its age. Experiments in reforestation have revealed that young forest plantations grow relatively well, but that with time the tips of the older shoots dry off and fresh shoots are then put out from below. Trees, therefore, develop abnormally as a result of the water shortage. If groundwater is available, however, healthy stands develop. Savanna-like communities are missing in the forest-steppe because individual deciduous species are unable to compete successfully with grasses. Only low shrubs such as *Spiraea, Caragana* and *Amygdalus* are common, although they generally occur on stony ground less suited to the dense root systems of the steppe grasses. The steppe component of the forest steppe, the meadow steppe, will be considered in the next section, which deals with zonobiome VII.

## Questions

1. What is the explanation for the wide ecological spread of beech (Fagus sylvatica)?
2. Is a long time without snow cover in the subnival region positive or negative for vegetation?
3. How do altitudinal belts of the north, central and southern Alps differ?
4. What are vicarious species?
5. For what process is the largest proportion of the incoming solar radiation used by the beech forest in Central Europe?
6. The alpine tree limit in side valleys is lower than on slopes, in valleys the polar tree limit moves further north. Why?
7. What are the significant ecological differences between the alpine and polar tree limits?
8. What percentage of the net primary production of a beech forest is eaten by herbivores?
9. Which leaves of beech possess the higher light compensation point – sun or shade?
10. Why are clonal plant species in the herbaceous layer of a beech forest advantaged?
11. What could be the ecological sense for the phenomenon of "the rich years" of many tree species in ZB VI?
12. What are the ecological advantages of old trees and old wood for a forest?
13. Which sequence is referred to with the terms plantation – managed forest – forest – virgin forest?
14. What is the explanation for the similar annual energy accumulation of a field, a meadow, a spruce forest and a beech forest?
15. What are snow valleys (schneetälchen)?

# VII Zonobiome of Steppes and Cold Deserts (Zonobiome of the Arid-Temperature Climate)

## Climate

In Eurasia, this continental zonobiome stretches from the mouth of the Danube across eastern Europe and Asia, almost to the Yellow Sea. In North America it occupies the entire Midwest, from southern Canada to the Gulf of Mexico. The degree of aridity varies considerably.

In ZB VII, the steppes, semi-deserts and desert regions with cold winters, six subzonobiomes can be distinguished:

1. semi-arid subzonobiome having a short period of drought (steppe and prairie vegetation, sZB VII).
2. arid subzonobiome with longer drought periods and winter rain [semi-desert, sZB VIIa(w)].
3. very arid subzonobiome [with a type VII climate (rIII)], i.e. with as little rain as the subtropical desert climate, but with winter rain (deserts).
4. arid sZB with longer drought periods and summer rain [semi-desert, sZB VIIa(s)].
5. very arid sZB but with summer rains (desert).
6. deserts of the cold mountainous plateaux (Tibet and Pamir) sZB VII(tIX).

The two semi-desert subzonobiomes are ecotones between steppes (Fig. 229, Tschakalow) and actual deserts. These arid ecotones of semi-deserts are characterised by the climate type VIIa (Fig. 229, Astrakhan). The semi-deserts (the sagebrush region in North America) are also more arid than the steppes, but less arid than the deserts and the vegetation is of a transitional type although there is a well-developed drought lasting about 4–6 months (Fig. 158).

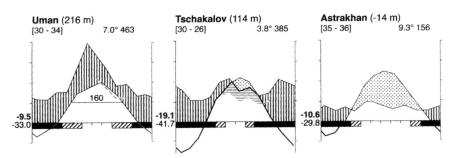

**Fig. 229.** Climate diagram from the forest steppe zone (with dry season), the steppe zone (with drought period and long dry season) and from the semi-desert (with long summer drought)

## 2 Soils of the East European Steppe Zone

The East European steppes are the cradle of the descriptive soil science, the foundations of which were laid by DOKUTCHAYEV (1898) and GLINKA (1914). There is no other region of comparable area where the parallel zonation of climate, soil type and vegetation can be seen so clearly. It must be added, however, that very little remains of the original vegetation. The conditions responsible for the clear zonation are the extreme uniformity of relief and the fact that the parent rock is to a large extent homogeneous (loess). The climate changes steadily from northwest to south-east, the summer temperatures and potential evaporation rising while the rainfall decreases, so that the aridity becomes more and more pronounced. Thus, over long distances there is an even gradient along ideal transects. The boundary between forest zone and forest-steppe zone coincides with the boundary between humid and arid regions. This means that to the north of this demarcation line the annual rainfall exceeds the potential evaporation, whereas to the south the latter is the higher of the two (Fig. 230) so that in depressions with no outflow (pods), saline soils form.

The distribution of the various soil types is depicted in a simplified form in Fig. 231. Humid regions have typical podsols and slightly podsolised grey forest soil, whereas arid regions have soils ranging from chernozem to chestnut and arid brown (burozem). The soil types are recognisable from their soil profiles which are shown in Fig. 232.

The chernozems are A–C soils, or pedocals, without a B horizon enriched in clay. The zones are subdivided as

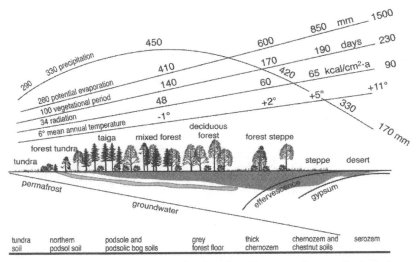

**Fig. 230.** Schematic climate, vegetation and soil profile of the eastern European lowlands from northwest to south-east (after SCHENNIKOV, from WALTER 1990). *Black* Humus horizon; *coloured* illuvial B-horizon; vegetational season in the tundra = daily mean above 0 °C, otherwise above 10 °C

follows: northern, thick, normal and southern chernozems. The humic A-horizon consists of a black $A_1$ layer, a slightly lighter $A_2$ layer and a loess layer slightly coloured by humus $A_3$. Below this is the C layer which is original unchanged prismatic loess. In the thick chernozem, the humus layer goes down to 170 cm, its thickness decreasing both to the north and south. Normal chernozem has the highest humus content, 7–8%, but in the eastern steppe regions the humus content is even higher. There is no deposition of clay in the chernozems, but in spring the downward flow of water from the melting snows carries with it calcium carbonate ($CaCO_3$) dissolved out from the upper horizons. If HCl is applied to soil from these leached upper horizons no effervescence occurs – only with soil from deeper layers is a positive reaction obtained. The more arid the climate, the nearer the effervescence level to the surface. Somewhat below this so-called effervescence horizon, the dissolved carbonates precipitate, usually in the form of very fine $CaCO_3$ threads reminiscent of mould (pseudomycelia). Further south, these carbonates also precipitate as small white nodules (bjeloglaski) and finally, only nodules are precipitated. The black humus-filled cross sections of abandoned burrows of ground squirrels (krotovinas) are recognisable in the soil

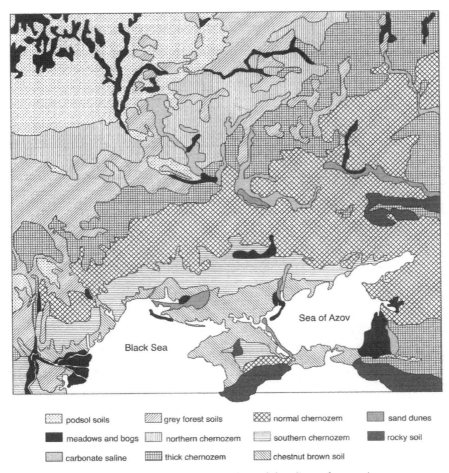

podsol soils     grey forest soils     normal chernozem     sand dunes

meadows and bogs     northern chernozem     southern chernozem     rocky soil

carbonate saline     thick chernozem     chestnut brown soil

**Fig. 231.** Soil map of the eastern European steppe region and the adjacent forest regions

Each soil type corresponds with a certain plant community.

profile. The changes in soil profile take place gradually from north to south, in conformity with the changes in climate reflecting the increasing aridity.

Beneath the forest of the forest-steppe zone, the upper soil layers remain wetter. The $A_0$ horizon is made up of litter which mixes only slightly with the mineral soil so that the humus horizon under the moist hornbeam forests (*Carpinus*) is light grey and that under the dry oak forests dark grey. The good friable structure is lost and the soil becomes laminated. Beneath the humus layer, there are

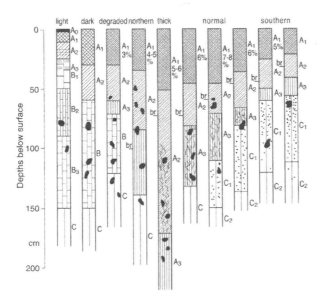

**Fig. 232.** Schematic representation of soil profiles of the forest steppe and steppe zones (west of the Dniepr), from north to south. *Percentages* Humus content of the $A_1$ horizon; *br* effervescence horizon; *wavy lines* pseudomycelia (CaCo₃); *small dots* CaCO₃ nodules; *large black spots* krotovinas (abandoned ground squirrel burrows); *horizontal dashes* laminated structure in forest soil

mealy, bleached sand grains and below this a compact B-horizon indicating the beginning of podsolisation. There is hardly a trace of this, however, in the degraded chernozem beneath the shrubby oaks which constitute the last outposts of the forest. Below the most humid parts of the meadow-steppes the soil is typical northern chernozem, with a very deep effervescence level and no CaCO₃ precipitation (see Figs. 231, 232).

On the basis of the surviving remnants of natural vegetation, it has been possible to show that every soil type has its corresponding plant community, as in the following summary:

| Soil type | Vegetational unit |
|---|---|
| Grey forest soil | Oak-hornbeam and oak-forest |
| Degraded chernozem | Oak-blackthorn (sloe) scrub (*Prunus spinosa*) |
| Northern chernozem | Damp meadow steppes with abundant herbaceous plants |
| Thick chernozem | Typical meadow steppe |
| Normal chernozem | Feather grass (*Stipa*) steppe with abundant herbaceous plants |
| Southern chernozem | Dry *Stipa* steppe, few herbaceous plants |

This scheme makes it possible to reconstruct the original vegetation using a soil map.

## Meadow Steppes on Thick Chernozem and Feather Grass Steppes

The word steppe comes from the Russian "stepj" and its use should be confined to those grass steppes of the temperate zone, such as the prairie and the pampas, that are comparable with the East European steppes. No steppes of this type occur in the tropics, where it is more appropriate to refer to "tropical grassland". The word steppe often conjures up a picture of dreary, poor vegetation although the very opposite holds true for the northern variant of the East European steppe. Nowadays these are the most fertile parts of Europe, with the best chernozem soils. In their natural condition they exceed even the lushest European meadows in the abundance of their colourful blossoms. Only in autumn do they give the impression of dryness.

The forest-steppe is a macromosaic of deciduous forest and meadow-steppe. The seasonal course of events is depicted in Figs. 233, 234.

When the snow melts the steppe soil is thoroughly wet, the temperature rises and a profusion of spring flowers develops. At the end of April the mauve blossoms of *Pulsatilla patens* appear, *Carex humilis* begins to shed its pollen and at the beginning of May they are joined by the large golden stars of *Adonis vernalis* and the pale blue inflorescences of *Hyacinthus leucophaeus*. By mid-May the steppe is verdant and *Lathyrus pannonicus*, *Iris aphylla* and *Anemone sylvestris* are in flower among the sprouting grasses. The most colourful stage is reached at the beginning of June when innumerable *Myosotis sylvatica*, *Senecio campestris* and *Ranunculus polyanthemus* are in bloom. At this point, the first plumes of *Stipa joannis* appear and, by early summer, the long feathery awns of the various *Stipa* species interspersed with the panicles of *Bromus riparius* (closely related to *B. erectus*) are swaying in wave-like motion in the wind. Intermingled with these plants are the blossoms of *Salvia pratensis* and *Tragopogon pratensis*. Toward the end of June the flowers of *Trifolium montanum*, *Chrysanthemum leucanthemum* and *Filipendula hexapetala* whiten the steppe, a colourful contrast being provided by *Campanula sibirica*, *C. persicifolia*, *Knautia arvensis* and *Echium rubrum*. At the beginning of July, when the pink flowering *Onobrychis arenaria* and the yellow *Galium verum* come into flower, the glorious colours begin to

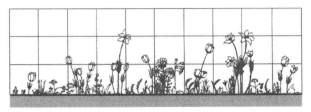

**Fig. 233.** Meadow steppe in spring (after POKROVSKAJA; from WALTER 1968). Vertical projection, quadrats in decimetres. *Top* Beginning of April, brown aspect with patches of mauve *Pulsatilla patens, Carex humilis* in pollen.
*Middle* End of April, yellow aspect due to *Adonis vernalis*, pale blue *Hyacinthus leucophaeus*. *Bottom* End of May, blue aspect due to *Myosotis sylvatica*, white *Anemone sylvestris*, yellow *Senecio campestris*, a few *Stipa* in bloom

fade. From mid-July onward plants begin to wither. Dark-blue panicles of *Delphinium litwinowi* and, later, the red-brown candles of *Veratrum nigrum* now make their appearance. From August onward, the steppes look dry and remain in this state until they are covered with snow.

This description shows that the dry meadows and steppe heath of central Europe merely represent poor extrazonal outposts, on dry shallow habitats, of the meadow steppes of humid climates. In floristic composition the two are very similar, except that in central Europe sub-Mediterranean elements (such as orchids), not found on the steppes, are also present. Further to the south of the meadow-steppes of the forest-steppe zone, the feather grass (*Stipa*) steppe appears on normal and southern chernozems. Various species of *Stipa* predominate and, faced with increasing dryness, the less drought-resistant

**Fig. 234.** Early summer aspect of the meadow-steppe (from WALTER & ALECHIN 1936). Many herbs in flower among the flowering feather grass *Stipa joannis* (those above 40 cm in height are: *Salvia pratensis, Hypochoeris maculata, Filipendula hexapetala, Scorzonera purpurea, Phlomis tuberosa* and *Echium rubrum*)

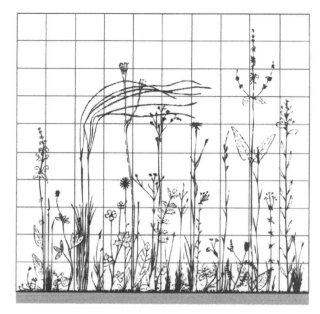

herbaceous species are incapable of successful competition and gradually recede. The density of the plant cover decreases to such an extent that the ground is covered in spring with the moss *Tortula (Syntrichia) ruralis* and the alga *Nostoc*. In spring, geophytes such as *Iris, Gagea* and *Tulipa* and some winter annuals (*Draba verna, Holosteum umbellatum*) are abundant. *Paeonia tenuifolia* is especially striking. Other herbaceous species make their appearance in summer (*Salvia nutans, S. nemorosa, Serratula, Jurinea, Phlomis* and others) and are joined later by Apiaceae (*Peucedanum, Ferula, Seseli, Falcaria*) and Compositae (*Linosyris*). Further to the south the density of the vegetation decreases still more. Apart from the feather grasses *Stipa capillata* and *Festuca sulcata*, herbaceous plants with very long taproots (*Eryngium campestre, Phlomis pungens, Centaurea, Limonium, Onosma*) are common. On the chestnut soils, the wormwood (*Artemisia*) species become more abundant and initiate the transition to wormwood semi-desert, as subzonobiome VIIa lying in between the steppe and the even more arid desert [VII (rIII)].

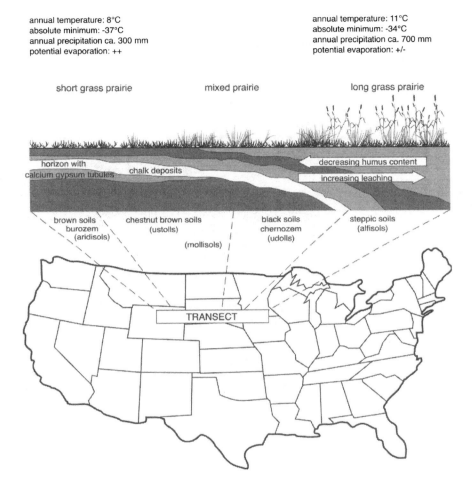

annual temperature: 8°C
absolute minimum: -37°C
annual precipitation ca. 300 mm
potential evaporation: ++

annual temperature: 11°C
absolute minimum: -34°C
annual precipitation ca. 700 mm
potential evaporation: +/-

short grass prairie          mixed prairie          long grass prairie

horizon with
calcium gypsum tubules     chalk deposits

decreasing humus content
increasing leaching

brown soils          chestnut brown soils          black soils          steppic soils
burozem                    (ustolls)                    chernozem            (alfisols)
(aridisols)                                             (udolls)
                              (mollisols)

TRANSECT

**Fig. 235.** Transect through the prairie of North America with vegetation and soil gradient (modified after BURROWS 1990)

## North American Prairie

Although the conditions prevailing on the prairies and on the steppes are very similar, the situation in the former is more complicated. Whereas the steppes stretch at a latitude of about 50° from the outposts of the Carpathians far beyond the borders of Europe to the east, the prairies,

The individual vegetational
zones, such as tall grass
prairie, mixed prairie and
short grass prairie, succeed
one another from east to
west with increasing aridity,
but within each zone there
is a floristic gradient from
north to south.

although beginning south of latitude 55° in Canada, extend in a north-southerly direction beyond latitude 30° and are succeeded by *Prosopis* savanna. Furthermore, the extensive plains of North America rise gradually from east to west to 1500 m above sea level (Fig. 235). Precipitation decreases from east to west, but temperature rises from north to south. This means that there is no clear-cut soil zonation, but rather a checkerboard arrangement of soil types (Fig. 237, p. 382).

*Andropogon* species, i.e. grasses of southern origin, are more common in the prairie than *Stipa*.

In North America, too, there is a transitional zone of forest steppe, in which the sides of the valleys and light soils are forested and the flat watersheds with heavy soils support grassland. Tall grass prairie corresponds to the northern meadow steppe on thick chernozem, but the prairie soils are wetter, the chalk is completely leached and there is no effervescence horizon. The question as to why no trees grow on the prairie has been settled experimentally by planting tree seedlings with and without the competition of grass roots. The results showed that if such competition is excluded, trees are, in fact, able to grow.

Wherever prairie fires do not occur the forest slowly encroaches upon the prairie, with a bush zone in the vanguard, at a rate of about 1 m every 3–5 years. Statistics show that, for 1965, an average of one fire caused by lightning occurred per 5000 ha of prairie and it is clear that such fires are a natural environmental factor favouring the grasses. It must also be borne in mind that, in earlier times, the prairie vegetation was much favoured by the large herds of grazing bison. A natural experiment was provided by the catastrophic drought of 1934–1941, the effects of which on the prairie vegetation were still evident in 1953. Such recurrent periods of drought every century are undoubtedly partially responsible for the absence of trees on the prairie.

Tall grass prairie is just as abundant in herbaceous plants as are the meadow-steppes and is floristically even richer. At the height of the flowering season in June, 70 species bloom simultaneously. The majority of the grasses (*Andropogon scoparius* and *A. gerardi*) are southern elements with C4 photosynthesis and do not flower until late summer; in normal years they are not troubled by the lack of water since the prairie soils are moist to a great depth. The grasses themselves are 40–100 cm tall, even 1–2 m with their inflorescences. In the mixed-prairie zone, apart from the tall grasses (*Andropogon scoparius, Stipa comata*), there are many short grasses (*Bouteloua gracilis, Buchloë dactyloides*). In the short grass prairie the latter

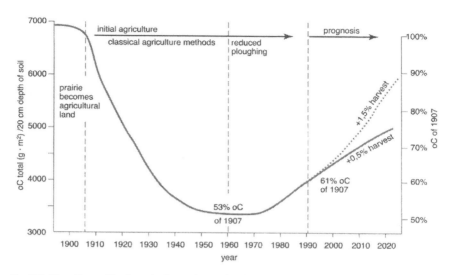

**Fig. 236.** The effects of land use in the prairie on humus content ($oC$ organic C) of the Great Plains in USA (after DONIGIAN et al. 1995)

assume the dominant role and herbaceous plants disappear. However, *Opuntia polyacantha* is abundant, particularly in overgrazed areas (KÜCHLER 1974). Grazing tends to alter the appearance of the prairie slightly in the direction of an apparently greater degree of aridity, the tall grass prairie turning into mixed prairie, which later becomes short grass prairie. The $CaCO_3$ deposits in the soil indicate an increasing aridity toward the west. In the short grass prairie, $CaCO_3$ nodules are found at a depth of only 25 cm, the humus horizon is very shallow and plant roots are shorter since they only have to penetrate a short distance into the horizon with the chalk deposits, these being an indication of the mean depth to which the soil contains moisture.

As part of the US/IBP, FRENCH (1979) published ten contributions in a volume containing ecological studies on production, consumers and grazing problems. Net annual production is approximately $2\ t\ ha^{-1}$ in the short grass prairie (stand about 65 cm high), $3\ t\ ha^{-1}$ in the mixed prairie (stand about 130 cm high) and $5\ t\ ha^{-1}$ in the tall grass prairie (stand up to 160 cm high).

The North American prairie has been made arable as have the steppes of eastern Europe. There are hardly any original prairies left. Ploughing has led to a large loss of humus, the percentage of soluble organic carbon (oC) in

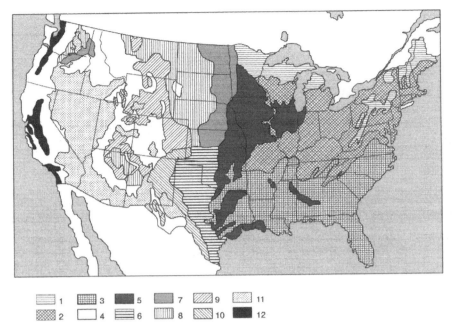

**Fig. 237.** Soil map of the United States (based on US Dept. of Agriculture map). *1* Podsol soils; *2* grey-brown forest soils; *3* yellow and red forest soils; *4* mountain soils (general); *5* prairie soils; *6* southern chernozem; *7* northern chernozem; *8* chestnut brown soils; *9* northern brown soils; *10* southern brown soils; *11* grey soils (serozems); *12* Pacific valley soils. Brown soils burozem. These soil types correspond to: 1 coniferous-forest zone, 2 and 3 mixed-forest and deciduous-forest zone, 5 tall grass prairie, 6–10 mixed and short grass prairie, 11 sagebrush (wormwood) semi-desert (in the north) and other desert types (in the south)

the soil has halved since ploughing started (Fig. 236). Not only the enormous changes in humus content caused by ploughing of the prairie into arable land, but also possible future developments aimed to secure the continuous sustainability of the land are shown (Fig. 236). Extensive research has shown how the loss of humus in the soil can be prevented by appropriate techniques, for example, by reduced intensity of cropping and by returning the litter.

Towards the north in the Canadian region and in the mountains, larger forest islands are increasing and demonstrate the transition to a forest steppe or to the forests of the taiga or montane forests. In the North American region *Populus tremuloides* is dominant. This species is able to form larger clones because of its extensive runners, one plant may even be able to form whole forests (Fig. 238).

**Fig. 238.** Small island-like groves of *Populus tremuloides* and other shrubs in the Wasatch mountains (Utah) (Photo: S.-W. BRECKLE)

**Fig. 239.** Open steppe forest with *Pinus banksiana* and *Populus tremuloides* in the Prince Albert Park (Saskatchewan, Canada) (Photo: S.-W. BRECKLE)

This species also occurs together with pines (*Pinus banksiana*) which are particularly liked by beavers for building dams (Fig. 239).

## Ecophysiology of Steppe and Prairie Species

The cold winter on the one hand and the drought of late summer on the other, limit the vegetational season of steppe plants. Only about 4 months of favourable growth conditions occur, in spring and early summer. Most of the species are hemicryptophytes and during this brief period they are obliged to build up a large productive leaf area at the smallest possible cost of material. Exact determina-

Production values for steppes: meadow steppe: phytomass, 23.7 t ha$^{-1}$ (84% underground); annual production 10.4 t ha$^{-1}$; feather grass steppe: phytomass, 20.0 t ha$^{-1}$ (91% underground); annual production 8.7 t ha$^{-1}$

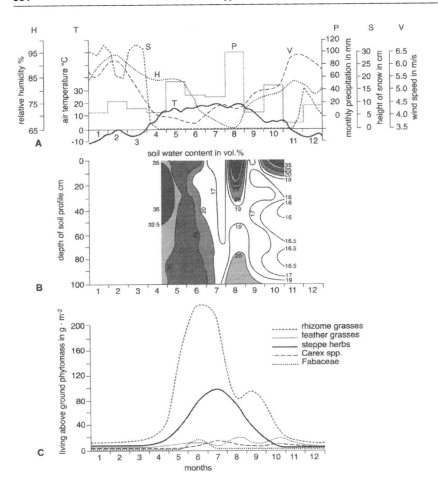

**Fig. 240.** Seasonal changes of abiotic (**A, B**) and biotic parameters (**C–H** see page 385) in (**A, B**) a meadow-steppe ecosystem in the largest central chernozem steppe reservation during 1957. **A** meteorological factors, **B** water content of soil, **C** aboveground phytomass. **D** Phenology of the vegetation; **E** dead aboveground phytomass; **F** invertebrate zoomass; **G** humus content; **H** number of rodents (predominant vertebrate group; from WALTER 1990)

tions of leaf area indices have not been made, but on the meadow-steppes the values are probably similar to those for deciduous forests. Nevertheless, the total leaf area varies greatly from year to year according to the rainfall. The figures for the above ground phytomass of the feather-grass steppe, which is poor in herbaceous plants, are

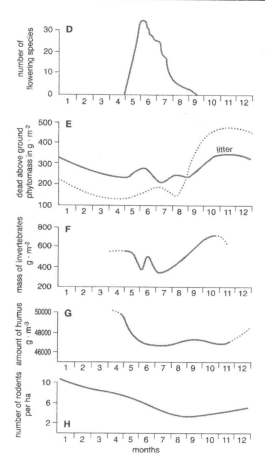

Fig. 240 D–H

4530–6250 kg ha$^{-1}$ in wet years and 710–2700 kg ha$^{-1}$ in dry years. This means that an insufficient supply of water is countered by a reduction in transpiring surface and, consequently, less productivity. The underground phytomass remains unchanged and is much larger than that above ground.

The aerial parts dying off each year form a litter layer on the ground (steppe felt) amounting to 8–10 t ha$^{-1}$ on the meadow steppe and 3 t ha$^{-1}$ on the dry steppe. When the underground parts die they are converted into humus by the soil organisms. In spring and summer the litter layer undergoes intense decomposition, with a minimum at the commencement of the drought period and a maximum at the beginning of winter. The seasonal changes of the environmental conditions and biotic parameter for the largest steppe reservation can be seen in Fig. 240.

In protected areas the accumulation of excess litter is detrimental to the regeneration of grasses. It is more difficult for grasses to rejuvenate and as a consequence the plant cover becomes patchy and weeds such as *Artemisia, Centaurea* and thistles become established. If the steppe vegetation is to be maintained in its original form, a certain amount of grazing is therefore indispensable. This was provided in earlier times by gazelles, saiga antelope, wild horses and donkeys, and above all, by the innumerable steppe rodents (ground squirrels and others) and grasshoppers. Earthworms and burrowing rodents contributed substantially to the mixing of the humus with the mineral soil. Occasional naturally occurring steppe fires led to the destruction of the accumulated litter. Nowadays, in the steppe reservations, the grass is mowed every 3 years for hay in order to reduce such accumulation. A similar state of ecological equilibrium exists between the grasses and herbaceous plants in the steppe as between woody plants and grasses in the savanna (see p. 170 f.). Grasses possess a very intensive, finely branched root system, whereas that of the herbaceous plants is extensive, often with very long taproots. Typical of the open spaces of the steppes are the so-called tumbleweeds *Eryngium, Falcaria, Seseli, Phlomis, Centaurea* and others). The rigid stem supporting the spherical dried-out inflorescence breaks off at the root collar and is rolled across the steppe in the wind, scattering its seeds as it goes. The rolling plants often get entangled with each other and form enormous masses which are driven at great speeds across the steppes by the wind.

The water relations of herbaceous steppe plants place them among the group of malakophyllic xerophytes. Their cell sap concentration is very low in spring. Temporary periods of dryness lead to wilting, accompanied by a sharp rise in cell-sap concentration. Late-flowering species that bloom when the drought commences cut down their transpiration losses by allowing their leaves to wither. The flowers and fruits require little water and receive the necessary building materials from the yellowing plant organs.

*Stipa* species regulate their transpiration by rolling up their leaves as well as closing their stomata, in this way also reducing photosynthesis. The distribution of the various species is determined by their adaptation to specific habitat conditions. The water relations of the steppe heath outposts of central Europe have been the subject of many investigations. This type of vegetation, an extrazonal relict of a xerothermic period of the post-glacial era, is confined to warm loess or calcareous slopes or to sandy soils and is made up of very hydro-labile, malakophyllic steppe species. The dryness of such biotopes in central Europe is due to the small field capacity of the soils and the high potential evaporation on southern slopes rather than to the climate itself. Instead of a long drought in the late autumn there are frequent but brief dry periods.

## Asiatic Steppes

After the interruption presented by the southern Urals, the East European steppe continues into the more continental climate of Asia, although east of Lake Baikal it is interrupted by numerous mountain ranges and is confined to the basins and wide valleys. Only in Outer Mongolia and Manchuria is the steppe once more recognisable as a distinct zone. The West Siberian steppe is similar in character to that of Europe, with certain floristic differences. *Lilium martagon* ssp. and *Hemerocallis* spp. frequently occur, but since the steppe climate east of Lake Baikal is extremely continental with very little snow in winter and a dry spring, the spring flora is missing. *Filifolium* (*Tanacetum*) *sibiricum* is well represented; its leaves turn a brilliant red in autumn. A further consequence of the extreme continental climate is seen in the large numbers of what, in western Europe, are alpine elements (species of the genera *Leontopodium, Androsace, Arenaria, Kobresia* and others; WALTER 1975 a).

Innumerable small lakes dot the very flat northern Siberian-Kasakhstan steppe; paradoxically these lakes result from the semi-arid climate. In a wet climate where every hollow overflows with water a river system can develop, whereas under semi-arid conditions the situation is different and each small hollow has its own catchment area. The hollows form wherever puddles accumulate after rain and the water seeps slowly into the ground. As a result, the soil particles become more compactly distributed and the soil volume decreases. This leads to settling of the soil surface and a deepening of the hollows. Such lake flats can be seen in other semi-arid regions from the air: in North Dakota (USA), in the pampas of Argentina and in western Australia. If such small lakes have subterranean drainage, however little, the water remains fresh, but if the water is only lost by evaporation, they become brackish (soda in slightly arid regions, otherwise chloride-sulphate).

In the eastern parts of the European steppe and in North Dakota, the edges of the small and usually round lakes are lined with aspen (*Populus tremula* resp. *P. tremuloides*) and the steppe appears to be dotted with small groves. Aspen groves of this kind (often birch in Siberia) constitute the forest element of the forest-steppe where the nemoral zone peters out in the continental climate and the steppe meets up with the boreal coniferous-forest zone (i.e. in zonoecotone VII/VIII), as in western Siberia and the Canadian steppe region.

## 7 Animal Life of the Steppe

Today, the fauna can only be observed in a few reserves and occasionally in the Siberian and Mongolian steppe, where nomads still graze their herds. Soil organisms play a very important role in the formation of the chernozems.

Just like the American prairie the steppe was originally **big-game** territory. Even in the eighteenth century, the tarpan was still extant (*Equus gmelini*), but in 1866 the last existing specimen was presented to the zoological garden in Moscow. Artiodactyls (cloven-hoofed animals) were well represented and it is assumed that the aurochs (*Bos primigenius*), which originally inhabited the steppe, gradually withdrew to the forests in the face of man's encroachment. The antelope (*Saiga tatarica*) survived longer and is even seen today in a few places (reserves near Astrakhan and elsewhere). Both red and roe deer were previously found in the forest steppe and the wild pig (*Sus scrofa*) visited the watering places and lived in reed stands. Light grazing is essential if the steppe vegetation is to be preserved but man has completely exterminated both big game and predators, leaving only the myriads of rodents.

A very important role in the formation of chernozems is played by soil organisms, above all **earthworms**. The larger of these (*Dendrobaena mariupolensis*) work through the soil in all directions and to a considerable depth. As many as 525 passages $m^{-2}$ have been counted in the uppermost metre and 110 $m^{-2}$ at a depth of 8 m. The smaller forms (*Allophora* spp.) tend to be confined to the upper soil layers. The soil is so thoroughly mixed by the earthworms that the lower layers are enriched with organic material from the surface; in addition, the earthworm passages make root penetration easier.

Next in importance are **ants,** which also stir up the soil, followed by **rodents** with underground burrows. The activities of the latter animals are well revealed by the "krotovinas", which are sections through tunnels that have been filled with humus soil from above and therefore show up in the profile as black circles in the loess soil. Rodents throw up earth from the deeper layers and 175 such heaps have been counted per hectare, occupying 0.5–2% of the area. In time, this activity leads to the formation of a microrelief with small successions of vegetation. Steppe shrubs (*Caragana frutex*, *Amygdalus nana* and others) frequently colonise the heaps because they are safe from the competition of grass roots. Rodents also help to loosen the soil.

Up to about 200 years ago the steppes were only lightly grazed since the nomads were constantly on the move with their herds and the vegetation remained almost unaltered. However, during the past two centuries the changeover to cultivation and the introduction of intensive grazing has destroyed the steppe ecosystem and, as a result, the fauna has suffered considerably. Only a few animals have adapted to the changed conditions. Rodents are a pest to farmers and insects that were harmless on the steppe now attack sugar beet and grain in large numbers and must be combated by chemical means. Cattle grazing has degraded the plant cover. The regeneration of steppe vegetation is a very slow process. It is impossible, however, to go into the question of the secondary successions here.

# Grass Steppes of the Southern Hemisphere

Compared to the area of the grass steppes of the northern hemisphere, the area occupied by those of the southern hemisphere is relatively small. The largest continuous area is the **pampas** in the eastern Argentinean province of Buenos Aires and parts of the neighbouring provinces. It may be considered a semi-arid variant of zonobiome V with relatively high precipitation during the hot summer. The pampas lies between the latitudes 32° and 28 °S and extends over about 0.5 million km$^2$, directly bordering the Atlantic coast. It is thus situated in a warm-temperate region and corresponds to the southernmost parts of the prairies of Oklahoma and Texas. Rainfall reaches 1000 mm in the north-east of the pampas and diminishes to 500 mm in the south-west, at its dry limit. Although these values appear high at first sight, it has to be remembered that temperatures and thus potential evaporation, are also very high (Buenos Aires: mean annual temperature, 16.1 °C). Despite this, the climate of the pampas has always been considered to be humid. The question has continually arisen as to why the pampas is bare of trees. The simplest explanation, as in all cases where no other explanation is available, is that the vegetation is of anthropogenic origin having arisen from an earlier forest vegetation as a result of fires set by man.

This assessment of the climate has, however, proved to be incorrect. Even in the wettest parts of the extremely flat pampas there are many shallow lakes with no outflow (locally known as lagoons) besides innumerable small pans which, although they contain water in spring, are dried out in summer. The water in the lagoons contains soda and is strongly alkaline. The soils surrounding the pans are alkaline (solonetz) and support the grass typical of saline soils (*Distichlis*). All of this points to a semi-arid climate such as has already been encountered in the forest steppe of eastern Europe. Measurements of potential evaporation have shown that, in the coastal regions of La Plata, evaporation and rainfall are equal, but evaporation exceeds rainfall in the pampas. In the more humid parts of the pampas, the negative water balance amounts to 100 mm and in its arid parts reaches 700 mm. In January and February rainfall is at a minimum and the potential evaporation is particularly high because daytime radiation is very intense and only in the evenings and at night do severe thunderstorms occur. Although well supplied with water in the spring, the vegetation is severely scorched by January.

On well-drained ground in a forest-steppe climate, a woodland vegetation is to be expected and, in fact, in the vicinity of the coast small groves of *Celtis spinosa* do occur on slight elevations with a porous limestone or sandy soil. On poorly drained ground there is grassy vegetation. Bushes grow on stony hillocks (FRANGI 1975). Almost nothing remains, however, of the original vegetation of the pampas. On the grazed areas European grasses have been introduced. They are softer than the pampas-grasses and are preferred by the European breeds of cattle. The many exotic trees which have been planted, whose roots are protected from competition by the intensive root system of the grasses, grow well everywhere.

Judging from small remnants occurring on ungrazed patches, it can be concluded that on the humid northeastern portion of the pampas, *Stipa-Bothriochloa laguroides* steppe composed of about 23 graminids and 46 herbaceous species originally prevailed (LEWIS & COLLANTES 1975). The soil profile beneath such a steppe has a thick humus horizon (1.5 m) and is reminiscent of thick chernozems or prairie soils. There are signs, however, of alternate high and low water content and the soils are a transitional form leading on to the subtropical grassland soils of southern Brazil. There is no indication that forests existed here previously. Where the groundwater table is high

**Fig. 241.** Southern tussock pampas with *Stipa brachychaeta* (central province of Buenos Aires) (Photo: E. WALTER)

there are stands of the dense tussocks of *Paspalum quadrifarium* which, with a very high groundwater table, are replaced by *Distichlis* on alkaline soils (pH = 8–9).

The dry south-western pampas was previously tussock grassland with *Stipa brachychaeta* and *Stipa trichotoma* and was almost entirely lacking in herbaceous plants. Tussock indicates a growth form which, although completely lacking in the northern hemisphere, is widespread in the southern hemisphere, with its mild winters. A tussock consists of bunch-like tufts, sometimes more than 1 m high, in which the hard, old, withered leaves are intermingled with the fresh, young, green leaves, thus providing the tussock grassland with its perpetual yellowish colour (Fig. 241). These grasses are of little grazing value and for this reason they are often ploughed under to give the European grasses a better chance to establish themselves.

Toward the west, where the rainfall has decreased to 500 mm annually falling mainly in the summer and the loess soil is replaced by light sandy soil, the pampas is replaced by xerophytic *Prosopis caldenia* woodland. As the rainfall decreases even further, the woodlands are succeeded by *Prosopis* savanna (Fig. 242) which strongly recalls the *Acacia* savanna of south-west Africa (zonoecotone II/III). At the same time, large stretches of saline soil are found bearing a halophytic vegetation. At a rainfall of less than 200 mm annually there is a **Larrea semi-desert** (see pp. 231, 392 f.) on stony ground, with many broomlike bushes belonging to various families (Caesalpiniaceae, Scrophulariaceae, Capparaceae, Asteraceae). With such a

**Fig. 242.** Tree savanna with *Prosopis caldenia* and a grass cover of *Stipa tenuissima* and *S. gynerioides* between Santa Rosa and Victoria (Argentina) (Photo: E. WALTER)

**Fig. 243.** Patagonian semi-desert with cushions of *Chuquiraga aurea* near Manuel Choique, province of Rio Negro (Photo: E. WALTER)

small transpiring surface and by drastically cutting down transpiration during the 6 months of drought, the semi-desert vegetation is able to survive with the meagre amount of soil water at its disposal. This amounts to 50–80 mm annually on flat ground, 25–55 mm on sloping ground and more than 140 mm in small valleys into which water drains.

The *Larrea* semi-desert extends along the eastern foot of the Andes to the northern part of Patagonia, where south of 40 °S strong west winds blow continuously across the Andes, which at this point are rather low (pass altitude about 1000 m). The wind is, however, of a föhn character, descending and dry. The eastern margins of the mountains have a rainfall of 4000 mm annually and support *Nothofagus* forests and are succeeded to the east by dry *Austrocedrus* forest and, following on these, a bushland of beautiful red-blossomed Proteaceae *Embothrium coccineum*. The woody plants then disappear and the Patagonian steppe commences. Only 100 km from the Andes the annual rainfall is 300 mm and, even farther away, diminishes to 160 mm. Apart from the true steppe on the westernmost margins of Patagonia where low tussock grasses are predominant (*Stipa* and *Festuca*), it is more correct to speak of the **Patagonian semi-desert** which is characterised by xerophytic cushion plants belonging to completely different families (Asteraceae, Apiaceae, Verbenaceae, Rubiaceae and others; Fig. 243). The ground is in many places 60–70% bare. The cushion-like form appears to be an adaptation to the constant strong wind (mean velocity 4–5 m s$^{-1}$); within the cushions a propitious microclimate can be achieved which is protected from the effects of the wind (HAGER 1987).

The Patagonian tussock grassland has much in common with that of Otago on the South Island of New Zealand, situated in the lee of the New Zealand Alps with a rainfall of 300 mm. Both lie south of latitude 40 °S. Low tussock grasses (*Festuca nova-zelandiae, Poa caespitosa*) predominate, but at an altitude of 750–2000 m where snow remains for 2–3 months of the year, they are replaced by taller tussock grasses, 1.5–2 m high (*Chionochloa* = *Danthonia*). Fire and grazing are partially responsible for the fact that tussock grassland has spread widely in places at the expense of the original *Nothofagus* forests. So far, no ecophysiological investigations of these grasslands have been undertaken.

## Subzonobiome of Semi-Deserts

Semi-desert is distinguishable from true desert by its diffuse vegetation although the ground is only covered to about 25%. In the true desert the density of the vegetation is still lower and at the same time a change from a diffuse to a contracted vegetation takes place. The plant cover of semi-deserts differs greatly. In the frost-free subtropics and tropics, plant cover consists mainly of woody plants and succulents and in the temperate zone, with cold winters, mainly of half-shrubs, especially the genus *Artemisia*.

**Fig. 244.** Dwarf shrub semi-desert covered with the wool-like grey *Ceratoides lanata* in the Rush valley near Toele (Utah, USA) (Photo: S.-W. Breckle)

This holds true for the semi-deserts of Eurasia as well as North America. Various species may dominate such semi-deserts, for example the genus *Atriplex* (*A. confertifolia*) in Utah and also *Ceratoides* (= *Eurotia; C. lanata*) in the Midwest of the USA (Fig. 244), in central Asia *C. latens* and other species. The characteristic cushion plants of windy Patagonia have already been mentioned.

Saline soils are widespread, as would be expected from the greater aridity of the semi-desert. This is particularly marked in eastern Europe where the broad expanses of the "sivash" to the north of the Crimea dry out in summer and are covered by a salt crust. The salt dust is blown north by the wind and deposited in the southern chernozem and **chestnut soil zones.** This input of Na salts causes **solonization** of the soil. In spring, the salt is washed out of the upper soil layers by water from the melting snows and the sodium-humus sol thus formed carries the sesquioxides ($Fe_2O_3$, $Al_2O_3$) along with it into the deeper soil layers. Here, precipitation occurs and a compact B-horizon of strongly alkaline reaction is formed (Fig. 245). The amount of salt deposited increases steadily toward the south. Humic material is entirely leached out of the A-horizon and the strongly alkaline B-horizon becomes harder and harder and, owing to the alternate swelling in the humid season and shrinkage in summer, assumes a columnar structure. This so-called **columnar solonetz** resembles the podsols in certain respects, although the latter are strongly acid in reaction, their acidification being effected by H-ions. Beneath the B-horizon of the solonetz soils, the very slightly soluble $CaCO_3$ precipitates as chalk nodules, followed by gypsum ($CaSO_4$) as tubular or druse-

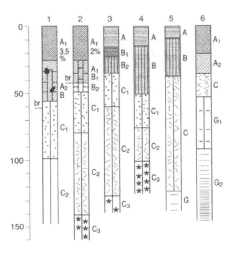

**Fig. 245.** Soil profiles in eastern Europe, weakly to strongly saline: *1* Slightly solonised southern chernozem, some compaction ($A_2B$); *2* dark-chestnut soil with B-horizon; *3* light-chestnut soil, strongly solonised (*A* poor in humus and laminated, *B* columnar and very compact); *4* typical columnar-solonetz soil; *5* solonetz changed by rising groundwater; *6* typical solonchak with high groundwater and $A_1$ rich in humus. Chalk nodules $C_1$, gypsum tubules $C_2$ in *2–4* and *C* in *5–6*, gypsum druses $C_3$, gley horizon (groundwater) *G*, $G_1$ and $G_2$

like deposits and the readily soluble salts are washed down into the groundwater.

If the groundwater rises, as is happening on the slowly sinking north coast of the Black Sea, a wet saline soil known as **solonchak** is formed. The groundwater is drawn to the surface by capillary forces and evaporates. This results in a horizon containing gypsum tubules above the gley horizon, followed by the humus horizon which bears a white salt crust in the dry season. Humic sols are not formed since the humus is precipitated in the presence of such a high salt concentration.

The steppe grasses recede on solonetz soils, to be replaced by *Artemisia maritima-salina* and *A. pauciflora*, as well as species of the genera *Camphorosma, Limonium, Kochia, Petrosimonia* and others (in North America, *Ceratoides lanata, Atriplex confertifolia, Kochia* spp. and others). Ground lichens (*Aspicilia*) and species of liverworts (*Riccia*) and *Nostoc* are also found.

On slightly elevated, non-saline ground, a **semi-desert arid brown soil** or **burozem** is formed. Its upper horizons contain only 2–3% humus and are brown in colour. The effervescence level is at a depth of 25 cm and plant cover amounts to less than 50%. The vegetation consists of *Festuca sulcata* and the low half-shrubs *Pyrethrum achilleifolium, Kochia prostrata* and *Artemisia maritima-incana*, which avoids saline soils. Only solitary individuals of *Stipa* species are seen, but in spring many ephemerals appear. In the Caspian lowlands the two associations often form a mosaic on the burozem and solonetz soils because of the nature of the microrelief. *Salicornia* and *Halocnemum* predominate on very wet solonchak and *Suaeda, Obione, Petrosimonia, Limonium caspica, Atriplex verrucifera* and others where it is less wet (see also LEVINA 1964; WALTER & BOX 1983).

After the Caspian Sea receded from the delta region of the Volga-Ural River system, the southern part of the Caspian lowlands was left covered with alluvial sand upon which *Artemisia maritima-incana, Agropyron cristatum, Festuca sulcata, Koeleria glauca* and others grew. However, the vegetation was destroyed by grazing, the sand became mobile and large bare wandering dunes or barchans, were formed. Now, whenever the sand becomes more stationary, a pioneer vegetation consisting of *Elymus giganteus* and *Agriophyllum arenarium*, a chenopod, can gain a foothold, followed by species of *Salsola* and *Corispermum*. In the dune valleys, *Aristida pennata* and *Artemisia scoparia* and others make their appearance and gradually the zonal vegetation is restored.

Sand dunes, particularly those devoid of vegetation, store water. Groundwater is always present beneath the dunes and this leads to the formation of small freshwater ponds in the dune valleys, around which *Elaeagnus angustifolia*, willows and poplars grow. Attempts to get willow (*Salix acuminata*) and poplar to grow on the sandy areas were initially successful. The plants developed well for the first 4 years at the expense of the water stored in the ground, but when this was exhausted they died. In Kazakhstan there are large areas of semi-desert between the southern Siberian steppe in the north and the desert in the south. It is comparable to the sagebrush zone of North America, with *Artemisia tridentata*.

A distinction is made between the middle Asiatic and the central Asiatic desert (Fig. 246). The former comprises the Irano-Turanian desert region occupying the southern portion of the Aralo-Caspian lowlands and the southern part of Kazakhstan, including Dsungaria. The central Asiatic desert comprises part of Dsungaria, the Gobi Desert, the western part of Ordos on the great bend of the Hwang-Ho, Ala-Shan, Bei-Shan and the Tarim basin (Kaschgaria), together with the Takla-Makan Desert and the more elevated Tsaidam basin.

## Subzonobiome of Middle Asiatic Deserts

This region lies in Iran north of the limit of date cultivation and is characterised by regular frosts.

The Tsaidam basin (Fig. 247) is succeeded by the high mountain desert of Tibet together with the Pamir in the extreme west.

In middle Asia, cyclonic rain is still received from the Atlantic Ocean, falling in the winter in the southern parts and mostly in the spring and summer in the north; in any case, the soil here is always wet in spring after the snows have melted. Rainfall diminishes from west to east. Floristically, the Irano-Turanian element is strongly represented. In contrast to the situation in middle Asia, the source of

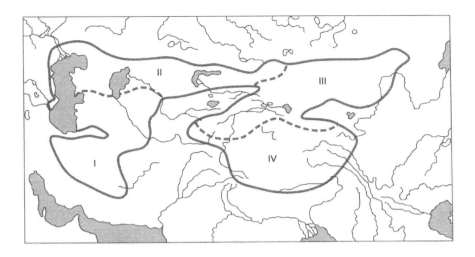

**Fig. 246.** Asiatic deserts of the temperate climatic zone (from WALTER 1990). Middle Asiatic deserts: *I* Irano-Turanian (in parts almost subtropical) and *II* Kazakhstan-Dsungarian deserts. Central Asiatic deserts: *III* in strict sense (hot summer) and *IV* Tibetan cold high-mountainous desert

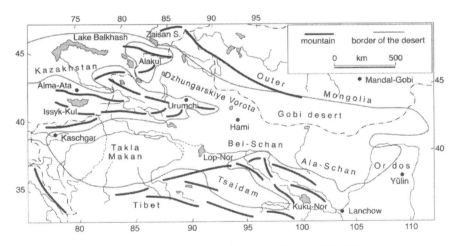

**Fig. 247.** Divisions of the central Asiatic desert regions (after WALTER 1990). Dzhungaria is a transitional region to middle Asia

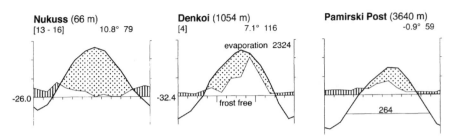

**Fig. 248.** Climate diagrams of Nukuss in middle Asia with winter rain, Denkoi in central Asia with summer rain and Pamirski Post (Murgab) in the cold desert (only 264 days with a mean temperature above −10 °C)

the moisture in central Asia is to be found in the extensions of the eastern Asiatic summer monsoons. Winter and spring are extremely dry and the aberrant rainfall distribution accounts for the predominance of East Chinese-Mongolian elements in the flora (see Fig. 248, Denkoi).

In these regions, the most detailed ecological investigations have been carried out on the vegetation of the Middle Asiatic desert in the Aralo-Caspian lowlands (formerly Turkestan). The rainfall in the entire region amounts to less than 250 mm. Since the winters are cold, evaporation at this time of year is very low. This is the reason why the annual evaporation at the Bay of Bogaz is only 1100 mm. The various types of vegetation are determined by the soils (biogeocene complexes):

The **ephemeral desert** is found on loess-like, salt-free soils that are very wet in spring, but dry from May onward. During the brief period of vegetation, lasting from the beginning of March until May, annual species and geophytes develop, the most common of which are *Carex hostii* (*C. stenophylla*) and *Poa bulbosa*. Here and there the 2-m-high *Ferula foetida* is encountered. The 40–50 annual species manage to produce ripe seeds within 30–45 days. In years with a good rainfall the desert presents the appearance of a meadow, producing a dry mass of 0.5–2.5 t ha$^{-1}$. It provides grazing for 3 months, but is completely lifeless for the rest of the year.

The **gypsum desert** is a stony desert (hamada) on the high plateaux of table mountains (mesas). The soil contains up to 50% gypsum which has the property of storing water. The situation is similar to that in the Sahara. Therophytes develop in the spring, but otherwise gypsum plants provide a ground cover of about 0.1%, except in the erosion gullies where plants are more abundant (Fig. 249). There are also a few halophytes.

**Fig. 249.** Desert almost bare of vegetation with hard cemented soils above gypsum layers in the Dasht-e-Margo in southern Afghanistan with small *Stipagrostis* clusters and dwarf shrubs of *Suaeda*, with fine material accumulating in their lee side (Photo: S.-W. BRECKLE)

**Fig. 250.** Sandy, loamy alluvial area in the semi-desert in Kazakhstan with dense growth of strong *Saxaul* bushes and small trees (*Haloxylon aphyllum*) (Photo: S.-W. BRECKLE)

The **halophyte desert** is found more extensively on soils with groundwater close to the surface, in the lower reaches of rivers, in hollows (shory), or around salt lakes. Most of the plants are hygrohalophytes (*Salicornia, Halocnemum, Haloxylon, Seidlitzia* and others; Fig. 250).

**Takyrs** are seemingly bare, clayey, flat expanses which are flooded in spring by surface water running down from the mountains, but which soon dry out again (Fig. 251). The shallow pools left behind warm up rapidly and harbour 92 species of Cyanophyta, 38 of Chlorophyta and other algae, producing in all 0.5 t ha$^{-1}$ dry matter with an N content of 4.5% (N fixed from the air). Lichens (*Diploschistes* and others) and (rarely) mosses colonise slightly higher areas. Flowering plants are rare.

**Fig. 251.** Dry takyr areas, with the ground split up into polygons. A *Tamarix* bush has established itself on drifted sand (Photo: P. Hanelt)

**Sand deserts** are particularly widespread: Karakum (black sand) between the Caspian and the Amu-Darya River and Kysylkum (red sand) between the Amu-Darya and Syr-Darya Rivers. The sandy soil favours the growth of a denser vegetation. Extensive ecological investigations carried out at the desert station Repetek since 1912 provide a basis for the following discussion of the region.

## 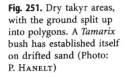 Karakum Desert

The Karakum Desert covers an area of 350,000 km$^2$ and occupies the southern part of the Turanian depression between the Caspian Sea in the west and the Amu-Darya River in the east. The latter arises at 5000 m above sea level in the mountains of Pamir and flows for 2600 km (Fig. 252).

This sand desert is a geographically well-delineated biome of the temperate-desert subzonobiome of zonobiome VII. It occupies a large basin that has been in the process of being filled with loose alluvial rocks by the Amu-Darya since the Tertiary. In the course of redeposition by wind, the dust particles have been deposited as loess on the Kopetdag slopes in the south and the sand has formed a dune region (Fig. 253).

The Amu-Darya originally emptied into the Caspian Sea, but was displaced to the east by the delta deposits of the Murgab and Tedzhen Rivers coming from the south so that it now empties into the Aral Sea. However, the river is still important to the region since its water infiltrates to feed a groundwater lake underlying the entire central Karakum with a slight inclination towards the Caspian Sea.

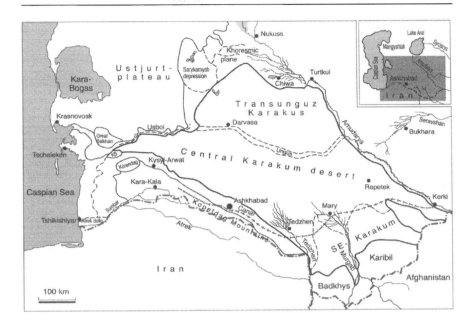

**Fig. 252.** Map of the Karakum Desert (*thickly outlined*). *KB* Lower Balkhan (from WALTER 1976)

**Fig. 253.** Karakum Desert with slightly raised dune relief and sparse growth of shrubs (*Haloxylon persicum* and others). Ground only covered in spring with ephemerals and ephemeroids (Photo: PETROV)

Salt pans form in the few places where this groundwater comes to the surface.

The water budget for the biome as a whole is given below:

| | |
|---|---|
| Groundwater infiltration from the Amu-Darya (mean) | 150 m³/s |
| Rainwater seepage in the barchan region | 30 m³/s |
| Infiltration from Murgab and Tedzhen | 21 m³/s |
| Subterranean inflow from Kopetdag (from the south) | 20 m³/s |
| Seepage from elevations and takyrs | 1 m³/s |
| Total inflow | 222 m³/s |
| Approximate losses are as follows: | |
| Evaporation from salt pans above high groundwater table (average) | 165 m³/s |
| Groundwater losses due to phreatophytes (i.e. plants dependent on groundwater) | 57 m³/s |
| Total losses | 222 m³/s |

Although the groundwater moves slowly from east to west, it has been proved that none of it flows into the Caspian Sea. The water is slightly brackish, although it is covered by a lens of fresh water beneath the bare dunes owing to rainwater seepage even in years when as little as 100 mm of rainfall is recorded. Wells sunk in such places yield good drinking water.

The climate of the Karakum can be seen on the climate diagram for Nukuss (Fig. 248). Fifty to 70% of the precipitation falls in spring which is the best season. The winter is cold, but no lasting snow cover forms. In the hot summer potential evaporation amounts to 1500–2000 mm, or 10–20 times the precipitation.

The ecological studies that have been carried out for many years in the protected area in the vicinity of the desert station Repetek will be very briefly summarised. The 34,000-ha area consists of 14,000 ha of bare barchans, 18,000 ha of plant-covered dunes and 2000 ha of dune valleys. The tree shrubs *Haloxylon persicum* and *H. ammodendron = aphyllum* are characteristic of the sandy desert.

Within the psammophyte complex, several distinct biogeocenes can be recognised:

Differences in vegetation and soil provide the basis for a distinction between the following biogeocene complexes in the Karakum: (1) psammophyte complex, covering 80% of the area, (2) takyr complex and (3) halophyte complex.

1. the biogeocene of Ammodendretum conollyi aristidosum on light, shifting sand, with a pioneer synusia of *Aristida karelinii* on the dune ridges and shrubs (*Ammodendron conollyi, Calligonum arborescens, Eremosparton* and others) on the upper slopes,
2. the biogeocene of Haloxyletum persici caricosum on stationary sand of the lower slopes and on sandy areas, with synusiae of spring and summer ephemerals and ephemeroids (143 species, 24 of which are common) and
3. the biogeocene of the deep dune valleys with groundwater at a depth of 5–8 m, which can be reached by the roots of the salt-tolerant *Haloxylon ammodendron*, a

tree that attains a height of 5–9 m and forms small woods. Here, too, a number of synusiae can be recognised in the undergrowth (ephemerals, halophytes).

The most widespread biogeocene is Haloxyletum persici caricosum (Fig. 253) with an open shrub layer 3–5 m tall (100–300 plants ha$^{-1}$), where aphyllic *Calligonum* species (Polygonaceae) are found. The age composition of the shrubs is as follows:

| Age in years | 1 | 2 | 3–5 | 6–10 | 11–15 | 16–20 | >20 | Dead |
|---|---|---|---|---|---|---|---|---|
| Percentage | 8 | 3 | 1 | 14 | 20 | 41 | 11 | 2 |

Ecological studies have revealed that the shrubs remain active throughout the year since the upper 2 m of sand always contains available water. The osmotic potential drops slightly during the drought, but water deficits are never very large; the otherwise very intense transpiration is reduced to half or a third in the drought period. Photosynthesis continues uninterrupted throughout the summer and is particularly intense in the ephemerals, which have only a short vegetational season.

In the Ammodendretum biogeocene on shifting sand, 83 mm of water was found in the upper 2 m of sand, although at the end of the vegetational season only 34 mm was left. This is a difference of 49 mm, of which 37 mm was used by the plants for transpiration and apparently only 12 mm evaporated.

Conditions on the fixed sand with denser vegetation are less favourable. It was found that 30 mm of water was transpired by the shrubs (*H. persicum* alone, 16 mm) and 17 mm by the dense undergrowth of *Carex physodes*, which gives a total of 47 mm. Water in the soil amounted to 62 mm in spring and 8 mm in autumn, which means that 54 mm was lost (thus 7 mm evaporated) in all.

In wooded, deep dune valleys situated above groundwater, transpiration amounted to a total of 149 mm (*H. ammodendron* 108 mm, other shrubs 30 mm, thin carpet of *Carex physodes* 11 mm), although the soil contained only 76 mm. Assuming that *H. ammodendron* takes the 108 mm that it requires from the groundwater, the 41 mm needed by the rest of the plants can be supplied by the water in the upper 2 m. In the case of *H. ammodendron*, 14% of the precipitation runs down its trunk, which makes it easier for the taproots to penetrate the soil to greater depths in the rainy years. Thus, despite the very high intensity of transpiration of Karakum shrubs per gram of fresh weight, the total transpiration per unit area is low owing to the small leaf area. The main adaptation of the plants is their aphylly and their ability, during drought, to shed any small leaves that may have formed in spring.

The aboveground phytomass was as follows: on shifting sand 80 kg ha$^{-1}$ (25% *Calligonum*, 12% *Aristida*) and on sand covered with vegetation, 2.4 t ha$^{-1}$ (85% *Haloxylon persicum*, 10% *Carex physodes*). The subterranean phytomass is much larger, but was only determined in the dune valleys with *Haloxylon ammodendron*, aboveground 6.4 t ha$^{-1}$ (82% *Haloxylon*) and subterranean, 19.4 t ha$^{-1}$ (49% *Haloxylon*). The annual primary production for *Haloxylon ammodendron* amounted to 1.17 t ha$^{-1}$ aboveground and 2.11 t ha$^{-1}$ belowground. These are high production figures for a desert and are only possible because of the groundwater.

The takyr biogeocene complex is found where water of episodic rivers flooding large plains has deposited a layer of clay, above which the water stands until it evaporates. Algal masses, chiefly cyanophytes (N-fixing as well), develop on the water and an algal skin is left when the water evaporates. Cracks formed on the dry ground rapidly close again when rainwater swells the soil. Lichens develop in places that are wet, but not flooded and ephemerals can be found further up the slopes. The phytomass (= primary production) in the three biogeocenes is as follows: algae = 0.1 t ha$^{-1}$, lichens = 0.3 t ha$^{-1}$ and ephemerals = 1.2–1.6 t ha$^{-1}$.

The halophyte biogeocene complex is found where salt pans have formed as a result of a high groundwater table. The biogeocenes, often consisting of only one species, are arranged concentrically around the central portion which has a salt crust. For the pioneer species *Halocnemum strobilaceum* a phytomass of 1.76 t ha$^{-1}$ (roots accounted for 1.04 t ha$^{-1}$) and an annual production of 0.5–0.7 t ha$^{-1}$ were found.

The delta region of the Amu-Darya supports luxuriant *Populus-Halimodendron* forests rich in lianas, as well as large expanses of reeds (*Phragmites*). The following figures were obtained for aboveground phytomass and primary production: floodplain forest 77.8 t ha$^{-1}$ aboveground phytomass and 11.4 t ha$^{-1}$ primary production; reeds 35 t ha$^{-1}$ phytomass and the extremely high figure of 18 t ha$^{-1}$ for primary production. Today, these stands have lost the groundwater because the Aral Sea is drying out (BRECKLE et al. 1997; KLÖTZLI 1997). These flood plain forests (tugai) which were once rich in species are nowadays almost completely destroyed.

Animals play a very important role. The herds of antelopes, donkeys and wild horses have been replaced by millions of karakul sheep which graze all year-round on the sandy desert and yield the Persian lamb skins. The animals prevent the sandy expanses from becoming overgrown with *Carex physodes* and the moss *Tortula desertorum*. At the same time, the sheep tread seeds from the shrubs into the soft ground facilitating germination. The soil is churned up by innumerable rodents and giant tortoises (100/ha) that feed on the ephemerals for 10 weeks to 4 months and spend the rest of the year asleep in the ground. However, the zoomass of all three biogeocenes is, as usual, not large: mammals $0.3–1.4$ kg ha$^{-1}$, birds $0.02–0.07$ kg ha$^{-1}$, reptiles $0.21–0.7$ kg ha$^{-1}$ (excluding tortoises), invertebrates maximum $15$ kg ha$^{-1}$ (above- and belowground).

Cycling of mineral elements was also studied, but the decomposers were only dealt with summarily. A detailed description of the Karakum Desert may be found in WALTER (1976) and WALTER & BOX (1983).

## Orobiome VII (rIII) in Middle Asia

The altitudinal belts of orobiome VII are particularly interesting in middle Asia, where the orobiome falls within zonobiome VII (rIII). The mountains, which rise to over 7000 m above sea level, are part of the Pamiro-Alai and Tyanshan systems. Almost every sequence of belts here exhibits peculiarities, depending upon the nature of the local ascending air masses. Nevertheless, two main types of sequences can be distinguished: (1) an arid sequence with no forest belt and (2) a more humid sequence with one to two forest belts (STANJUKOVITSCH 1973).

In the most extreme cases in central Tyanshan, the semi-desert belt is followed by a mountain steppe belt from 2000–2900 m above sea level with an admixture of alpine elements such as *Leontopodium alpinum, Polygonum viviparum, Thalictrum alpinum* and others from 2600 m upward in the subalpine belt. Steppe elements also extend into the lower alpine belt (*Kobresia* swards), up to 3500 m, above which they disappear. The transition from high-montane to subalpine and alpine belts is a gradual one. The explanation for this remarkable mixture of steppe and alpine elements is that the steppe plants require 4 months of favourable conditions for vegetation and, on account of the intense radiation, these are found up to 3500 m above sea level in the arid mountainous climate. The remaining 8 months of the year may be arid or cold. The alpine elements, however, can manage with a

**Fig. 254.** Steppe forest in the Transilii-Alatau south of the Almaty (Kazakhstan) near Medeo at 2000 m above sea level, exclusively with *Picea schrenkiana*, lower belt with Malus (Photo: S.-W. BRECKLE)

shorter period of vegetation, but are able to go on growing as long as conditions are moist enough for them to compete successfully with the steppe plants (WALTER 1975a).

In less extreme situations, on the northern slopes in subalpine belts, stands of *Juniperus* trees occur. These take on an espalier form on the ground in the alpine belt proper.

Of special interest are humid belts above the semi-desert, with xerophilic trees (*Pistacia, Crataegus*), and mountainous steppe are succeeded by a **deciduous tree belt with wild fruit trees** such as *Amygdalus communis, Juglans regia, Malus sieversii* with large fruits and species of *Prunus* and *Pyrus*. The capital of Kazakhstan is Almaty (formerly Alma-Ata) which means apple father, because the *Malus* belt is very well developed above the city, in the Transili-Alatau. When the fruit is ripe in the *Malus* belt the entire population gathered around camp fires to preserve apples either as stewed fruit or dried apples. The deciduous belt is followed by a coniferous belt (Fig. 254) consisting of *Picea schrenkiana* and seldom *Abies semenovii* in places. This is succeeded by the alpine belt.

**Afghanistan**, with the Hindu Kush Mountains and the influence of monsoon winds in the east (FREITAG 1971b; BRECKLE 1971, 1973, 1974, 1983), provides another example of similar vegetation. The biodiversity and number of endemic plants is much larger in Hindu Kush than in Alatau. Nuristan in the east of Afghanistan is an example of an area which is particularly rich in number of species (Fig. 255) and in the eastern part floristic elements from the Himalayan region become evident. The peaks of the Hindu Kush rise up to above 7000 m and even above 5000 m about 40 species of higher plants can be found

**Fig. 255.** Steppe forest in the deep valleys of the eastern Hindu Kush (upper Nuristan valley) with *Pinus gerardiana* on the slopes and shrubs in the valley (Photo: S.-W. BRECKLE)

**Fig. 256.** Mountain peak (Kohe-Baba Tangi, 6800 m above sea level) in the north-eastern Hindu Kush (Wakhan, Afghanistan) with glacial slopes and stony area with open alpine nival vegetation above 5000 m above sea level (Photo: S.-W. BRECKLE)

(BRECKLE 1971, 1973, 1974). Despite the summer drought the alpine nival belt profits from the large snow and ice reserves from the glaciers (Fig. 256).

The succession of altitudinal belts in the Front Range of the Rocky Mountains near Colorado Springs provides an example of an orobiome in the semi-arid climate region VII of North America: the short grass prairie at the foot of the mountains is succeeded at 1500 m above sea level by a belt with long grass prairie and then by a belt only 50 m wide of deciduous shrubs, with *Pinus edulis* and *Juniperus* (pinyon belt). This is followed by the forest belts, in which *Pinus ponderosa, Pseudotsuga menziesii* and *Picea engelmanii,* successively achieve dominance. The tree line is reached at 3700 m above sea level, followed by a narrow subalpine belt with dwarf *Picea* and *Dasiphora (Potentilla) fruticosa* bushes and finally the alpine belt.

## Subzonobiome of the Central Asiatic Deserts

As already mentioned, the last traces of the Chinese monsoons or low pressure weather front coming from the east are still noticeable in this region, which explains why rains fall in summer and rainfall diminishes from east (Ordos 250 mm) to west (Lop-Nor depression 11 mm). Winter and spring are dry and the spring ephemerals, so typical of middle Asia, are entirely absent in central Asia. The flora is poor, with shrubby psammophytes (*Caragana, Hedysarum, Artemisia* and others) predominating among the East Chinese-Mongolian elements. *Stipa*, too, is represented by central Asiatic species. Buckthorn (*Hippophaë rhamnoides*) and the tall grass chii (*Lasiagrostis splendens*) are widespread. Apart from *Populus diversifolia* and *Elaeagnus*, the floodplain forests contain *Ulmus pumila*. Among the halophytes, *Nitraria schoberi* and species of *Zygophyllum, Reaumuria, Kalidium* and *Lycium* deserve mention. The character of the deserts is influenced by their geological structure and the nature of the rock. The geographical position of the deserts is shown in Fig. 247.

1. **Ordos.** This is the region lying in the bend of the Hwang-Ho to the north of the Great Wall of China, which runs along the edge of the wandering dune region. It joins up with the steppe region of the loess plains of the upper Hwang-Ho, nowadays cultivated and very deeply dissected by erosion gullies. The Ordos is a *Stipa* steppe differing greatly, however, from that of eastern Europe because of the dry spring. The underlying rock of the true Ordos region is soft sandstone which has given rise to large expanses of sand and dunes with widespread *Artemisia ordosica* semi-desert vegetation with *Pycnostelma* (Asclepiadaceae) (cover, 30–40%). In the central undrained parts, there are lakes containing $Na_2CO_3$ and NaCl.

2. **Ala-Shan.** This is a desert consisting largely of sandy wastes with barchans. It lies to the west of the Hwang-Ho and stretches as far as the Nan-Shan Mountains in the south. To the north it borders the Gobi Desert near the Gushun-Nor. Rainfall decreases from 219 mm in the east to 68 mm in the west with the potential evaporation increasing from 2400 to 3700 mm. The rainfall maximum occurs in August, the mean annual temperature is 8 °C and the minima range from –25 to –32 °C. Groundwater is present in the dune region. The encircling mountains have a higher rainfall. Above the desert and steppe altitudinal belts at an altitude of 1900–

**Fig. 257.** Tarim basin, river-seepage region with nebka landscape (dunes heaped around the *Nitraria schoberi* bushes (Photo: PETROV)

2500 m, mesophytic shrubs appear including *Lonicera, Rosa, Rhamnus* and *Dasiphora (Potentilla) fruticosa* and others. Above this, and up to 3000 m, there is coniferous forest with *Picea asperata, Pinus tabulaeformis* and *Juniperus rigida*, succeeded by subalpine shrubs and alpine mats.

3. **Bei-Shan.** This region is west of the Ala-Shan and is an ancient elevated block rising from 1000–2791 m above sea level. It is bounded on the west by the Lop-Nor depression and Hami. Rainfall amounts to 39–85 mm and potential evaporation to 3000 mm. The vegetation is low shrub desert consisting of central Asiatic species with a few halophytes. *Picea asperata* is found growing on the highest points.

4. **Tarim basin with Takla-Makan.** The basin is 1300 km long and 500 km wide and surrounded on three sides by high, snow-covered mountains. It is the most arid part of central Asia with hot summers and cold winters (minimum, −27.6 °C). Despite this, it is well provided with groundwater fed by the mountain rivers. The 2000-km-long Tarim River has an average flow of 1200m$^3$ s$^{-1}$ and forms wide flood plains. In its lower reaches the river continually changes its course and its water seeps far into the central sandy desert (Fig. 257). Lop-Nor is sometimes a salt lake 100 km in diameter and is sometimes completely dried out. The sandy desert Takla-Makan is devoid of vegetation, but water can readily be obtained by sinking wells in the dune valleys.

5. **Tsaidam.** This is an elevated basin at an altitude of 2700–3000 m, completely surrounded by much higher mountains from which it receives its water. It is cut off from the Lop-Nor depression by the Altyn-Tag. The

mean annual temperature is approximately 0 °C, the
minimum being below -30 °C. The central part of the
basin was a large lake in the Pleistocene, but is a barren salt desert today. *Artemisia* semi-desert is found on
the sandy soil at the foot of the mountains. Tsaidam
forms a transition to still higher Tibet.

6. **Gobi** (Mongolian = desert). This region is north of the
above-mentioned deserts and covers the entire southern
part of Outer Mongolia. It is separated from the forests
and steppes to the east by the Khingan Mountains. In
the west it touches on Dsungaria which, thanks to rain
originating in Atlantic cyclones, is middle Asiatic in
character. The Gobi is gradually replaced to the north
by the Mongolian *Stipa* steppe with *Aneurolepidium
(Agropyron)* and *Artemisia* species. Saline and gypsum
soils are common in the desert. The central areas are
devoid of vegetation and covered by a stony pavement;
elsewhere the plant cover is sparse with a dry-mass
production of scarcely 100–200 kg ha$^{-1}$ compared with
400–500 kg ha$^{-1}$ in the northern steppe areas. On low,
brackish ground, *Nitraria sibirica, Lasiagrotis, Kalidium*
and other halophytes are found and on areas covered
by drift sand, *Haloxylon ammodendron* is seen. Nowhere in the entire western Gobi does groundwater
come to the surface. There are some oases in the east.
The Mongolian Altai mountain range extends into the
Gobi from the north-west, continuing as the Gobi Altai.
In the latter, only a steppe altitudinal belt is reached,
whereas in the former a coniferous belt with *Larix* is
present on northern slopes, albeit completely Siberian
in character.

## 14 Subzonobiome of Cold High Plateau Deserts of Tibet and Pamir (sZB VII, tIX)

Lying between the high mountain barrier of the Himalayas
to the south and the Kwen-Lun and Altyn-Tag to the north
is Tibet, the largest highland mass of the world, with an
average height of 4200–4800 m above sea level. It extends
2000 km from east to west and is 1200 km wide from north
to south and consists of debris-filled basins, encircled in
turn by mountain ranges which are 1000 m higher still
(Fig. 258). Water from melting snow forms swampy, frost-
debris areas with the cyperaceous *Kobresia tibetica* and
there are occasional "salt lakes" and even sand dunes.

The monsoon still exerts an influence in the southern
and eastern parts and, in the deeply incised valleys forming the upper reaches of the large southern and eastern

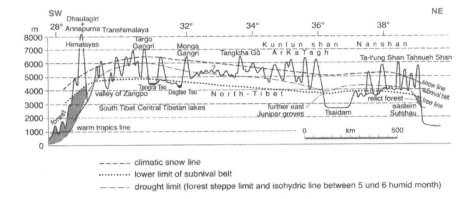

----- climatic snow line
············ lower limit of subnival belt
— — — drought limit (forest steppe limit and isohydric line between 5 und 6 humid month)

**Fig. 258.** Climatic profile through Tibet from south-west to north-east and the somewhat lower lying Tsaidam Desert. The determination of the forest line in Tibet and Tsaidam by temperature is only theoretical as the forest line is actually determined by drought. Forest is found only on the south face of the Himalayas and only as a narrow high zone in the Richthofen mountains (Nanshan) (from WISSMANN 1961)

Asiatic river systems, south-east Chinese and Himalayan forest elements appear.

The larger western and central area, the Chang Tang Desert, is characterised by the most extreme type of climate. The annual mean temperature is −5 °C and only July has a positive mean of +8 °C. Daily temperature variations of as much as 37 °C can occur, but the rainfall seldom exceeds 100 mm. The flora is poor and very young, having developed after the Ice Age. It consists of central Asiatic elements (*Ceratoides, Kochia, Reaumuria, Rheum, Ephedra, Tanacetum, Myricaria* and others).

At the western end of the highland plateau from which the high mountain ranges originate, is **Pamir**. At the Pamir Biological Station, 3864 m above sea level, Russian scientists have carried out ecophysiological investigations. The mean annual rainfall at this point totals 66 mm, most of it falling between May and August. The air is dry and solar radiation totals 90% of the solar constant so that the ground warms up to 52 °C in summer, although there are only 10–30 nights in the entire year when there is no frost (see Fig. 258, Pamirski Post). There is no closed snow cover. The soils are so dry that they do not freeze.

Dwarf shrubs, 10–15 cm high, grow on the desert-like habitats: *Ceratoides papposa, Artemisia rhodantha, Tanacetum pamiricum* or *Stipa glareosa* and the cushion-like *Acantholimon diapensioides*. Along the streams in the valleys, however, there are alpine meadows (Fig. 259).

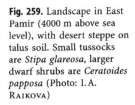

**Fig. 259.** Landscape in East Pamir (4000 m above sea level), with desert steppe on talus soil. Small tussocks are *Stipa glareosa*, larger dwarf shrubs are *Ceratoides papposa* (Photo: I. A. RAIKOVA)

Growth in the dry habitats is extremely slow. *Ceratoides* flowers only after 25 years, but lives for 100–300 years. The root systems are strongly developed and their mass is 10–12 times that of the shoot system. Most of the roots are found in the uppermost 40 cm of the soil, i.e. in the layers which warm up to more than 10 °C in summer. Roots extend more than 2 m laterally. The reserves of water in the upper meter of the skeletal soil amount to 26 mm at the most and 5 mm at the least. This is very little, but it nevertheless suffices for the scanty vegetation, even though the transpiration rate is quite high. Photosynthesis is intense only during the morning hours and the daily production is given as 25 mg dm$^{-2}$ leaf area. Respiratory losses are slowed by the low night temperatures.

Three biogeocenes were studied (see Table 22): one desert-like in which *Ceratoides* predominates on talus soil; a second, more steppe-like, on clay soil, with *Artemisia* and *Stipa glareosa*; and a third with low herbaceous cushions on stony ground with relatively good water conditions in gullies.

It is obvious that despite the relatively high intensity of transpiration the water requirements of the biogeocene are so small that they can be met by the precipitation. Only herbaceous, cushion-like plants in the vicinity of streams receive additional water from inflow. In general one-half to one third of the precipitation is used to meet losses due to transpiration.

The situation is complicated in the orobiomes by the fact that the sequence of altitudinal belts is highly dependent upon precipitation. In regions with less than 100 mm

**Table 22.** Phytomass and water consumption in East Pamir

| Biogeocene | Desert | Steppe | Cushion plant |
|---|---|---|---|
| Plant cover (%) | 5–18 | 15–20 | 15–30 |
| Phytomass (t ha$^{-1}$) | 0.14–0.54 | 0.09–0.48 | 0.4–0.89 |
| Transpiration (g g$^{-1}$ fresh wt. h$^{-1}$) | 0.3–0.9 | 0.1–0.7 | 0.1–0.19 |
| Water consumption during growing season (mm) | 8–40 | 6–87 | 25–446 |

of precipitation there is no true snow line, since even at more than 5500 m above sea level the small quantities of snow evaporate owing to the strong radiation. The desert continues to the upper limits of vegetation, whereas in other regions alpine steppe would occur at such altitudes or, where precipitation is above 500 mm, even alpine meadows could be expected (Western Pamir). Cushion-like plants usually play a leading role in the upper alpine belt.

# Man in the Steppe

Steppes cover large areas in the northern hemisphere and feed numerous wild grazing animals, buffalo in the prairie of North America, wild cattle and horses in Eurasia and probably many other herbivores. In central Asia the Bactrian camel (two-humped) plays a large role. Asiatic horse-riding people were, in previous millennia, able to penetrate far into Europe because of their nomadic life style and the possibilities of long **migrations.** Steppes and prairies have only in recent times been used almost exclusively as arable land. **Dust erosion** of the prairies in the 1930s and in Kazakhstan in the 1960s have caused great damage and desertification.

The **availability of water** is decisive for life in these areas, either as natural sources (and thus oasis) or as deep wells, which nowadays are drilled to supply water reservoirs. Furthermore, the scarce vegetation is used by herds of animals (usually sheep and a few goats, but also cattle) requiring large spaces. The grazing of steppes and semideserts has become an acute problem because of overgrazing, degradation and erosion of the soil. A solution which might ease the problem would be the replacement of domestic animals by wild animals (CAMPBELL 1985). Steppe herbivores are excellently adapted to their environment and might be able to re-create the equilibrium in the easily destroyed ecosystems. They could provide more meat than domestic animals on the same area if correctly managed. Cattle should only be used in temperate climates on areas which can cope with intensive grazing.

## 16    Zonoecotone VI/VIII – Boreonemoral Zone

Unlike the zones discussed so far, the boreal zonobiome VIII of the northern hemisphere with coniferous forests and cold-temperate climate encircles the entire globe through northern Eurasia and North America (interrupted by oceans). To the south, where an oceanic climate prevails, the boreal zone borders on the nemoral, deciduous forest zone, but in the regions with a continental climate it adjoins arid steppes or semi-desert (Fig. 260). There is no sharp boundary between deciduous forest zone and coniferous forest zone; instead, a transitional boreo-nemoral ecotone VI/VIII is intercalated between the two. This consists of either mixed stands of coniferous (mainly pine) and deciduous species or a macromosaic-like arrangement, with pure deciduous forest on favourable habitats with good soil and pure coniferous forest on less favourable habitats with poor soils. In eastern North America, different species of *Pinus* represent the conifers in the mixed stands, mainly *Pinus strobus* in the neighbourhood of the Great Lakes (although *Tsuga canadensis* is also found, as well as *Juniperus virginiana* in the south-east). Pine trees are often the pioneer woody species following forest fires or occurring on abandoned arable land. Since pines grow more rapidly than deciduous species on poor soils, they constitute the upper tree stratum, but their regeneration in such mixed stands is problematic if there is dense deciduous undergrowth. For this reason, pine trees are only successful where fire plays a recurrent role. It has been shown that fires caused by lightning are a common occurrence in such forests, particularly on sandy soils where there is a dry litter layer in summer. In Europe the situation is much simpler. On the poor fluvio-glacial sands extending as wide belts in front of the end moraines in central and eastern Europe, pure pine forests can be found (Pinetum) in regions which belong, climatically speaking, to the deciduous-forest zone. In eastern Europe, they are called "bor". On rather better, loamy-sandy soils, there is an additional lower tree stratum of oak and the forest is then called "subor" (Querceto-Pinetum).

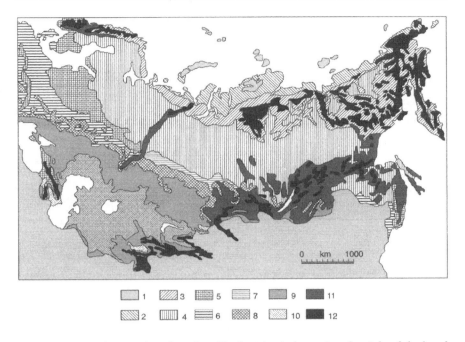

**Fig. 260.** Vegetational zones of Northern Euro-Siberia: *1* Arctic desert; *2* tundra; *3* dwarf shrub and forest tundra; *4* boreal coniferous forest zone; *5* mixed forest zone; *6* deciduous forest zone; *7* small-leaved deciduous forest zone; *8* forest steppe; *9* grass steppe; *10* semi-desert and desert; *11* mountainous coniferous forest; *12* alpine zone

On loamy soils, hornbeam (*Carpinus betulus*) occurs as well and the forests, now with three strata, are termed "sugrudki" (Carpineto-Querceto-Pinetum). Finally, on loess there are the zonal deciduous forests known as "grud", with oak in the upper stratum and hornbeam in the lower stratum (Querceto-Carpinetum). Such forests have been drastically altered by human interference. Forest fires and felling of deciduous trees for fuel have encouraged the growth of pines, whereas the removal of pines for use as valuable building material has resulted in the formation of pure deciduous forests. Yet another disrupting factor is the practice of forest grazing. In central Europe, extensive pine forests have arisen as a result of forestry activities in what were formerly pure deciduous forest regions, for example in the Upper Rhine Valley. Further north (in southern Scandinavia and in central-eastern Europe), spruce (*Picea abies*) and oak (*Quercus robur*) are more common. These species form a macromosaic but do not mix with each other (KLÖTZLI 1975). Since the better soils, once the site of oak forests, are now mostly cultivated, the proportion of remaining spruce forests has risen. Furthermore, spruce is encouraged by the forestry industry. In central Europe, the **spruce forests** at lower altitudes have all been planted by man and more and more spruce is appearing in the landscape because of its economic value. Damage to spruce, as well as to fir and deciduous trees, however, has increased noticeably in the past several years as a result of "acid rain" and other pollutants and the additional depletion of the soil.

The boundary between the boreo-nemoral and the true boreal zone coincides, in Europe, with the northern distribution limit of the oak. It runs along a latitude of 60°N through southern Sweden, extends along the southern coast of Finland, thence to the middle Kama River, where the steppe borders on to the boreal zone.

## Questions

1. Why is it important to distinguish between grassland and steppe both in definition and structure?
2. What different effects do summer and winter rain maxima have on vegetation?
3. How is the equilibrium of competition between grasses (steppe) and herbaceous plants and dwarf shrubs (semi-desert) described?
4. Why is the regeneration of the steppe after cultivation almost impossible?
5. What are the significant differences between the deserts of ZB VII (rIII) compared to the deserts of ZB III?
6. Why is there no zone of deciduous forests (ZB VI) in Mongolia between the zones of semi-desert/steppe (ZB VII) towards the north of the taiga (ZB VIII)?

# VIII   Zonobiome of the Taiga
## (Zonobiome of the Cold-Temperate Boreal Climate)

### Climate and Coniferous Species of the Boreal Zone

The true boreal zone (Fig. 260) commences at the point where the climate becomes too unfavourable for the hardwood deciduous species, i.e. when summers become too short and winters too long. This is recognisable in the climate diagram as the point where the duration of the period with a daily average temperature of more than 10 °C drops below 120 days and the cold season lasts longer than 6 months (Fig. 261). The northern boundary between the boreal zone and the Arctic is where only approximately 30 days with a daily mean temperature above 10 °C and a cold season of 8 months are typical of the climate.

Nevertheless, in view of the large distances over which this zone extends, it would be incorrect to speak of a uniform climate. Rather, a distinction should be made between a cold oceanic climate with a relatively small temperature amplitude and a cold continental climate with, in extreme cases, a yearly temperature span of 100 K (from a maximum of +30 °C to a minimum of –70 °C). Temperature conditions also change from north to south.

Zonobiome VIII is characterised by long, cold winters and short summers with extreme temperature fluctuations during the year. ZB VIII extends over enormous areas in Eurasia which can be differentiated into several subzonobiomes.

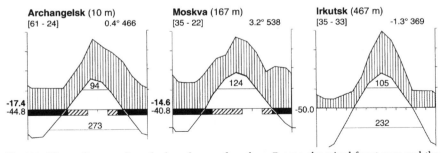

**Fig. 261.** Climate diagrams from the boreal zone of northern Europe, the mixed-forest zone and the boreal zone of Siberia. *Values* on the upper and lower horizontal lines indicate the number of days with a mean temperature above +10 °C and above –10 °C, respectively

Because of the climatic conditions several subzonobiomes can be distinguished: northern, central, southern
(with evergreen conifers) and extreme continental (with
summer green *Larix*). Oceanic subzonobiomes with birch
are also found in north-west Europe and north-east Asia.
The taiga in Canada and Alaska is similarly divided, but
does not extend over such a large area (WALKER 1998). In
the oceanic region, species of birch (*Betula*) are very important (AHTI & JALAS 1968). The floristic composition of
the tree stratum also changes across the large distances of
this zone. Coniferous forests in North America and eastern Asia contain a large number of different species,
whereas those in the Euro-Siberian region contain very
few. The vegetational history responsible for this is the
same as in zonobiome VI. Many species of the genera *Pinus, Picea, Abies, Larix*, as well as *Tsuga, Thuja, Chamae-
cyparis* and *Juniperus*, are found in North America,
although the last four belong in fact to the transitional
zone. The specific representatives of these genera on the
Pacific Coast are different from those in the east and only
one species, *Picea glauca*, extends from Newfoundland
across the Bering Strait. Apart from the species mentioned, *Picea mariana*, found on poor soils, occurs at the
forest line towards the Arctic and *Larix laricina* in the
continental regions. *Abies balsamea, Thuja occidentalis*
and *Pinus banksiana* also occur. *Pinus banksiana* is found
on sites previously laid bare by fire. The coniferous belt
in the mountain regions contains widely differing species.

In contrast, only two species are of any importance in
the boreal zone of Europe, spruce (*Picea abies*) and pine
(*Pinus sylvestris*). Only in the eastern regions is the European spruce replaced by the closely related Siberian species, *Picea obovata*, with the addition of other forest species (*Abies sibirica, Larix sibirica* and *Pinus sibirica*, a
subspecies of *Pinus cembra*). The proportion of spruce
gradually decreases, until in the continental parts of eastern Siberia spruce is entirely absent. At the same time,
*Larix sibirica* is replaced by *L. dahurica*. Larch forests
alone cover 2.5 million $km^2$ in Siberia. In northern Japan
the number of coniferous species greatly increases again.

In the northern European zone traces of the ice ages
are everywhere. The change in landscape caused by large
inland ice masses and their melting is particularly obvious in the European taiga (Scandinavia) and is obvious
in the development of the Baltic Sea (Fig. 262), which
took its present form only a few 1000 years ago. In the
surrounding landscape, glacial traces can be seen in the
formation of numerous "dead ice" lakes (Fig. 263) and the
various deposits (moraine deposits, glacial soils).

**Fig. 262.** Late glacial and Holocene development of the Baltic Sea. **A** Baltic ice lake with fresh water (10,200 years ago); **B** Ancylus lake with fresh water (8000 years ago); **C** Litorina sea (5000 years ago) (from DIERSSEN 1996)

**Fig. 263.** Formation of water bodies after melting of compressed ("dead") ice under the deposits from the advancing ice (*left*) and in the moraine deposit (right)

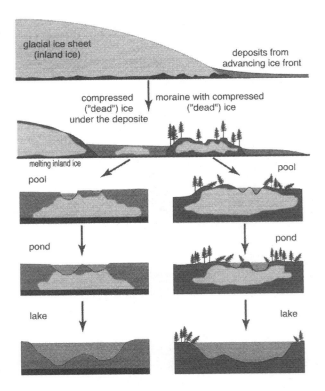

## 2 Oceanic Birch Forests in Zonobiome VIII

The large differences in the continental climate across the taiga zone of Eurasia are expressed in significant floristic differences, as mentioned previously. In both oceanic climatic zones on the Atlantic in Norway and on the Pacific in Kamchatka are light forests of birch and pine. Spruce is missing almost completely in Norway, thus a taiga zone is not developed. In northern Scandinavia the forest limit is formed by *Betula tortuosa* (closely related to *B. alba*). *Betula tortuosa* is a low growing tree with an irregular bent stem ("drunken trees"). Similar light forests occur in **Kamchatka** where the polar forest line is formed by *Betula ermannii* and, in parts, by *Pinus pumila*. *Betula ermannii* dominates; it has a very dense wood (specific mass >1) and can become many hundreds of years old and is widely distributed, even as far as Japan and Korea. Occasionally, other birches are

found (*B. japonica, B. middendorfii*) and *Larix* species
(*L. gmelinii, L. kamtschatica, L. cajanderii*). Strictly speaking, the forests of Kamchatka do not belong to the taiga.
Spruce forests only rarely occur (with *Picea ajanensis*).

# European Boreal Forest Zone

Typical of zonobiome VIII in northern Europe is the dark
spruce forest known as **taiga**, which occurs on podsol
soils with a raw humus layer, bleached alluvial horizon
and compact B-horizon. Soils of this kind are formed
from every type of parent rock in the humid boreal zone,
but the fewer bases contained in the rock, the better developed is the podsol soil. Litter from spruce does not decompose readily and lies above the $A_0$ horizon, which
consists of an organic mass of interwoven rhizomes and
roots of the dwarf shrubs as well as the mycelia of fungi.
This is the raw humus layer and it can readily be lifted off
the underlying $A_1$-horizon (humus rich mineral soil). For
this reason it is also called 'dry turf'. Humic acids formed
in this layer are carried down in rainwater and completely
leach out the bases and sesquioxides ($Fe_2O_3$, $Al_2O_3$), leaving nothing but fine, bleached quartz sand in the $A_2$ horizon (bleached horizon). At the point where the underlying
soil is not yet bleached, the humus and sesquioxides are
precipitated because of either decreasing acidity or removal of water by the roots of the trees. This is the origin
of the B-horizon, which can be either dark brown (humus
podsols) or rusty red (iron podsols).

Apart from the tree stratum of the **spruce forests** (Piceetum typicum), a herbaceous stratum and a closed
mossy stratum may also be present. Predominating in the
herbaceous layer are cowberry (*Vaccinium myrtillus*), as
well as red bilberries (*Vaccinium vitis-idaea*) in the drier
forests, or very commonly in the southern zone, wood
sorrel (*Oxalis acetosella*). The following species are also
very characteristic: *Lycopodium annotinum, Maianthemum
bifolium, Linnaea borealis, Listera cordata, Pyrola (Moneses) uniflora* and others. Wherever the groundwater table
is high, more raw humus accumulates and peat is formed
and this in turn leads to the formation of raised bogs,
where at first *Polytrichum* dominates in the mossy stratum, but is later ousted by peat moss (*Sphagnum*). If well-airated, flowing groundwater is present, spruce forests are
replaced by floodplain forests.

Apart from spruce forests, the proportion of **pine forests** (Pinetum) in the boreal zone is always very high, with

*Pinus sylvestris* displacing spruce in dry habitats. The herbaceous stratum of these thin forests consists of heather (*Calluna vulgaris*) and cowberries. Many lichens (*Cladonia, Cetraria*; see Fig. 268, p. 427) are found in the mossy stratum. Characteristic species in the herbaceous layer are *Pyrola* species, *Goodyera repens, Lycopodium complanatum* and others. Pines are often found in habitats favourable for spruce, but after forest fires which may be caused by lightning, masses of *Molinia coerulea, Calamagrostis epigeios* or *Pteridium aquilinum* spring up, appearing in the order mentioned, according to increasing dryness of the habitat.

On such burned sites birch and aspen are the first tree species to grow; they are later ousted by pine, beneath which spruce grows more slowly. In northern Sweden the birch stage lasts for 150 years and the pine stage for 500 years, but the fact that fire usually recurs before the zonal vegetation has arrived at the spruce stage explains the high proportion of pine that is found. Pine is only absent in moist habitats where there is very little danger of fire. Corresponding types of forests are found in North America; although they are floristically somewhat richer in species.

## Ecology of Coniferous Forests

The denser the stand, the less sunshine can penetrate to the forest floor. Soil beneath a spruce forest is 2 K colder than that in the open. The snow covering, too, is thinner in the forest so that the ground freezes to a greater depth during the long winter. The frost depth beneath a dense forest stand in which the frost remained until the beginning of August was found to be 85 cm, as compared with 50 cm beneath a thinner stand from which the frost had disappeared by the beginning of June. The roots of the spruce are very shallow, usually being confined to the upper 20 cm of the soil or even less, if the groundwater table is high. For high productivity, spruce forests require a continuous supply of water and a medium-depth groundwater table, whereas pine which roots more deeply is not so sensitive to a dry soil. The total water lost annually by a typical spruce forest amounts to 250 mm in the northern taiga, 350 mm in the central taiga and 450 mm in the southern taiga. The mean annual production of organic mass is 5.5 t ha$^{-1}$ and wood production is 3 t ha$^{-1}$ (the latter can reach 5 t ha$^{-1}$ in the southern taiga). The largest annual increment is achieved in the north after 60

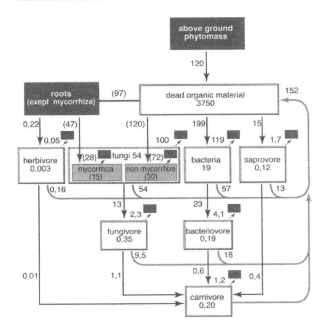

**Fig. 264.** Carbon supply (in *box* as g C m$^{-2}$) carbon flux (*arrows* in g C m$^{-2}$ a$^{-1}$) in a pine forest in central Sweden. The mineral soil up to 30 cm deep is incorporated. *Arrows* to the black boxes: respiration loss. *Grey line* indicates transport via dead organisms or excrement. *Values in brackets* are uncertain (from DIERSSEN 1996)

years, but in the south after only 30–40 years. The phytomass of the tree stratum of pine forests reaches a maximum of 270 t ha$^{-1}$ and that of the undergrowth in old stands is 20 t ha$^{-1}$. Comparable values from a pine forest in central Sweden are given in Fig. 264.

The quantity of litter produced by older stands on their way to maturity can exceed 1000 t ha$^{-1}$. This is not accumulated, however, but is continuously decomposed until a state of equilibrium between addition and loss of litter is reached at a litter mass of 50 t ha$^{-1}$. On very wet habitats the organic matter accumulates in the form of peat. Under such unfavourable conditions the annual increase in dry mass of the tree stratum is often less than that of the other strata. The figures are 850 kg ha$^{-1}$ for the tree stratum (total 1906 kg ha$^{-1}$) in the herbaceous type of spruce-swamp forest and 104 kg ha$^{-1}$ (total 1780 kg ha$^{-1}$) for the tree stratum in pine-raised bog. The leaf area index is relatively high because the trees bear at least two years of needles. In pine forests of the boreo-nemoral zone, the LAI is 9 to 10 and in the spruce forests of the taiga it exceeds 11.

Conifers invariably possess **ectotrophic mycorrhiza**, the fungal hyphae greatly enlarging the range of the root system and rendering the nutrients contained in the raw humus layer more easily available to the trees. Plants in the herbaceous stratum are exposed to severe competition from the tree roots. On shallow granitic soil all of the available water may be used up by the pine trees so that a herbaceous stratum is completely lacking and the ground is covered with lichens (Fig. 268, p. 427). Even the young pine saplings are unable to mature in the face of such root competition and, in fact, only succeed in growing in places where an old tree is dying (Fig. 267, p. 427) and competition from its roots is therefore lacking. Where the soil is wetter the roots of the trees utilise the nitrogen in the soil to such an extent that only dwarf shrubs with extremely low nutrient requirements (*Vaccinium myrtillus*) can grow under the trees. However, if the roots of the trees are severed in order to exclude them from competition and conditions of illumination remain unchanged, other more demanding species take hold. Examples are provided by *Oxalis acetosella* or even the nitrophilic raspberry (*Rubus idaeus*) which is otherwise only found in clearings away from the competition of the tree roots. Thus, it is more often the quantity of nutrients available to the plants than the amount of light which determines the composition of the herbaceous stratum (see also pp. 30, 69).

Water relations of a spruce forest in Sweden have been analysed:

A large part of the rainwater is withheld by the crowns of trees (interception) (50% as compared to 30% in the thinner east European stands). Moss and litter layers retain a further portion of the water and in the end, only about one third of the total rainfall reaches the roots. In the summer months, this was found to amount to 90 mm and for the rest of the year 202 mm, which gives a total of 292 mm which is almost completely lost by transpiration in a 40-year-old stand. In wet habitats, as much as 378 mm is lost to the atmosphere by transpiration so that a part of the water must be drawn from the groundwater. Although most ecophysiological investigations have been carried out in the spruce belt of the Alps, the situation is probably analogous to that in the boreal zone.

Active transpiration is paralleled by equally intense photosynthesis. Spruce possesses two kinds of needles, sun needles and shade needles. The situation is reminiscent of that in beech, but with the difference that the active period for the evergreen spruce begins very early in spring and continues into the autumn, until the onset of occasional frost. The seasons of low nocturnal temperatures and small respiratory losses are particularly favourable for the net gain of dry matter. Nevertheless, after a night of frost, photosynthesis is temporarily inhibited, although it is not until the beginning of the cold season proper that the spruce falls into a state of dormancy in which it does not even assimilate on sunny days. At the same time, respiration sinks to such a low level that it can hardly be measured and accounts for only negligible material losses. Needles lose their fresh green colour at this time and chloroplasts are difficult to recognise under the microscope. After a long period of cold it takes a little time for photosynthesis to regain its normal level in the spring, since the photosynthetic apparatus must be reactivated. Young *Pinus cembra* saplings in the mountains have been shown to spend the winter beneath the snow with green needles and recommence $CO_2$ assimilation immediately in the higher spring temperatures.

The transition to winter dormancy is accompanied by a process of "hardening," or in other words, a great increase in resistance to frost (see p. 311). Conifers of the boreal zone undergo the same processes as deciduous trees. Whereas in their non-hardened autumn condition spruce needles are killed by frost at a temperature of $-7\,°C$, they are capable of enduring a temperature of $-40\,°C$ in the winter without suffering any damage. Young spruce buds are quite vulnerable to light frosts in spring and may, therefore, be easily damaged by late frosts. **Frost resistance of the needles** can be artificially changed by the influence of low temperatures in late autumn and spring and the dehardening process can be affected by the influence of normal room temperatures, especially in December and late winter. Hardening prevents the occurrence of frost damage in coniferous trees in their natural habitats even at temperatures as low as $-60\,°C$, such as occur in Siberia. Thanks to the state of winter dormancy these trees are in a position to survive the complete darkness of the polar winters. The varying degree of adaptation achieved by the different species is reflected in their distribution. Only a few species can tolerate the extreme continental Siberian winter, the deciduous Siberian larch better than the evergreen species. Variations within a species also occur, depending upon their provenance. Spruce from the Alps behave differently to members of the same species taken from the northern boreal zone, or again, spruce from the upper tree limit behave differently to spruce from lower altitudes. The more extreme the conditions, the more pointed the crowns of the trees, showing that the growth of lateral twigs is more strongly inhibited than that of the main shoot. In polar regions, the same phenomenon is also observable in pine.

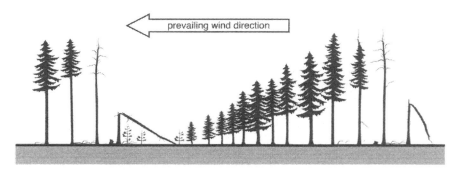

**Fig. 265.** Schematic transect through a forest with *Abies balsamea* (Maine, north-eastern USA) showing a regeneration wave similar to the Shimagare phenomenon in Japan (see Fig. 266) (modified from SPRUGEL 1976, from BURROWS 1990)

**Fig. 266.** *Abies veitchii* forest from the Shimagare slope (Japanese Alps) with a wave of dying trees. On the *left* the regeneration wave is starting (Photo: S.-W. BRECKLE)

Whether or not this shape results from a selection of mutants better able to withstand the weight of snow is unknown, but the same phenomenon has been observed in fir trees at the lower, dry limit in Albania where snow is not an important factor. The most likely explanation is that whenever the general situation is unfavourable, growth of lateral twigs is inhibited before the main shoot suffers (the reverse is true if light conditions are poor). On dry slopes in Utah (North America), *Picea*, *Abies* and *Pseudotsuga* have extremely conical crowns on dry slopes, whereas on the valley floor their crowns are rounded. The danger due to the weight of snow is the same in both situations.

**Fig. 267.** Pine-spruce taiga south of St. Petersburg with trees blown down by wind with upright root pans (as an indication of scale: Professor Okmir Agachanjanz). *Vaccinium* species dominate in the herbaceous layer, lower sites have become boggy; they are small *Sphagnum* depressions (Photo: S.-W. BRECKLE)

**Fig. 268.** Dry lichen-pine forest in central Norway near Grimsdalen (350 mm annual precipitation). The herbaceous layer consists almost exclusively of a dense carpet of fruticose lichen (*Cetraria, Cladonia, Alectoria* etc.) growing up to 25 cm high (Photo: S.-W. BRECKLE)

Another phenomenon worth mentioning is that of **waves of regeneration**: in pine forests in Japan it has been observed that wide strips of dead trees gradually move along and are replaced by a new wave of young trees of almost the same age. This so-called **Shimagare phenomenon** (Figs. 265, 266) only occurs in monospecific forests with trees of the same age usually growing very close together as a natural monoculture. The **natural dying of forests** presupposes not only certain rare events (storms, fire strips) leading to a synchronisation of trees of the same age, but also spatial self organisation in which cohorts of trees of the same age die synchronously and young trees follow on synchronously (MUELLER-DOMBOIS 1987; JELTSCH 1992). Examples of pine forests are shown in Figs. 267, 268.

## Siberian Taiga

Pine (*Pinus sylvestris*) in its several forms occurs in the whole of the Eurasian taiga zone. However, pine does not form a zonal vegetation and only fills gaps, for example fire gaps, on poor sandy soils and on boggy soils. *Picea obovata*, which is closely related to the European *P. abies* occurs as the dominant species and forms the **dark taiga** together with the stone pine (*Pinus sibirica*, closely related to the alpine *P. cembra*) and with *Abies sibirica*. Pure stands of *Abies sibirica* also occur and are called **black or**

**Fig. 269.** Virgin pine-birch taiga between Irkutsk and Kultuk, with *Rhododondron dahuricum* and *Ledum palustre* in the lower shrub layer (Photo: U. KULL)

**lightless taiga.** However, in extreme continental eastern Siberia *Larix gmelinii* (= *L. dahurica*) occurs in pure stands forming the **light taiga.** In the more northern zone all taiga types also contain *Larix sibirica*, which forms the polar forest limit in western Siberia and is replaced by *L. gmelinii* in eastern Siberia. Birch also occurs in open gaps in the Siberian taiga (Fig. 269) and not only forms the pioneering transition stage in areas of fire and storm, but also survives for long periods in undisturbed stands.

## Extreme Continental Larch Forests of Eastern Siberia with Thermokarst Formations

Shady coniferous forests of western Siberia, with *Picea obovata, Abies sibirica* and *Pinus sibirica* (dark taiga) differ significantly from those in the light forests of the deciduous *Larix gmelinii* (light taiga) in eastern Siberia. The latter is a vast subzonobiome with an extreme continental boreal climate (absolute temperature fluctuations per year of as much as 100 K) and can be seen on the climate diagrams in Fig. 270. In North America, a similar less extreme climatic region is found around Fort Yukon in Alaska (Fig. 282).

Precipitation is very low in this region (less than 250 mm) but the slow thawing of the upper soil layers compensates and there is sufficient melt water for forests to thrive. These larch forests usually have an undergrowth of dwarf shrubs (*Vaccinium uliginosum, Arctous alpina*), the drier soils with *Vaccinium vitis-idaea* and *Dryas crenulata* and the wetter soils with *Ledum palustre*. In the very dry type of forest the ground is covered by lichens.

With annual mean temperatures as low as −10 °C, the ground in eastern Siberia is permanently frozen down to a depth of 250–400 m; it is a **permafrost soil**. In the relatively warm summers only the upper 10– 50 cm thaw out, although as much as 100–150 cm thaw in places where drainage is good.

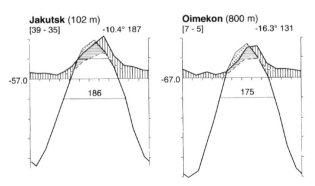

**Jakutsk** (102 m)
[39 - 35]                    -10.4° 187

-57.0

186

**Oimekon** (800 m)
[7 - 5]                    -16.3° 131

-67.0

175

**Fig. 270.** Climate diagrams for the extreme cold-continental region of eastern Siberia. Oimekon is the cold pole of the northern hemisphere

**Fig. 271.** Forest tundra in the Cherski Mountain country of eastern Siberia. Larch forest (*Larix dahurica*) on the slopes, moss tundra in the cold valleys, with dwarf birch (*Betula exilis*) and *Rhododendron parviflorum* (Photo: V. N. PAVLOV)

Further north, the thin forests change and are gradually replaced by scattered trees (redkolesye) and then by a dwarf shrub tundra with *Betula exilis* (knee-high) and *Rhododendron parviflorum* (Fig. 271).

The thermokarst conditions of this coldest part of the northern hemisphere are especially impressive.

B. FRENZEL wrote on the thermokarst:

"The permafrost of Siberia and most probably that of Alaska has existed since the early ice ages. Each ice age contributed to its expansion, while each warm interglacial period reduced its territory and thickness. In this type of landscape, however, even the warmer climates are suitable for the new formation of permafrost, although the depth of the soil subject to annual thawing is greater than during colder periods or ice ages. The appearance and dissolution of permafrost are related phenomena. Ecologically and geomorphologically, these processes are especially pronounced on fine grained sedimental formations."

During the ice ages, loess (an aeolian sediment) and its derivatives were deposited on large areas of the former climatic zones with extremely cold winters. In the present high continental climates of the boreal coniferous forest zone, these soils are filled up to 80% of their volume with the ice of permafrost. Local perturbations in the radiation balance and heat flux between the atmosphere and soil caused by, for example, natural forest fires which recur in

**Fig. 272.** Destruction of a *Larix dahurica* forest by thermokarst on the Aldan River. The upper 5–8 m of the soil on which the forest grows thaws and the flowing mud, supersaturated with water, transports the fallen trees to the river (Photo: B. Frenzel)

a given area every 180–240 years, or erosion by rivers, etc. result first in an increase in the depth of thawing in summer. Since the soil was previously supersaturated with ice, the thawed layer slides downhill if situated on a slope. In Siberia this is referred to as the "Jedom" series, meaning that the loose soil is "eaten away". The more intense summer thaw of the top soil results in a decrease in soil volume. This is generally designated as **thermokarst** (Fig. 272). On horizontal surfaces, the soil reacts by collapsing in on itself so that depressions of several kilometres in length without drainage are formed. In these so-called **alasses** the groundwater level rises, thereby drowning forests located within them (Fig. 273). These phenomena were described by travellers through Siberia as early as in the 17th century, a clear indication that the formation of alasses is a natural process and may be traced back to the end of the late glacial period (approximately 12,000–10,000 years ago). At present, alass development is being increased by clear-cutting and building projects.

Alasses are especially common in the Vilyuy basin of central and eastern Yakutia, with its highly continental climate, and in the adjacent regions. Although the centres of the alasses are at first usually filled with water, the steep slopes around the edges, which may be up to 50 m high, are better drained, receive more solar radiation and are covered with a colourful steppe vegetation. Approximately one third of the 900 species of higher plants in Yakutia belong to such plant communities, although their total surface area covers only a few percent of the country's total area. If the soil of the alasses sinks slowly and retains

The permafrost region, in which all life appears to be extremely limited as a result of the winter cold is extremely dynamic.

**Fig. 273.** An alass forming in the larch taiga of central Yakutia. Thawing of the upper 15–20 m of the permafrost causes the soil to sink to the same depth and the absence of drainage in the basin causes drowning of the *Larix dahurica* forest (Photo: B. FRENZEL)

**Fig. 274.** A bulgunnjachi in central Yakutia, situated in an alass with natural meadow community. The slopes of the broken, therefore decaying, pingo are covered with a grass steppe vegetation and the top is still covered with trees (Photo: B. FRENZEL)

only a small amount of water, the dead larch or pine forests are replaced by natural meadow communities, which today are of great importance for cattle raising. The number of species is often much lower in brackish locations. Alass lakes may eventually become land again. This results in a changed heat flux and expansion of the permafrost. Since alass lakes contain a great amount of water, large mounds with centres of ice, known as **"bulgunnjachi" or "pingos"**, develop (Fig. 274). They increase in height, either until radiation in summer inhibits the formation of the ice-filled centres, or until their growth causes them to break open so that the summer heat is able to penetrate

**Fig. 275.** Grass and herbaceous steppe on the south-exposed banks of the Aldan River in central Yakutia. The fine-grained ice age sediments are filled up to 80% (vol) with the ice of permafrost, especially in the large polygonal ice wedge systems. When the ice wedges are dissected by the river, the purer ice melts first. Starting from the higher surfaces, these are then colonised by forest as a result of the improved drainage conditions. The drier mineral soil between the ice wedges is covered with grass steppes on the upper slopes and with herbaceous steppes on the middle and lower slopes (cf. Fig. 276) which slide onto the melting base (Photo: B. FRENZEL)

deeply into the bulgunnjachi, thereby leading to their decay. The lifetime of such mounds varies from between a few decades to several thousand years. In any case they always contribute to the variety of biotopes, since their well-drained steep slopes with the increased solar radiation often offer advantageous opportunities for the settlement of colourful steppe communities.

The occurrence of steppe communities in Yakutia is typical on all dry steep southern slopes, which are very warm in summer, as for example on the slopes along the large rivers (Figs. 275, 276).

In general, the climate of Yakutia should at least be designated as semi-arid as clearly illustrated by the climate diagram in Fig. 270. Even in the euclimatotopes potential evaporation is higher than the low annual precipitation. This also explains the treeless areas, called **"tscharany"** (charani), within the larch forests, in which the high evaporation results in the concentration of salts. Halophytes, such as *Atriplex litoralis, Spergularia marina* and *Salicornia europaea*, which are also native to the seacoasts, grow on such brackish, solonised soils. On the wet salty soils the grasses *Puccinellia tenuiflora* and *Hordeum brevisubulatum* are also found (WALTER 1974).

Since steep southern slopes in the far north are struck at noon by the low-lying sun at a right angle, some typical steppe species are also able to grow on Wrangel Island at 71 °N (JURTSEV 1981), these include: *Ephedra monostachya, Stipa krylovii, Koeleria cristata, Festuca* spp. and other grasses, *Pulsatilla* spp., *Potentilla* spp., *Astragalus* and *Oxytropis* spp., *Linum perenne, Veronica incana, Ga-*

**Fig. 276.** Herbaceous steppe on the south-exposed slope of a small side valley of the Lena within the dominant *Pinus sylvestris* taiga. The candle-shaped inflorescences of *Orostachys spinulosa* (Crassulaceae), dense fructescences of *Alyssum* sp. (*lower right*) and *Ephedra monosperma* (not shown) characterise these stands. This type of steppe flourishes on permafrost (Photo: B. FRENZEL)

*lium verum, Artemisia frigida, Leontopodium campestre, Aster alpinus* (typical of Siberian steppes). These steppe islands also exist extrazonally on warm southern slopes as relicts of the zonal steppes of the glacial periods, when the climate was more extremely continental. At that time, descending föhn-like winds coming off the great ice masses and warming in the process were deflected towards the east across the ice-free periglacial surfaces and deposited the deep loess sediment. Summers were apparently so hot that large dry cracks formed in the loess, in which freezing of permafrosts led to the accumulation of ice (Jedome series). Based on recent Russian research, it must be assumed that such periglacial steppes existed zonally across all of Eurasia and North America with an abundant steppe fauna of rodents, antelopes and wild horses, as well as the woolly rhinoceros and the mammoth. These large mammals only disappeared with the advent of man.

Tundra vegetation probably only grew in boggy or swampy areas around lakes (therefore, as pedobiomes) and was restricted to deeper parts of the relief. The plentiful presence of *Ephedra* and *Artemisia* pollen in the pollen spectra of peat samples from the glacial period indicates that these species grew in the surrounding cold steppes which today are only found in a few locations, particularly in Yakutia.

In the post-glacial period the entire air circulation was altered when the ice melted; the ice sea developed, the land bridge between eastern Asia and Alaska was severed and the Gulf Stream brought warm water to the newly formed polar ice sea. The climate of the northern latitudes was determined by the westerly currents flowing from the Aleutians on the one side and from Iceland on the other. It became humid and the western flanks of the continents were given an oceanic nature. The tundra vegetation expanded its range into the northern region of the former periglacial steppes and colonised the areas which had just become free of ice. Forest vegetation followed, emerging from refuges, until tundra and forest zones assumed their present situations. Periglacial steppe vegetation receded into the arid regions of today's continental steppes, along with its typical fauna. Animal species which were not able to keep up with the changes, or perhaps even because of the arrival of man, became extinct. This was especially true for the largest animals such as mammoth, woolly rhinoceros, giant elk and others (see WALTER & BRECKLE 1991).

These facts have been presented in order to demonstrate that today's zonal tundra vegetation and the boreal coniferous forest zone with its numerous mires are recently developed phenomena in their present form. Elevated bogs also probably did not exist earlier. Certain relicts of the periglacial steppes are found on calcareous cliffs in Central Russia. *Carex humilis*, with its scattered distribution in the recent steppe heaths, is also to be considered a periglacial relict. Many species occurring in the alpine mats belong genetically to typical steppe genera, such as *Astragalus, Oxytropis, Potentilla, Pulsatilla, Festuca, Avena* and especially *Artemisia*, the edelweiss (*Leontopodium*) and *Aster alpinus*.

## Orobiome VIII – Mountain Tundra

There are only a few altitudinal belts in these northern latitudes of zonobiome VIII and the tree line is soon reached. The latter is formed by *Picea, Pinus sibirica*, or *Larix*, depending upon the geographical situation. However, the tree line is succeeded by mountainous tundra in the alpine belt and not by typical tundra or alpine mats in the Alps (STANJUKOVICH 1973). In the Alps, the first snow falls on unfrozen ground and a thick covering of snow keeps the temperature of the soil at about 0 °C throughout the winter. The perennial herbaceous plants are thus exposed neither to severe frost nor to frost desiccation and the vegetation consists of dense alpine mats. The situation in the mountainous tundra is different. The ground is already frozen when the first snow falls and the snow covering is thin and is blown away from the summits. Permafrost, which is not found in the Alps, prevails. Winter storms are violent and the weathering due to frost action is very intense. The debris gradually moves further and further down the slopes (**solifluction**) and the fine earth is blown away. All of these factors are responsible for the bareness of the summits in the mountainous tundra and thus for the description **"golzy"** (Russian *golyj*= bare). They are covered by lichens and a few mosses with isolated dwarf shrubs between the rocks. The situation is very reminiscent of the windswept ridges of the Alps, with *Loiseleuria* and the same lichens (see p. 357 f). Conditions in the subalpine or **"podgolez"** belt are rather better because drifting snow can accumulate. Mountainous tundra is encountered in the continental climate region as far south as 50 °N and is still found in Altai. In the oceanic region of the boreal zone (Scandinavia, Kamchatka) there is no mountainous tundra and the alpine belt is similar to that in the Alps. Snow in winter is plentiful and the tree line consists of birch (*Betula* spp.).

## Mires of the Boreal Zone (Peinohelobiomes)

Extensive areas of the boreal zone are not covered by the true zonal vegetation, which is coniferous forest, but by mires (moors, bogs).

The boreal zone has a humid climate with rainfall exceeding potential evaporation, which means that the water balance is positive. If surplus water is prevented from draining into the rivers, the groundwater table rises and mires are formed. Since the soils of the boreal zone are poor and acid (podsols), the groundwater is acid in reaction and has a low mineral content. It is usually brown in colour owing to humus sols. The situation only differs if the underlying rock is limestone. Large expanses of the boreal zone in Euro-Siberia and North America are very flat so the groundwater table is high. As long as it remains more than 50 cm below the ground for the larger part of the year, tree growth is possible, otherwise the growth of trees is inhibited and forests are replaced by mires.

In large areas of Finland, more than 40% of the total land is covered by mires, in places even 60%. The same holds true for the boreal zone of eastern Europe and especially western Siberia which is entirely covered by swamps and mires except in the vicinity of the rivers. In Kamchatka, Alaska and Labrador as well as the regions to the south of Hudson Bay, the situation is similar in places. For this reason, the pedobiome of the mires has to be dealt with after the coniferous forests. The dividing line between the two is often difficult to establish. In the spruce forests already mentioned, with *Polytrichum* and *Sphagnum*, peat formation is well developed (Fig. 267, p. 427).

In the geological meaning of the word, a mire or bog ("moor" in German) must have a peat layer at least 20–30 cm deep. If the peat layer is thinner, or if its content of combustible material falls below 15–30%, then in German, the term "anmoor" is applied. In the ecological sense, mires are plant communities which are dependent upon a high groundwater table, but are independent of the thickness of the peat layer upon which they grow. On account of the poor aeration of the soil, the roots of the plants remain near the surface and therefore only the consistency of the uppermost layer is important.

Three types of mires can be distinguished according to the origin of their soil water:

1. **Topogenous mires** are associated with a very high groundwater table and for this reason occupy the lowest portions of the relief or occur wherever spring water is available. Many widely differing types of fens belong to this group.

According to the origin and the nature of the water in the soil of mires, they are divided into topogenous, ombrogenous and soligenous mires.

The low-nutrient, i.e. oligotrophic, mires which are only found in cool to cold climates are all **peinohelobiomes**. According to their structure and topography, several distinct types can be recognised which are bound to certain climatic conditions (Fig. 277): blanket bogs – raised bogs – aapa mires – palsa bogs.

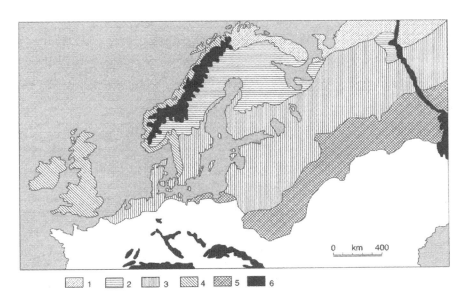

**Fig. 277.** Distribution of the various types of mires (bogs) in northern Europe. *1* Palsa bogs; *2* aapa mires; *3* typical raised bogs; *4* blanket bogs; *5* forest-raised bogs; *6* mountain bogs. The *white patches* in the southern regions have mainly topogenous fens

2. **Ombrogenous mires or raised bogs** are higher than
   their surroundings and are exclusively watered by the
   rainwater falling onto them.
3. **Soligenous mires** are also watered by rain, but because
   they are not higher than the surrounding country they
   also receive water draining in from surrounding slopes
   when the snow melts.

If the groundwater of the topogenous mires or fens contains mineral substances and is rich in nutrients, the
mires are termed eutrophic or minerotrophic. Rainwater,
on the other hand, is very pure and is poor in nutrients
so that the ombrogenous mires are said to be oligotrophic
or ombrotrophic. The run-in water received by soligenous
mires contains rather more nutrients, unless it comes entirely from melting snow. Therefore they may be oligotrophic to minerotrophic. Since the groundwater of the
boreal zone is poor in mineral salts, it is not easy to distinguish between fens and raised bogs and it is more
usual to speak of mesotrophic transitional bogs. If the
water contains less than 1 mg $l^{-1}$ Ca, less demanding species typical of the oligotrophic mires are to be found.

Eutrophic mires or fens in which *Carex* spp. play a
leading role occur in the temperate zone if the groundwater contains calcium but is not saline, regardless of climate. All of these pedobiomes are helobiomes.

1. **Blanket bogs.** These have already been encountered in
   the extreme oceanic climate of the Atlantic heath region
   of the British Isles and along the west coast of Scandinavia (p. 319). They cover the entire terrain.
2. **Raised bogs.** These are typical of the rather less oceanic, north-west corner of central Europe with its heath
   regions, the entire boreo-nemoral zone and the southern part of the boreal zone. In their typical form they
   are devoid of trees, but as the climate becomes drier
   and more continental, pine grows out into the bogs to
   form the so-called forest-raised bogs. The entire southern margin of the boreal zone consists of such bog forests (Figs. 277, 278). They occur in north-east Germany,
   northern Poland and occasionally in the foothills of the
   Alps.

The raised bogs of Eurasia and North America are con-
fined to the oceanic regions, but aapa and palsa bogs are circumpolar in distribution.

3. **Aapa mires or string bogs.** These are found north of
   the raised-bog zone, but above all in Fennoscandia and
   western Siberia. They are gently sloping soligenous
   mires with slightly raised ombrotrophic ridges, or
   strings, running at right angles to the slope. In between
   the ridges are elongated hollows filled with minerotrophic water (Finnish *rimpis*, Swedish *flarke*). The entire bog presents a descending, terraced aspect reminis-

**Fig. 278.** Bog lakes and taiga alternate. Central Sweden near Sunnersta (Photo: S.-W. BRECKLE)

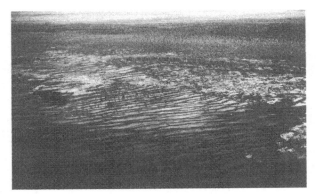

**Fig. 279.** Endless expanses of the western Siberian bog region. String bogs with bog lakes on the horizon. Aerial photograph taken from a Zeppelin on 17 August 1929 (Photo: Archives of the Zeppelin Museum, Friedrichshafen)

cent of terraced rice fields. The ridges are in part the result of the lateral pushing effect of the ice covering the rimpis in winter (see Fig. 279).

4. **Palsa bogs of the peat-hummock tundra.** These begin beyond the boreal zone in the forest tundra in regions where the annual mean temperature is below $-1\,°C$. Ice in the ground itself is partly responsible for the formation of the peat hummocks, which may be as much as 20–35 m long, 10–15 m wide and 2–3 m (up to 7 m) high. On slightly elevated ground the snow cover is thinner and the frost can penetrate more rapidly into the peaty soil. Layers of ice are formed which attract water from the unfrozen surrounding peat. As the ice thickens, it pushes up the peat and at least part of the hummock remains since not all of the ice melts in summer. As a result the snow covering in the following

**Fig. 280.** Palsa or peat hummock bog in northern Finland (Photo: E. WALTER)

year is still thinner and the soil freezes even more rapidly. From year to year the ice becomes thicker and the peat hummocks with their core of ice grow higher and higher. In summer the structure as a whole sinks into the ground somewhat, giving rise to a ditch-like depression filled with water, in which dwarf birch (*Betula nana*) and cotton grasses (*Eriophorum*) grow (Fig. 280). The tops of the peat hummocks (palsen) dry out in summer, crack and undergo wind erosion. It can be assumed that the majority of palsen are subfossil structures from a period with a colder climate and are now in a state of disintegration. These may be considered thermokarst phenomena on a smaller scale.

## 9 Ecology of Raised Bogs

Peat mosses (*Sphagnum* species) play the largest part in the formation of raised bogs. They contain large numbers of dead cells which easily fill up with water by capillary action, so that the cushion-like plants are of a spongy nature and contain many times their own dry weight in water. As the plants grow upward their lower ends die off and are converted into peat. As the cushions grow larger and larger they merge with one another, until finally the entire area presents the watch-glass appearance of a typical raised bog. Because peat mosses cannot tolerate drying out they require uniformly damp, cool summers. Since they colonise only poor, acid soils, podsols are well suited to their requirements. For this reason raised bogs usually originate in boreal coniferous forests which have gradually become wetter.

In a large, growing, raised bog it is possible to distinguish a very wet, only slightly convex high area, a better drained and relatively steep marginal slope and a surrounding minerotrophic fen, termed a **"lagg."** The high area is not absolutely flat, but consists of small hummocks or **"bults"** which extend above the mossy areas and "schlenken" or small hollows in the mossy carpet which are filled with water and in which such hygrophilic peat plants as *Carex limosa* or *Scheuchzeria* grow. Several schlenken may unite to form small bog-pools known as "blänken" or "kolke" (Fig. 278). They are usually 1.5–2 m deep and are filled with soft detritus. Surplus water drains off the high areas in small gullies called "rüllen".

Only a few flowering plants are able to survive in the raised bogs and they have to be species that are undemanding as regards nutrients: *Eriophorum vaginatum, Trichophorum caespitosum*, as well as the dwarf shrubs *Andromeda polifolia, Vaccinium oxycoccus, V. vitis-idaea, V. uliginosum, Calluna vulgaris* and *Empetrum*. In Atlantic regions, *Narthecium* is found; in the east *Ledum palustre* and *Chamaedaphne calyculata* and in the north *Rubus chamaemorus, Betula nana* and *Scheuchzeria palustris*.

The second ecological factor affecting the survival of other plants, apart from the **scarcity of nutrient materials**, is the danger of their being overgrown by the peat mosses. The viable tips of the latter are the substrate upon which the flowering plants germinate. Depending upon the amount of water available, peat mosses grow 3.5–10 cm annually, and if the flowering plants are not to be smothered by the mosses, they are obliged to raise their shoot bases by this amount, either by elongating the rhizome or by putting out adventitious roots. In places where the peat mosses grow more slowly, such as on relatively dry hummocks or well-drained slopes, it is easier for other plants to survive and here, dwarf shrubs are usually seen. Definite zonation can be distinguished on the individual hummocks (bults): *Eriophorum vaginatum* and *Andromeda* at the base, *Vaccinium oxycoccus* further up and other dwarf shrubs at the top. The top of the hummock (bult) is often so dry that *Sphagnum* is replaced by other mosses (*Polytrichum strictum, Entodon schreberi*) and even by lichens (*Cladonia* species, *Cetraria*).

Trees (pine, spruce) confronted with *Sphagnum* moss are at a great disadvantage since not only is the base of the tree trunk fixed, but growth is also extremely slow on such a poor substrate. Often not more than the topmost twigs are higher than the hummocks. Bog forest occurs only in places where the growth of the peat mosses is limited by a drier climate. Once a bog is drained, peat moss ceases to grow and the area is rapidly converted into heath in which dwarf shrubs assume a leading role, soon to be joined by trees such as birch, pine or spruce. This is the state of the majority of contemporary bogs in central Europe and the effect is increased by the additional N supply from the atmosphere.

Even though the water is just as poor in nutrients as the rest of the bog, species typical of minerotrophic soil often grow along the runnels or on the edges of the pools. It appears that running water or water agitated by waves provides the plants with more nutrients than stagnant water in which only diffusion of nutrients takes place. Because of their high water content, bog soils warm up very slowly and are therefore cold habitats, which explains the presence of northern Arctic floristic elements including **relicts of the glacial period**. Furthermore, on a raised bog, these relicts do not have to compete with more rapidly growing and demanding species.

With the exception of *Drosera* species, which supplement their nitrogen supplies considerably by digesting the insects caught on their leaves, all of the other plants are xeromorphic, despite there being excess water at their disposal. This is ascribed to a lack of nitrogen. It has generally been observed that "xeromorphism" occurs whenever the growth of the plant is inhibited by a deficiency, for example, lack or excess of water (leading to poorly oxygenated soil), low soil temperatures which hinder the uptake of nitrogen or direct nitrogen deficiency. These "xeromorphoses" are a symptom of deficiency and it is therefore more accurate to use the term **peinomorphoses** (Greek *peine*, hunger). A survey of the mires of north-western Europe was published by OVERBECK (1975).

## The Western Siberian Lowlands – The Largest Bog Region of the Earth

Of the entire peat deposits of the earth 40% is located in the West Siberian Lowlands; these together with more than 100,000 bog lakes, store a volume of water said to be the equivalent of twice the annual drainage of the immense Ob-Irtysh River system.

This region, comprising the Ob-Irtysh Basin, constitutes a peinohelobiome of almost inconceivable extent. It stretches over 800 km from the forest tundra in the north to the steppes in the south and as much as 1800 km from the Urals west of the Yenisey River in the east. Settlements are found only along the rivers which also serve for communication. A closer study of the bog region proper began only a few decades ago (POPOV 1971–1975). Topography, climate and the hydrological situation can be considered as the factors responsible for the formation of the bogs, which have replaced the dark taiga.

This vast basin has a foundation of Meso-Neocenozoic layers. The ice ages of the Pleistocene left no marks except that alluvial and, in places, water-impermeable sediments were deposited which promoted the retention of water. Peat mosses (*Sphagnum* species) which initiate peat formation, readily took hold in the wet and poor-quality podsol soils. With an annual precipitation of 500 mm the climate is humid; evaporation accounts for only 240–300 mm and drainage for a further 127–270 mm. As far as temperature is concerned, the climate is very continental with 174 days free of frost and 100 days on which the daily mean is above 10 °C. The summers are thus relatively warm and as a result plant production and peat growth are considerable. There are even brief periods of drought so that the danger of forest fires cannot be excluded. Burned areas rapidly turn into bog.

The hydrological conditions of the region are of very special importance. The rivers have not cut deeply into the ground and are extremely sinuous, so that drainage is poor. The spring high water begins in the upper reaches of the Ob and Irtysh 1.5 months before the snow melts in the lower reaches, i.e. when the rivers are still covered with ice in the north. When the ice begins to break it forms high walls which dam the water on the upstream side. Since the sources of the Ob are fed by the glaciers of the Altai range, summer high water soon follows. Thus the rivers run high (12 m above low water) for almost the whole of the short Siberian summer and the low watersheds are flooded and unite with the bog lakes to produce one enormous expanse of water.

*The rivers of western Siberia do not drain the region. On the contrary, their waters become dammed and promote bog formation.*

Bog formation began as far back as the subarctic period of the post-glacial era. The starting point was provided by the wide, shallow depressions containing water poor in mineral salts, in which *Scheuchzeria* bogs could develop, with *Eriophorum vaginatum* and *Sphagnum* species. Mesotrophic *Scheuchzeria* peats corresponding to this period are found at the base of the oldest peat profiles at a depth of 4–7 m. The oligotrophic phase is indicated by the appearance of the most important peat moss species, *Sphagnum fuscum*, beginning in the mid-post-glacial. At this time the bogs bulged upward, the groundwater table rose and as a result the adjacent forests became water-logged. *Sphagnum* species took hold beneath the dying trees and the bogs rapidly spread horizontally in all directions. In all younger bog profiles (the majority of bogs are young) the peat is 3–4 m thick and the lowest horizon invariably includes pinewood and the remains of bark. The oligotrophic phase, with *Sphagnum fuscum* peat, commences immediately above this.

The bogs of western Siberia are **string bogs** (Fig. 279) with a mean gradient of only 0.0008–0.004. On the more or less wide strings *Pinus sylvestris* (in the stunted form *P. willkommii*) and *Ledum palustre* and the dwarf shrubs *Chamaedaphne calyculata*, *Andromeda polifolia* and *Oxycoccus microcarpus* as well as scattered individuals of *Rubus chamaemorus* and *Drosera rotundifolia* grow. The moss layer consists of *Sphagnum fuscum*. Patches of lichens (*Cladonia* spp., *Cetraria*) are rare. Growing in the hollows (wet depressions between the strings or ridges or schlenken) is *Eriophorum vaginatum* with *Sphagnum balticum* or *Scheuchzeria* (or *Carex limosa*) with *Sphagnum majus*, although *Rhynchospora alba* with *Sphagnum cuspidatum* can also be encountered.

String bogs tend to regress on the wet watersheds resulting in formation of bog lakes, particularly where recent tectonic movements involve subsidence. Newer aerial geological surveys have revealed a subsidence of about 0.07–0.25 mm/year in some regions. This is sufficient to disrupt the very labile equilibrium between strings and hollows and bring about the accumulation of increasing amounts of water. This surplus of water initiates the phenomena connected with regression. Oxygen deficiency ensues even in the uppermost peat layers and methane gas is formed. If drilling is done in such places, fountains of liquid peat are forced out by the escaping methane gas. Naturally escaping gas kills off the vegetation and pools form on the dead patches. These pools gradually unite to form larger ones which again grow in size as wave action causes the banks to collapse. Bog lakes of all sizes, together with the wet hollows, form a hydrological system which represents an ecological unit known as a peinohydrobiome (Fig. 279) on account of its low nutrient content (ash content of only 2–4%; WALTER 1977). In places where an independent drainage system develops, the string bogs may become dry. Where the bog runnels cut into the peat, the banks are better drained and a narrow strip of forest consisting of *Pinus sylvestris*, *Betula* and *Pinus cembra* ssp. *sibirica* may develop. The above description of raised bogs applies to the taiga zone. Further north, the thickness of the peat layers decreases due to the shorter vegetational period and smaller production.

Different types of bogs are found to the south of the taiga. In the forest-steppe region, which is zonoecotone VII/VIII in Siberia with birch-aspen forests, the calcium content of the groundwater is already high. Slightly domed eutrophic *Hypnaceae* mires with *Carex* (sedge) species predominate and peat growth is inhibited by the greater dryness of the climate. On such mires islands of oligotrophic forest bogs ("ryami") can develop. In the southern areas, mires are only found in the lowest-lying places, chiefly in the wide river valleys. The ash content of the peat may be very high (19%). "Hummock and hollow" fens are common, the hummocks ("bults") consisting of old clumps of *Carex caespitosa* and *C. omskiana*. Such fens are helobiomes.

Still further south, in the northern steppe zone, the climate is semi-arid. Instead of a river system, innumerable small undrained lakelets are present in the Baraba depression, a situation similar to that in the pampas (p. 389). Some of these lakes become brackish. They are surrounded by eutrophic mires or even halophytic swamps with halophytes and thus represent a transition to a halo-helobiome (zonoecotone VIII/VII).

## Man in the Taiga

In this century the enormous expanse of the taiga has not saved it from being cut up in many places and in others, completely destroyed over large areas by drilling for gas and oil, pipelines and by mineral extraction and the required space intensive infrastructure. For a long period the Siberian taiga was an area which was difficult to access. Fur hunters and individual settlers moved through the area which had been settled in many places by ethnic tribes. However, these settlements were in small areas with large distances between. The original use of the taiga by nomads and widely distributed individual settlements was in every aspect **sustainable.** It should also be mentioned that many large mammals survived all ice ages and became extinct only in the late glacial period. Much evidence supports the theory that this was the consequence of the destructive influence of mankind.

In places, human intervention and the extent of the destruction of the Siberian taiga surpass the destruction of the tropical rain forests.

## Zonoecotone VIII/IX (Forest Tundra) and the Polar Forest and Tree Line

Just as the forest-steppe forms the zonoecotone VI/VII between forest and steppe proper, a macromosaic arrangement of forest and tundra, zonoecotone VIII/IX the **forest tundra,** provides a transition from the boreal forest zone to the treeless tundra. The first sign of a transition is the occurrence of scattered treeless patches, usually on raised ground, within the forest region. These become more frequent towards the north until only scattered islands of forest remain, finally consisting of low, stunted, malformed specimens. Whereas this zone of stunted trees is quite narrow in the mountains, it may extend for hundreds of kilometres on flat land. The tree most typical of oceanic regions is birch (Fig. 281); of the extreme continental regions it is larch (Fig. 282); and elsewhere, spruce. The polar forest line can be assumed to be determined by factors

**Fig. 281.** The Arctic polar forest limit with *Betula tortuosa* above the Torneträsk (oceanic influence) in northern Sweden (Photo: S.-W. BRECKLE)

**Fig. 282.** The Arctic forest limit with *Larix gmelinii* (extreme continental influence) in north-east Siberia, light forest tapering off in the Moma valley of the Cherski mountain chain (Photo: O. AGACHANJANZ)

similar to those governing the tree line in the Alps. Frost desiccation is accentuated by winter storms and the forest extends further north on slopes of the river valleys in places with snow and sheltered from wind, where well-drained soils thaw out to a greater depth in summer or where the rivers flowing from the south carry warmer water. However, failure to regenerate is also held to be one of the factors determining the forest line in polar regions. At their northernmost limits of distribution, trees rarely produce viable seeds; those that are produced are often eaten by animals or swept by storms over the smooth, snowy surfaces far up to the north where they can no longer develop. Only a few of them germinate in warm niches. The thick blankets of moss and lichens found in forest-tundra do not provide a very suitable substrate for the germination and establishment of tree seedlings. Man and his accompanying herds of reindeer play an important role in this part of the world, both on account of the damage done by the animals and, more importantly, as a result of wood utilisation since the natural growth rate of the trees is extremely slow. As a rule, a tree seedling only succeeds in establishing itself if the temperature has been especially favourable for 2 years in succession. Even so, its further growth is very slow and after 20–25 years the tree may scarcely be taller than the plants in the herbaceous layer, the annual increase in height amounting to only 1–2 cm. The size of annual rings of the stem is closely correlated with the July temperatures. The northernmost original forests, taiga with 2–5 m high trees of *Larix gmelinii* is found nowadays on the Taimyr peninsula in Amyras, next to the mouth of the Chatanga at 72°30′N. Growth is very slow, for example a 104-year-old stem had a diameter of 9.5 cm (WALTER & BRECKLE 1990). Open areas in the forest-tundra are usually occupied by dwarf-shrub tundra which also constitutes the southernmost true tundra (Fig. 260). During the warm period of the post-glacial era the forests extended considerably further to the north, as shown by tree stumps found in the peat of the present tundra.

## Questions

1. Where are the large area (zonal) forests with dominant pines?
2. Why do some plants standing almost in water (Calluna in bogs) have a xeromorphic appearance?
3. Why is the growth of forests possible in permafrost soils?
4. What promotes the damming of the water upstream of Siberian rivers and helps to form enormous raised bogs in West Siberia?
5. What do the terms "light", "black or lightless" and "dark" taiga mean?
6. Is it possible for birch forests to form a zonal vegetation? Under what conditions?
7. Which ecological conditions in the taiga are advantageous to form a "herbaceous" layer of lichens?

# IX  Zonobiome of the Tundra
# (Zonobiome of the Arctic Climate)

## Climate and Vegetation of the Tundra

Zonobiome IX comprises two partial areas, situated at great distances from each other, which are very different because of the very different distribution of land and sea masses in the northern and southern hemispheres and the corresponding latitudes. The continental feature of the climate decreases significantly from south to north.

The largest tundra region completely devoid of forest is an area of 3 million km² in northern Siberia. At most, there are 188 days in the year with a mean temperature above 0 °C and sometimes as few as 55. This is caused by the low solar angle during the summer and darkness in winter. The low summer temperatures are partially due to the large amount of heat required to melt the snow and thaw out the ground. Winters are rather mild in the oceanic regions but extremely cold in the continental regions (Fig. 283). However, the cold pole still falls within the forest region at Oimekon near Verkhoyansk, although the mean annual temperature at this point is –16.1 °C and the

**Zonobiome IX** comprises the areas around the poles of the earth. In the northern hemisphere, it is the tundra regions and cold deserts north of the Arctic forest limit. In the southern hemisphere, south of the Antarctic forest limit tundras occur only in small areas and on some islands. The Antarctic is covered by an enormous ice desert.

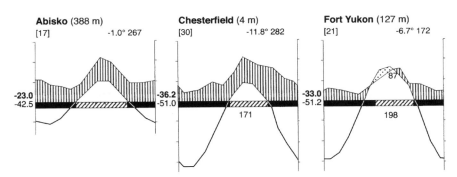

**Fig. 283.** Climate diagrams from the forest tundra of Sweden (oceanic;. *left*), the tundra of North America (*centre*) and the extreme continental boreal region of Alaska (*right*; cf. Fig. 5, Verchojansk and Fig 270)

permafrost extends far down into the ground (Fig. 270). The depth to which the ground freezes in winter has no influence upon the vegetation, since the growth of plants depends only on the thickness of the upper soil layer thawing out in summer.

In the southern tundra the **growing season** commences in June and lasts until September. The wind is of great importance here because it causes irregular drifting of snow, which, in turn, is responsible for the mosaic-like arrangement of the vegetation. In winter, storms can reach a wind velocity of 15–30 m/s. Precipitation is slight, often being less than 20 mm/month. Since potential evaporation is also very low, the climate is humid. Surplus water is unable to seep into the ground because of the permafrost and thus, extensive swamps are formed. However, the amount of peat formed is negligible because plant productivity is low. Snowfall amounts only to 20–50 cm annually, the raised ground being blown free of snow so that the abrasive action of snow and ice are decisive mechanical factors influencing the vegetation.

The steep, stony southern slopes that warm up relatively well in summer when the sun is low in the sky often look like **"flower gardens"**. Together with the banks of streams and rivers, these slopes constitute the most favourable habitats in this zone. Flat, raised ground with stone nets (polygonal soil) and gentle slopes subject to solifluction are, on the other hand, sparsely colonised. Vast stretches of land are covered with dwarf birch and dwarf willows and with *Eriophorum* and *Carex* species. The drier soils support a pure lichen tundra, whereas on wet ground mosses predominate, but there are no *Sphagnum* species. As far as air temperature is concerned, meteorological measurements carried out at a height of 2 m are no indication of the temperature of the low plant cover. By the time the air temperature has risen to $0\,^\circ$C, the ground has usually already thawed out to a depth of 0.5 m and the development of the vegetation is in progress. In fact, the daytime temperature of the plants is often $10\,^\circ$C higher than that of the air, but in spite of this the summer is often too short for the seeds to ripen. In an attempt to overcome this problem, half of the plant species in Greenland, for example, produce their flower buds in the preceding season so that no delay is involved in their coming into flower at the beginning of the following warmer season. Buds and leaves usually spend the winter safely beneath a covering of snow. However, open flowers die. Of particular interest are the **aperiodic species**, such as the tiny Brassicacea *Braya humilis*. Their development is protracted over several years and can be broken off temporarily for the winter at any stage. These species, therefore, are unaffected by the shortness of the summer and may flower either at the beginning of the vegetational season or at a later date, the buds having been formed as much as 2 years previously.

Dispersal of seeds and fruits is carried out by the wind skating them over the snow in 84% of the species and by water in 10%; berries are found only in the forest tundra. The size of the seeds reflects the low productivity of the tundra, 75% of the species produce seeds weighing less than 1 mg. The majority of seeds in the tundra require the influence of the low winter temperatures before they are capable of germination, but in spring they are then in a position to germinate rapidly and have time to accumulate some reserves before the autumn. Some 1.5% of the total species are viviparous including various grasses, as well as *Polygonum, Stellaria* and *Cerastium* species. Open spaces, such as occur on the lower Lena, are colonised rapidly because of the extremely prolific seed production. The majority of tundra species are hemicryptophytes or chamaephytes. The brief period of vegetation at low temperatures is unfavourable for annuals or therophytes (in contrast to conditions in the desert), which are therefore represented only by *Koenigia islandica,* three species of *Gentiana, Montia lamprosperma,* two species of *Pedicularis* and a few others. Most of the plants develop thick roots which serve as storage organs. Individual plants, including herbaceous specimens, may live more than 100 years with dwarf shrubs reaching an age of 40–200 years.

**Nitrogen supply** presents a problem of great importance in this region. Mineralisation and nitrogen uptake are very inhibited because of the low temperatures. Leguminosae (*Oxytropis, Hedysarum, Astragalus*) possess root nodules lying immediately beneath the surface, where the soil warms up in summer. It has recently been shown that *Dryas drummondii,* a pioneer species growing in Alaska, has nodules similar to those of *Alnus* and that the nitrogen content of the soil rises from 33 to 400 kg ha$^{-1}$ during the pioneer stage of colonisation. Animal dung is of great value as a source of nitrogen. If there is no nitrogen in the soil, then only mosses and lichens are found.

In some of the trough valleys in the interior of Peary Land (northern Greenland) at a latitude of 80° the climate is completely different from that of the rest of the Arctic. As a result of the descending winds blowing from the interior there is no rain in summer and desert-like conditions prevail, with salt efflorescence and an alkaline soil supporting a few halophytes. Apart from these plants, vegetation is not completely lacking because the snow drifting down from the mountains in winter provides a source of water when it melts in the spring. Because the soil thaws out to a depth of about 1 m, water can seep down and is then available for plants such as *Braya purpurescens* which can develop taproots more than 1 m in length. Fifty-nine days of the year are frost-free and the mean July temperature is 6 °C.

## Ecophysiological Investigations

The temperature of low plants and the soil is relatively constant during the polar day with 24 h low sun, however, the direction of radiation still has some effects. Differences between plants and air temperature can become very significant during fine days (Fig. 284) and adequate temperatures are the precondition for plants to have an active metabolic exchange.

Arctic plants have well-balanced water relations and their cell sap osmotic potential lies between –0.7 and – 0.2 MPa. The fact that some Arctic plants exhibit xero-

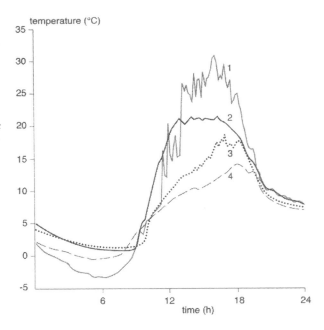

**Fig. 284.** Temperature changes during the day at the soil surface in a catena of Carici rupestris Dryadetum (*CD*) to Salicetum polaris on 29 August 1990, a fine day at Liefdefjord in north-west Spitzbergen (90 m above sea level; from Dierssen 1996). 1 *CD/Carex nardina*-Fazies; 2 *CD/Dryas*-Fazies; 3 *CD/Carex misandra*-Fazies; 4 Salicetum polaris

morphic features can be attributed to an inherited peinomorphosis, caused by nitrogen deficiency, similar to that
seen in plants of raised bogs. A low soil temperature renders the uptake of nitrogen more difficult.

Photosynthesis and the resulting production of organic
matter are of vital importance. The maximum $CO_2$ assimi-
lation does not exceed $12 \, \text{mg} \, \text{dm}^{-2} \, \text{h}^{-1}$. On cloudy days
the $CO_2$ uptake curve temporarily sinks below zero. How-
ever, since photosynthesis can usually be carried on
around the clock with a minimum at midnight when illumination is poor, uptake of $CO_2$ on a summer's day
amounts to $100 \, \text{mg} \, \text{dm}^{-2} \, \text{day}^{-1}$ which corresponds to
60 mg of starch.

Such quantities suffice for the accumulation of adequate reserves in the summer. The primary production of
the plant cover in 1 year in the subarctic regions of Swedish Lapland near Abisko (growth season, 111 days)
amounts to $2500 \, \text{kg} \, \text{ha}^{-1}$, in Alaska (growth season,
70 days), $830 \, \text{kg} \, \text{ha}^{-1}$ and in the high Arctic (growth season, 60 days), only $30 \, \text{kg} \, \text{ha}^{-1}$. In an Arctic willow scrub
in Greenland the phytomass was $5.5 \, \text{t} \, \text{ha}^{-1}$. The "tundra
biome" (zonobiome IX) was intensively studied within the
framework of the I.B.P. (see BLISS & WIELGOLASKI 1973).

## Animal Life in the Arctic Tundra

The extensive tundra of Siberia provides us with one of
the few remaining opportunities of encountering a fauna
in its rather original state and studying its influence upon
the vegetation. The majority of larger vertebrates leave the
tundra in winter, birds migrate to the south and only lemmings and ground squirrels are left behind. The polar fox
and the snowy owl leave the northernmost regions when
prey becomes scarce.

**Lemmings** neither hibernate nor lay up stores for the
winter. Rather, they remain active beneath the hard covering of snow and exist mainly upon the young buds of Cyperaceae. Although weighing only 50 g, a lemming requires 40–50 kg of fresh plant material annually. As a rule,
it colonises well-drained southern slopes and builds a nest
of Cyperaceae shoots in the vicinity of its feeding area,
which for one family, usually covers $100–200 \, \text{m}^2$. An entire
colony occupies about 1–1.5 ha and destroys 90–94% of
the entire vegetation. In such places, *Eriophorum angusti-
folium* does not succeed in flowering. The number of lemmings reaches a maximum, on an average, every 3 years.
The dried-out vegetation is not eaten, but forms a kind of

hay in spring $(1-1.2 \text{ t ha}^{-1})$, which is then washed together and forms peaty hummocks (bults). When they abandon their winter quarters, the lemmings proceed to build on higher ground, throwing up as much as $250 \text{ kg ha}^{-1}$ of earth. On disrupted habitats such as these a characteristic plant community is found which is the beginning of a secondary succession. The same situation is produced by the burrowing of ground squirrels and is comparable to the formation of molehills in European meadows. In this manner, the plant cover is held in a **continuously dynamic state**. Flocks of water birds, mainly geese, arrive in spring and destroy 50–80% of the plant cover by pecking off the young shoots of *Oxytropis* and pulling up the starchy rhizomes of *Eriophorum*. On the bare ground solifluction sets in until a thick covering of moss develops. Nesting and flocking places of these birds are well manured and support nitrophilous species such as *Rhodiola*, *Stellaria*, *Polemonium*, *Myosotis*, *Draba* and *Papaver*.

Although **reindeer** only remain in the tundra in winter if large snow-free areas are available for grazing, they too must be considered as belonging to these regions. In summer, the animals graze singly and have little effect on the vegetation, but in the autumn when they gather together in large herds their trampling is very obvious. Lichens and dwarf shrubs are eaten and destroyed so that grassy communities consisting of *Deschampsia* and *Poa* gain ground. The eating pattern of reindeer is very adaptable according to the supply of food (Fig. 285).

The number of wild reindeer is decreasing nowadays in favour of the domesticated reindeer. In the tundra reindeer are the most important herbivores; in the North American tundra it is the caribou (*Rangifer caribou*), whilst in the Euro-Siberian tundra the reindeer (*Rangifer tarandus*) occurs. Predatory animals such as the polar fox and the occasional bear and lynx have only a small indirect effect upon the plant cover.

It is now known that during the last 20,000 years numerous species of large animals became extinct (MARTIN 1984; SIMMONS 1996). It is even considered that up to 200 genera of large mammals and birds disappeared by the end of the last ice age. In North America about two thirds of the large mammals documented at the end of the last ice age (thus about 13,000 years ago) became extinct. Amongst these animals are 3 forms of elephant, 15 species of hoofed animals (ungulates), many large rodents and predators and 6 species of Edentates (including giant sloth, armadillo and anteater). There are no indications of similar rates of extinction in earlier ice age periods to suggest that extinction was caused by climatic changes.

residence time (%/t)

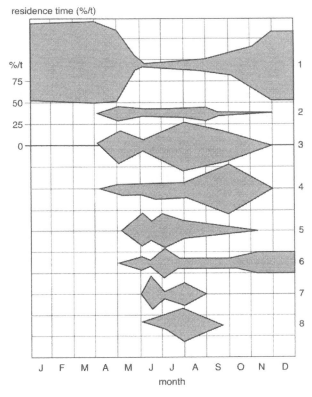

**Fig. 285.** Eating habits of wild reindeer during the year in the Hardangervidda (after SKOGLAND 1983). **1** Loiseleurio-Diapension; **2** Cladonio-Juncetum trifidi; **3** Phyllodoco-Vaccinion myrtilli and Potentillo-Polygonion vivipari; **4** Nardo-Caricion gigelowi; **5** Adenostylion alliariae; **6** Caricion nigrae; **7** Ranunculo-Salicetum herbaceae; **8** Cassiopo-Salicetum herbaceae

Rather, just at the beginning of the last ice age, large waves of Indians crossed the Bering Strait and were the main reason for this mass extinction of fauna. The conquest of the continent from Canada to Mexico probably took place in 350–500 years. Within the period of 500 years most of the species became extinct. Similar times for extinction have been reported from New Zealand, Madagascar and Java, always after the advent of man. The rate of extinction was not quite as dramatic in Eurasia. Whilst in North America at least 24 genera disappeared, probably 9 species disappeared in Eurasia, amongst those the mammoth (Mammonteus primigenius = Elephas p.), woolly rhinoceros (*Coelodonta antiquitatis*), giant elk (*Megaloceros giganteus*), musk ox (*Ovibus moschatus*), steppe bison (*Bison priscus*), a species of buffalo (*Homoioceros antiquus*), as well as three carnivores

(SIMMONS 1996). The musk ox has been re-introduced in some locations. Besides the better hunting techniques developed by prehistoric man, the rapid warming and the advance of forest vegetation is probably also responsible for the sudden rate of extinction of large animals.

##  Man in the Tundra

Annual movement of reindeer herds has greatly influenced the hunting behaviour of people in the tundra. Domesticated reindeer herds, for example of the Tungus, move in summer from the taiga into the tundra and in early autumn back into the taiga. Several Eskimo tribes adjusted to the Arctic conditions and live in the tundra. These tribes trade amongst each other, some predominantly hunt whales and seals, others are active further inland as caribou hunters, but also catch mountain sheep, elk, beaver, bear, blue hare, ducks and geese. Meat, fat and whale oil is traded for caribou meat and berries (CAMPBELL 1985). The lifestyle of Eskimos in northern Alaska and particularly their style of housing in characteristically round huts is an example for the possible lifestyle of man during the ice ages in Europe, for example, in the Magdalenian period. Huts are built half underground and were well packed with grass and covered by a roof of whale ribs with animals hides which provides excellent insulation. Similar building styles are known from many excavations from the Magdalenian period with the first documentation of Cro-Magnon man 30,000 years ago and the main activity in southern France (Dordogne) 19,000–13,000 years ago. Reindeer was the main meat source of Cro-Magnon man, but remains of bison, mammoth, horse and wild cattle have also been excavated from homesteads. It is probable that the systematic exploitation of the rich availability of wild animals accelerated with Cro-Magnon man. This form of food provision, mainly from wild animals, was well established at the end of the upper Pleistocene. Even today, it is similar for the tribes of the tundra, however, with improved technological methods and the availability of dogs, boats and sledges. Stone age man in the tundra of the ice ages in central Europe as well as Eskimos had highly developed technical aids: well insulated huts, clothing, traps etc. and even simple machines such as harpoons and catapults. Today, Western civilisation has caused far-reaching changes. Alcohol is a massive problem. Snowmobiles replace dog drawn sledges and hunters use guns. Today, a few Eskimos can decimate a large caribou herd within one day. Hunting has become very simple and it is no longer important that all parts of a slain animal are used; only the best parts are taken. Too many wild animals are hunted in order to be sold.

## The Cold Arctic Desert and Solifluction

The cold Arctic desert is the northernmost of the three subzonobiomes of zonobiome IX. Here, too, oceanic and continental regions can be distinguished (ALEKSANDROVA 1971). In the cold desert days on which the temperature rises twice above zero are very frequent; this causes soil movement, a phenomenon known as **solifluction.** In the tundra itself, as in the alpine and nival altitudes of mountains, local ice formation leads to an increase in volume of the wet soil under the plant cover and to the formation of peat hummocks and frost hillocks or bult tundra. Even on slopes with a very slight gradient the soil is pushed downward stepwise, giving the impression of cattle paths, shallow terraces running parallel to the contours; Fig. 286 shows a cross section through one of these steps. Earth movements of this kind become increasingly pronounced towards the north.

In autumn, where a wet unfrozen layer is sandwiched between the permafrost layer below and a freezing layer above, the ground may burst through the upper frozen crust in places and cover the vegetation with a layer of liquid clay. A bare patch several centimetres higher than the surroundings is thus formed (Figs. 287, 288); this landscape is called "patchy tundra" (Fig. 290).

Another effect of frost is that stones gradually work up to the soil surface. Figure 289 illustrates this process. As

From south to north the three following subzonobiomes can be distinguished in the Arctic tundra:
1. dwarf shrub tundra in the region covered by forest in post-glacial times,
2. true moss and lichen tundra,
3. cold desert, commencing where plant growth becomes sparse.

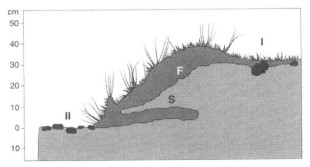

**Fig. 286.** Flow of soil on a gentle slope in the Arctic (Alaska). The fibrous peat layer (*F*) with the living plant cover, has moved about 30 cm, from *I* to *II*. A fold has formed in the process, into which the free silt soil is trapped (*S*) (after WALTER 1960)

**Fig. 287.** Tundra with larger soil patches and block polygons in Sognefjell (Jotunheimen, central Norway) (Photo: S.-W. BRECKLE)

**Fig. 288.** Patchy tundra with a vertical section through one of the patches. The patches develop because the liquid clay between the permafrost soil below and the freezing layer above is pressed up and flows over the surface forming a slightly raised clay patch a few centimetres higher, without vegetation (after WALTER 1990)

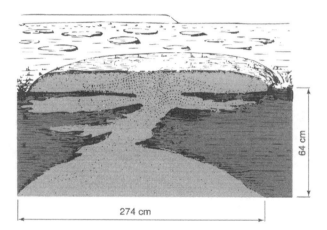

the upper layer of the soil freezes it draws up water from below and increases in volume, at the same time carrying stones of the freezing layer upward. Beneath each stone, a cavity is left in which fine sand collects, with the result that when the ground thaws the stone is higher than before. If the process is repeated time and again on days when the temperature rises above zero and then sinks again, the stone finally reaches the surface. The soil usually freezes from a number of locations separated by distances of one or more metres. Stones are not only lifted out of the soil, but are also displaced laterally. The result is a network of stones between the freezing locations, known as **polygonal or stone-net soil** (Fig. 289). Solitary

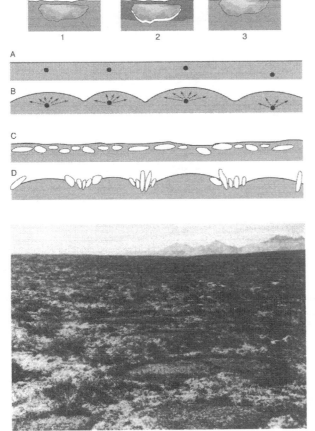

**Fig. 289.** *Top* Schematic representation of the processes involved in freezing and thawing of the ground: *1* before freezing; *2* soil frozen on top, but stone is lifted; *3* after thawing, the stone moves to the surface. *Below* Stone-network formation. **A** At • freezing centre; **B** *arrows* indicating the direction in which the stones move; **C** original position of stones in the ground; **D** final position, after formation of the frost-stone network, or polygonal soil (cross section) (from WALTER 1960)

**Fig. 290.** Patchy soil tundra in the Cherski mountains of eastern Siberia at 1100 m above sea level. *Betula exilis* and *Rhododendron parviflorum* predominate. Frost patches are clearly recognisable in the foreground (Photo: V.N. PAVLOV)

plants take refuge between the stones where movement is at a minimum. If this takes place on a slope, the stones are also pushed down the gradient giving rise to a "stone stream", or striped soils. Earth movements of this nature mean that the Arctic vegetation is constantly disturbed and suffers accordingly, as can be observed in Iceland (LÖTSCHERT 1974) and even more clearly on Spitzbergen. Solifluction plays a similar role in mountainous regions particularly in the upper alpine and nival belts, although the areas involved are not as large as in the Arctic. As regards the composition of Arctic vegetation, floristic variation around the North Pole is relatively small.

**Fig. 291.** Antarctic cliff ice-
desert with thin cover of lichens on the cliff (Admiral-
ity Bay, King George Is-
land). On the flat terrace
banks, penguins gather in
breeding colonies (Photo:
L. KAPPEN)

## Antarctic and Subantarctic Islands

Only two flowering plants have been found on the edges
of the ice-covered Antarctic continent: *Colobanthus crassi-
folius* (Caryophyllaceae) and a grass, *Deschampsia antarctica*. In recent times *Poa pratensis* has been imported, but
otherwise only mosses, terrestrial algae and lichens are to
be found, in total some 100 species. Plants are confined to
places on the coast that are sometimes free of snow
(Fig. 291), such as steep cliffs and talus, but their phytomass is quantitatively negligible. Bacteria and fungi have
been found in soil samples. Penguins and many other animals occurring at times in the coastal regions of the Antarctic find their food in the sea. Whether there are invertebrates living on the lichens or the algae on stones is not
known.

The sea surrounding the Antarctic, with its continuous
westerly storms, is scattered with many tiny islands, most
of them south of the 50th parallel. They are bare of trees
since the summers are cool, although the winters are not
cold. Conditions are almost isothermic on the islands, e.g.
the temperature fluctuates between 2.8 and 7.7 °C on the
Macquarie Islands (at 54°3′S) over the entire year. Driz-
zling rain and fog are typical of the weather. The Antarctic islands are sometimes referred to as wind deserts since
only in sheltered places is there a somewhat more luxuriant vegetation. The commonest plant on the Kerguelen
Islands is the dense, cushion-shaped *Azorella selago*
(Apiaceae). In earlier times seafarers were aware of the
anti-scorbutic action of the large-leaved Kerguelan cab-
bage, *Pringlea antiscorbutica* and used it as a fresh vegeta-

ble. *Acaena* species are common on all of the islands and also many mosses, ferns and lichens occur, together with tussock grassland (*Festuca* and *Poa* species). As in all very windy habitats a variety of cushion-plants are characteristic of the subantarctic. Terrestrial conditions in the Antarctic have been dealt with by HOLDGATE (1970, vol. 2, parts XII–XIII).

## Questions

1. How long must the vegetational period last so that phanerogam vegetation can occur?
2. Why do continental characteristics increase strongly between 60° to 80°S and decrease strongly from 60° to 80°N?
3. What are the effects of inhibited mineralisation of organic substances for the tundra?
4. How does the typical podsol soil of ZB IX differ from that of ZB VIII?
5. What is typical for a wind desert and where does it occur?
6. What are the preconditions for the occurrence of solifluction processes?
7. How can the development of the polygon soil be explained?
8. Which higher plants comprise the flora of the Antarctic?

# Summary and Conclusions

## Phytomass and Primary Production of the Various Vegetational Zones and of the Entire Biosphere

The geo-biosphere covers the earth's surface as a thin layer, indeed the thinnest skin. It includes the upper horizons of the soil, the atmosphere near the ground, insofar as organisms penetrate this space and all the surface waters. In this space all **biological metabolism** occurs. More than 99% of the earth's biomass is phytomass and discussion is limited to its distribution. Amounts of phytomass are distinctly related to zonobiomes.

Accurate determination of phytomass and primary production is difficult. BAZILEVICH et al. (1970) published calculations based on the rapidly accumulating literature for the various thermal zones and bioclimatic regions of the earth. These authors calculated mean phytomass and mean annual primary production for the various regions as dry mass (in metric tonnes) per hectare (t ha$^{-1}$). On the basis of measurements of the areas covered by individual regions, excluding rivers, lakes, glaciers and permanent snow, total phytomass and total annual primary production for the various regions were obtained. The sum of these figures is the phytomass and annual production of the land surface of the earth. In addition, Table 23 gives corresponding data for the waters of the earth. The values involved are potential, i.e. they are based on natural vegetation not influenced by man.

BAZILEVICH et al. (1970) distinguish five thermal zones: (1) polar (Arctic), (2) boreal, (3) temperate, (4) subtropical and (5) tropical. The first two zones have humid climates; the three others are subdivided into three regions: humid (h), semi-arid (s) and arid (a) (see Fig. 292 and Table 23). This structure differs from the division into zonobiomes as the following shows:

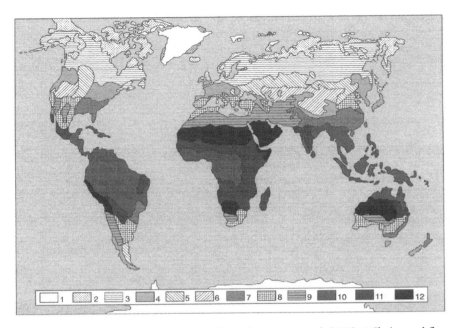

**Fig. 292.** Thermal and bioclimatic zones according to BAZILEVICH et al. (1970). *1* Glaciers and firn regions; *2* Arctic zone; *3* boreal zone; *4–6* temperate zone: *4* humid region; *5* semi-arid region; *6* arid region; *7–9* subtropical zone: *7* humid regions; *8* semi-arid regions; *9* arid regions; *10–12* tropical zone: *10* humid regions; *11* semi-arid regions; *12* arid regions

| Thermal zones and climatic regions | Zonobiome |
|---|---|
| Zone 1 | ZB IX |
| Zone 2 | ZB VIII |
| Zone 3 h, s, a | ZB VI and VII |
| Zone 4 h, s, a | ZB V, IV and III (outside the tropics) |
| Zone 5 h, s, a | ZB I, II and III (within the tropics) |

Comparison of terrestrial production with that in the oceans shows that the latter, with $60 \times 10^9$ t equals only about one third that of the land, although the surface area of the oceans is almost three times larger. It is remarkable that the phytomass in the oceans is minute in comparison with the 300 times greater primary production. This is understandable since plankton plants are single-celled and divide and multiply continuously (see p. 101). In contrast, terrestrial primary production is 7% of the corresponding phytomass.

Adding up the total supply of atmospheric active carbon of the earth yields about 748 Gt C (mostly as $CO_2$) in

**Table 23.** Distribution of potential productivity of the earth

| Climatic zone | Area $(10^6 \text{ km}^2)$ | Phytomass | | Primary production | |
|---|---|---|---|---|---|
| | | Total $(10^9 \text{ t})$ | Average $(\text{t ha}^{-1})$ | Total $(10^9 \text{ t a}^{-1})$ | Average $(\text{t ha}^{-1} \text{ a}^{-1})$ |
| Polar | 8.5 | 13.8 | 17.1 | 1.33 | 1.6 |
| Boreal | 23.20 | 439 | 189 | 15.2 | 6.5 |
| Temperate | | | | | |
| Humid | 7.39 | 254 | 342 | 9.34 | 12.8 |
| Semi-arid | 8.10 | 16.8 | 20.8 | 6.64 | 8.2 |
| Arid | 7.04 | 8.24 | 11.7 | 1.99 | 2.8 |
| Subtropical | | | | | |
| Humid | 6.24 | 228 | 366 | 15.9 | 25.5 |
| Semi-arid | 8.29 | 81.9 | 98.7 | 11.5 | 13.8 |
| Arid | 9.73 | 13.6 | 13.9 | 7.14 | 7.3 |
| Tropical | | | | | |
| Humid | 26.5 | 1166 | 440 | 77.3 | 29.2 |
| Semi-arid | 16.0 | 172 | 107 | 22.6 | 14.1 |
| Arid | 12.8 | 9.01 | 7.0 | 2.62 | 2.0 |
| Geo-biosphere | | | | | |
| Landmass | 133 | 2400 | 180 | 172 | 12.8 |
| Glaciers | 13.9 | 0 | 0 | 0 | 0 |
| Hydro-biosphere | | | | | |
| Lakes and rivers | 2.0 | 0.04 | 0.2 | 1.0 | 5.0 |
| Oceans | 361 | 0.17 | 0.005 | 60.0 | 1.7 |

(after BAZILEVICH et al. 1970)

the atmosphere, about 2000 Gt C are bound to the land surface and 38,000 Gt C are bound in the oceans, thus about 50 times more than in the atmosphere (T. Boden, pers. comm.). For the mass of consumers and decomposers on all the continents, a figure of only $20 \times 10^9$ t dry mass is given, less than 1% of the phytomass. In the oceans, on the other hand, values of $3 \times 10^9$ t dry mass are calculated, which is more than 15 times the phytomass in oceans. In contrast to single-celled plants, consumers in oceans include large animals which are exploited for human consumption. In contrast, various examples have already been given to illustrate the comparatively low zoomass of large terrestrial consumers. The phytomass on land is mostly wood in forests. This is 82% of the total phytomass on all the continents, although forests cover is only 39% of the land area. The principal part of the forest phytomass, about 50%, is found in tropical forests, about 20% in the boreal zone and about 15% in each of the subtropical and temperate zones. These figures should be memorised for "global change" discussions. Phytomass in deserts is only 0.8% of the land total, a very small amount considering that deserts occupy 22% of the land area.

The average phytomass in forests of the humid regions rises continuously with increasingly favourable temperature conditions, from 189 t ha$^{-1}$ in the boreal zone to 440 t ha$^{-1}$ in the tropics. On the other hand, the average phytomass in the tropical arid regions is the smallest with 7 t ha$^{-1}$. Drought combined with continual high temperatures is particularly unfavourable for plant growth.

The total annual potential primary production of the biosphere on land, in the oceans and in lakes and rivers is 233×10$^9$ t. Land masses contribute 172×10$^9$ t, lakes and rivers 1×10$^9$ t and oceans 60×10$^9$ t.

As for the average annual primary production, on land it is 12.8 t ha$^{-1}$, i.e. more than seven times that of the oceans and about 2.5 times that of the lakes and rivers with their aquatic and swamp vegetation.

Approaching the equator, the primary production of the humid land regions increases. It doubles from the boreal to the temperate zones and doubles again from the temperate zones to the subtropics, but increases little from the subtropics to tropics. Differences between the humid and semi-arid regions are not so great as those between the respective amounts of phytomass since the wood in the forest does not contribute to production and it is the leaf area that is important (see comparison of meadows and forest in Solling, p. 350). The relatively high production of 13.8 and 7.3 t ha$^{-1}$ in the subtropical semi-arid and arid regions, respectively, is noteworthy. Production depends on the often very luxuriant and productive ephemeral vegetation which can develop during the favourable, cooler rainy part of the year.

LIETH & WHITTAKER (1975) give rather different figures. Instead of calculating the potential production, they worked out the actual production, also considering the areas under cultivation. Their values for terrestrial production are thus lower. The most exact figures available for primary production are given by these authors, who arrive at the figure of 121.7×10$^9$ t dry mass for an area of 149×10$^6$ km$^2$ of land.

At the time when the human population was 3 billion with a biomass of 0.2×10$^9$ t, it can be assumed to have

consumed approximately the total agricultural product of
0.7% of the primary production of the biosphere. Energy
consumption is estimated to be $2.8 \times 10^{18}$ cal since only a
part of the energy taken up as food is utilised. These figures do not seem particularly high, but it can be assumed
that, since then, the rapid population explosion has resulted in a significant increase in consumption.

## Conclusion from an Ecological Point of View

This volume is a summary of the great natural, ecological
relationships of the entire geo-biosphere, comprehension
of which is essential for an accurate prediction of the dangers arising from increasing human interference in natural
systems. The effects of this intervention are so numerous
and so profound that they cannot be discussed within the
framework of this review. Thanks to human mental capacities, humankind has created, side by side with the natural world, its own **apparently independent world**, one of a
technically oriented world economy. As a result of increasing urbanisation, humans have become more and more
estranged from nature. Humankind is losing the ground
beneath its feet, considers everything technically possible
and believes in unlimited economic growth.

Based on detailed objective investigations, the Club of
Rome warned as early as 1972 of the Utopian nature of
this attitude and predicted an economic crisis unless
counter-measures were taken immediately (see also
GRUHL 1975). Nothing was done. However, the crisis occurred locally a long time ago and regionally in many
places. Still we are being told to watch for the rosy glow
of economic growth somewhere on the horizon. Although
everyone speaks of "ecology", a basic change in our way
of thinking has not taken place. The so-called economic
forces still have priority over everything else. Destruction
of the environment, on which the existence of mankind is
dependent, continues across the entire globe almost unre-
lentingly. Local damage is simply hidden by cosmetic
measures, although the problems are on a global level.
The two greatest dangers to be mentioned here are:

1. population explosion
2. excessive technological developments.

These were discussed in detail by WALTER in a different
publication (WALTER 1989). Considering this publication,
one has to ask how it will be possible, with sustainable
land use, to maintain the basis of life for humans for

many generations of mankind (thus for centuries to millennia).

## Population Explosion in Developing Countries

AURELIO PECCIO, president of the Club of Rome, warned emphatically in the 1981 German edition of his publication, *The Future in our Hand* that the earth's population was increasing at such an alarming rate that something must be done about it immediately. According to PECCIO, 223 children were born on the earth each minute, that is 321,000 each day and 120 million each year. This number was significantly higher in 1999. However, the exponential increase has slightly dropped in recent years.

At present, the population of the earth is increasing by 1 billion in less than 15 years. If it were possible to keep all new-born children alive, which is obviously our goal, in the next 10 years there would be 1.2 billion children on earth under 10 years of age to be provided for and sent to school. In another 10 years jobs would have to be created for them and they would, in turn, begin to have their own children. The population explosion proceeds with an **exponential increase** (Fig. 293). Two thousand years ago the earth's population may be estimated to have been approximately 200–300 million. Up to 1800, the population increased only slowly.

At about that time the population started to increase rapidly. T. R. MALTHUS warned as early as 1798 of the danger. However, the industrial age was just beginning in western Europe and workers were needed. Thus, MALTHUS's warning was not heeded. Especially in industrialising countries with their medical advances, the population in the cities rose significantly. More recently, as a result of a higher standards of living and the increased desire for a more comfortable life, the number of children per family has decreased so rapidly that the population of the industrial countries has now almost ceased to grow.

Colonialism and communism as well as capitalism with the dogma of exponential economic growth are perhaps nothing else but an "interglacial error" (after Succow, pers. comm.).

However, the situation is almost catastrophic in all other countries. Illness and epidemics were successfully curbed in these countries as well, so the death rate is declining, but the birth rate is hardly falling. As a consequence, the population has and is increasing with great speed; it has happened within a few decades. This is only a moment in time in the development of human history. The situation in these countries in which millions of people are undernourished or starving, is due to the population explosion and must be met with counter-measures

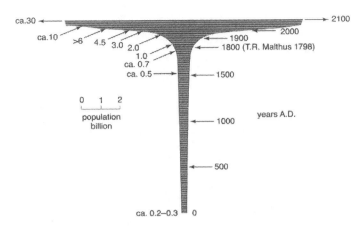

**Fig. 293.** Growth of the earth's population from 0 a.d. to the present in the mushroom shape of an atomic bomb explosion. It took millions of years from the first appearance of man on earth for the human population to reach 1 billion in the middle of the nineteenth century. One century later the population had reached 2 billion; in 37 more years, 3 billion; and in only 13 more years, 4 billion. It passed the 4.5 billion mark in 1970, exceeded the 6 billion at the turn of the millennium and may reach 10 billion in about 2030 a.d. It is not possible to depict at this scale the population level in 2100 a.d. because by this time the top of mushroom cloud will have doubled in width to over 20 billion. According to recent calculations, the number should level off at 11–14 billion (Birg, pers. comm.), whilst the population in Europe should shrink considerably from 2005 onwards unless there is strong immigration

immediately. Hunger is only a symptom. It is the natural result of such uncontrolled growth, a phenomenon which holds true for all living beings in ecological systems, including mankind. The unlimited exploitation by one species at the expense of all others cannot be upheld. No foreign aid programmes are capable of annulling this law of nature. Humans, with their earthly bodies which must be nourished, are a part of nature. Although well-intentioned, food-aid programs are especially dangerous, since they only tend to provoke the population to increase more; it is as if one were trying to extinguish a fire with oil.

A foreign graduate of the agricultural school in Stuttgart Hohenheim in Germany, who later became a professor of agricultural sciences in his own country, warned in a radio show: "Keep your hands off the developing countries! They must solve their problems on their own. Foreign aid programs only inhibit this."

Notably honest in this respect is also the statement of the foreign aid consultant of the World Church Council, JONATHAN FREYERS, which is based on his experience. However, it caused outrage among a wide range of uninformed individuals. According to a newspaper article, he believes that the food delivery programmes are causing terrible damage. Since the food is distributed among the poorest, the farming population no longer makes an effort, thereby also becoming a welfare recipient. Domestic production then collapses and the number of welfare recipients increases even more in an endless vicious cycle.

More recent mottos, such as "Help them to help themselves" consider neither the lifestyles nor the way of thinking of the native populations. Their ways of life have been determined by strict cultural codes over thousands of years. Their cultural practices were optimally adapted to their environments, otherwise they could not have survived over the centuries.

Even colonial rule did not change much in this respect. The power struggles between individual tribes were stopped and unsettled land was converted to farmland or plantations on which workers were able to earn money. This began a gradual inclusion in the European economic system. Sudden independence and the attempt to develop unified countries with democratic rule within the borders of the former colonies resulted in chaos and tribal warfare everywhere. The uneducated masses were unable to compensate and fill the void for which Europeans required several thousand years of development. The Europeans of 2000 years ago would certainly also not have been able to accomplish this successfully. The strict cultural rules restricting sexual relationship were also no longer effective and the door was opened to an uninhibited reproductive drive, followed by an enormous increase in the birth rate. Private possessions in the European sense had been unknown, everything belonged to the tribe and was regulated by the tribe and, therefore, the drive for private initiative was lacking.

Most foreign aid workers return home deeply disappointed. As long as a project is directed with the necessary leadership, the work is accomplished willingly and diligently even in the missions. As soon as the guidance ceases nothing else is undertaken and only the outer façade is kept intact, which, however, is insufficient. Those who never worked practically in developing countries themselves often voice the argument, "But something must be done; we have to help the developing countries." It should not be forgotten, however, that these are sovereign countries which are very distrustful and consider any fundamental advice as an attempt at neocolonialism. This is especially true for advice regarding the control of the population explosion. As long as this is not achieved, however, any form of help is senseless or even detrimental.

Of course, it is true that the development of commercial industries, SOS childrens' villages, care for the blind etc. are of great help for the people they serve and must be praised, but they do nothing to change the basic catastrophic situation which continues to grow worse. Even the more or less secondary argument that **foreign aid** is also helping to open up new markets for Western industrial products, for which the developing countries represent an unlimited demand, is not based on reality.

The products can only be delivered on credit which cannot be repaid and on which interest cannot be collected. Such countries as Brazil and Mexico, although rich in natural resources, exemplify this fact. In addition, it must be recognised that an economic system is being forced on the developing countries, but it is unclear if it is capable of guaranteeing mankind a lasting existence, or whether it will itself someday pop like a shiny soap bubble. In the past, every civilisation which had become estranged from nature eventually collapsed and was replaced by "barbarians" whose way of life was more in unison with nature. However, nowadays the estrangement and the problem can no longer be solved regionally, but must be solved globally.

## Excessive Technological Developments in Industrial Countries

Technological advances have made it possible to continually raise the standard of living in industrial countries, which is regarded as progress. This progress is measured by the gross national product (including for example all car repairs following accidents) or by the average per capita income (between $ 250 old age pension and millions of pounds for consultant fees). A further desired goal is to shorten the number of working hours as much as possible and thereby assure the longest possible amount of leisure time so that everyone is able to have the possibility for "self-realisation", because they are worth it.

This ideal has already been achieved, not in one of the industrialised countries, but in the smallest island nation on the coral island in Nauru in the Pacific Ocean (approximately $2\,°S$ and $164\,°E$). There, one should expect to find the happiest people on earth. The average per capita income is above that of the wealthiest industrialised countries. The weekly working hours number zero and the amount of free time is unlimited (report in IWZ, 8 to 14 January 1983). Children are born as pensioners. The island is $21.4\,km^2$ in size and rises to $60\,m$ above sea level.

The population numbers 4000 Nauruans. A deposit of fossil guano several feet deep, the purest phosphate deposits known, is located on the island. This was discovered by the German Colonial Office in 1900, which soon began exploitation. After World War I, Great Britain, Australia and New Zealand subsequently continued the exploitation with various degrees of intensity. In 1968, the Nauruan chief, Hammer de Roburt, was able to negotiate the island's independence within the British Commonwealth and in 1979 the phosphate reserves became the property of the Nauruans. Since then Nauruans no longer have to work. This is done by foreign "guest workers" from Australia, New Zealand, Hong Kong, Taiwan and elsewhere, who are not allowed to obtain Nauruan citizenship. An annual yield of 2 million t of phosphate is being sold on the world market.

The main pastimes of Nauruans is sleeping, eating (corpulency is considered beautiful) and watching television (favourites are Mickey Mouse and wild-west films as well as Australian advertisements). Their obesity makes sports too strenuous. They have the most modern automobiles for driving along the 18-km-long road around the island. Empty beer cans litter the countryside. One hobby is fishing in high-horsepower motor boats. They have provided themselves with the luxury of "Air Nauru" airlines, which runs on a deficit with six jets flown by Australian pilots to Melbourne, Hong Kong, Manila and Samoa, as well as a luxury ship line.

The future is being secured by saving two thirds of the income in the Nauru Royalties Trust and securely investing it in real estate, hotels and businesses in foreign countries. The "Nauru House" in Melbourne, with its 50 floors, is the highest commercial building in Australia. There is a "but" in all this, however. It is estimated that the phosphate reserves are only enough to last a few more years. All that will remain is a sterile coral landscape with 10–20 m high tooth-shaped cliffs. If asked why they do not use their resource more frugally, Nauruans reply that they are no different from the rest of the world and they love money just as much as the Europeans and Americans do. They just live from day to day as long as they have enough. The industrialised countries are actually behaving similarly. All warnings that the natural resources will soon be depleted have not changed anything. More important are the next elections and unpleasant decisions are postponed for later, even the increasingly urgent ecological problems.

It cannot be denied that most people in the industrialised nations do not know how to utilise their **leisure** time; it has to be organised and commercialised at the same time. The leisure industry has become a lucrative business, exemplified by the numerous travel agencies and mass holiday quarters in the rapidly growing domestic and foreign holiday resorts with their amusement centres. Holiday guests does not have to worry about anything; they can passively let everything happen and only have to pay the appropriate price. In foreign countries, they live in a ghetto, just as if they were at home, even though the desperation of the developing countries cannot be overlooked.

What is the benefit of this mass tourism besides the consumption of photographic material? Other than that, it is only a passive assimilation similar to the flood of manipulated information from the mass media. It comes up on the observer so fast that there no chance of it being subject to serious thought. This also holds true for education in the schools, as well as in the universities. The amount of information to be conveyed increases constantly and there is no time left for the critical discussion of problems.

Independent thinking is not promoted. Many believe that thinking can be left to computers. Science has been divided into innumerable specialised disciplines and is approaching the danger of becoming a Tower of Babel. The **masses** have rendered fruitful discussions in small circles impossible. A mass lecture is not much different from a television programme. Students passively absorb the information and only begin to study a few weeks before the examinations. This type of knowledge does not last very long. Understanding is lacking.

It has been warned that people are developing an increasingly hostile attitude towards technology. The real problem, however, is an increasingly inhuman technology. Technology, which is supposed to help mankind by making his life more comfortable and pleasant, has developed its own independent dynamics and is forcing the human masses more and more into their own mould and into dependency. It should also not be forgotten that the main purpose of technology has always been the development of weapons. Wars have always given technology its greatest impulses towards further advances. New discoveries were always used immediately for **weapons technology**. If it had not been for the two World Wars, technology and mass production would not yet have attained their present levels. Although the weapons arsenal is enough to extinguish humankind ten times over, the armament race goes on with no end in sight. Unfortunately, experience has taught that new weapons are always put to use.

The inhumanity of technology is also expressed by the destruction of the environment. While in central Europe large areas of forest are sacrificed to technology every year, the environmentally conscious corporations in Japan are trying to increase the amount of forest. All steel works of the Nippon Steel Coop., all plants and research institutes of Honda Motors Co. and of Topay Industries, the power plants of the Tokyo Electric Co. and of the Kansai Electric Co. and others reforest the areas around their industrial complexes to have them serve as air filters and recreational areas. Indigenous tree species have already attained a height of 10 m (MIYAWAKI 1983). Wood can be imported from Borneo.

In Europe, however, the buildings are merely cement blocks surrounded by parking lots or naked lawns at best. The rest of the environment is poisoned. Although the permissible values for the individual poisons may not be exceeded, no one knows whether these values are effective for the accumulation of a number of different poisonous substances. The increase in allergies and the accumulation of pollutants or heavy metals in arable land may be mentioned. Pollutants in human milk were very high in the 1970s, but breast feeding was not discouraged because substitutes contained just as many pollutants and the advantages of breastfeeding also had to be considered. In the 1990s the pollution in human milk has considerably decreased because of better nutrition and lower permissible values for pollutants.

The increasing use of **technology in agriculture** is an especially serious matter. Formerly, the large, mostly autonomous farms which ran without external sources of energy, were the only industries existing in a kind of harmonious equilibrium with the environment. These are now being replaced by agricultural factories with huge amounts of organic refuse which has become difficult to dispose of. Cleared, monotonous landscapes have developed. One of the serious consequences is the accelerated rate of decline in biodiversity since 1960. "Red lists" indicate not only an increasing danger to rare, usually specialised, species but also to previously widely distributed and common species (RUCKDESCHEL 1996). Not only agriculture, but also forestry and hunting, tourism, open cast mining, industry etc. are the main cause of the drastic reduction of the diversity of small habitats (see Figs. 211, 215).

Farms have been absorbed into the whirlpool of mainstream world economy and have lost their ability to withstand economic crises. Technology, therefore, is undermining mankind's natural basis of life! For this reason, ecologists cannot be expected to accept this excessive technology with open arms. It is their duty to warn repeatedly of the approaching danger.

People can do without very much and they can live on very little if they have to, but they need clean air to breath, clean water to drink, food free of poisons and the natural use of their body strength. Products of technology are not objects of necessity. Instead, they only serve for greater comfort or prestige. Worldwide **propaganda** and insistent **advertising** is used to stoke demand artificially. Everyone should be able to have everything. The main purpose of technology is not to serve human interests, but rather to generate profit and serve economic interests, especially those of large corporations. Rationalisation in industry (robots, micro-electronics) is increasingly forcing human workers out of the production process and is degrading them to mere consumers of the mass-produced wares. But how can the masses buy industrial products if their earnings are not assured because they have become unemployed? The economic force of competition is brought as an argument – an endless spiral and vicious circle!

A new modesty is required. Better to be poor and healthy than rich and half-dead.

Technology has made people neither happier nor healthier. Civilisation illnesses of physical and psychological nature are constantly on the rise. The increase in the average life span can only be attributed to the greater amount of medicine, the costs of which are rising astronomically. Walter expressed his opinion on this aspect as follows: "Looking back over the past eight decades of my own life, it is difficult to find criteria according to which the merits of technology should be judged. Any judgement is certainly subjective. Stress was definitely not a problem in earlier times. Even the ocean crossings for research excursions were a pleasant recovery before and after the work was done and allowed for a gradual adaptation. This does not hold true for modern trips by air; even the transition to the new climate, the new environment and new time zone is much too abrupt."

"Mass processing" would not be possible without technology. It has led to the increasing number of unemployed, who have become a heavy burden on the future. And the solution to the problem is so easy, requiring only a little more solidarity and less egoism. Even the most optimistic economists no longer reckon with economic growth as it was a decade or two ago. Even the population of Europe is too large.

## Sustainable Land Use

**Sustainable land use** is only possible in an area where, over many generations, the use of land, settlements and living from the land are in harmony with the existing vegetation and fauna. This includes maintaining the productiveness of the soils over long periods.

Each living being is influenced by its environment and conversely, also influences its environment. The latter becomes more obvious, the larger the population density. In the meantime, mankind has achieved a frightening population density and the effects on the environment increase exponentially (see Figs. 293, 294).

Land use always changes the soils and increases erosion. The formation of soil, however, is a lengthy process. Soil erosion destroys the most valuable resources for centuries or thousands of years. This, however, differs in the individual zonobiomes.

Sustainable land use can only be achieved if the population density (including towns) does not exceed a certain value and if the methods of land use for agriculture, husbandry and forestry are related to natural processes, resulting in the introduction of a **cyclic economy** at all levels. This necessarily also includes industrial processes.

Meanwhile global change is noticeable everywhere. The effects are shown in Fig. 294 including particularly the changes in chemical composition of the atmosphere. An increase in $CO_2$ and other trace gases ($CH_4$, $N_2O$, CFC and others) will lead to a changed equilibrium of the irradiation of the earth. Seasonal swings during the year in the northern hemisphere, shown in the so-called Mauna-Loa curve, have been known for a long time (Fig. 295).

**Fig. 294.** Components of global change (after VITOUSEK 1994). *Thick arrows* indicate strong effects

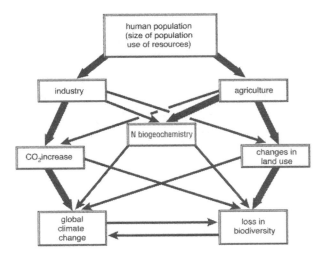

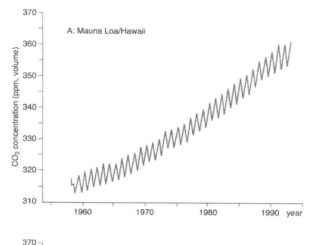

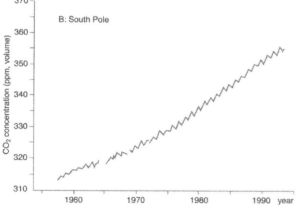

**Fig. 295.** $CO_2$ concentration in the atmosphere on Mauna Loa in Hawaii (**A**) and on the South Pole (**B**). Annual oscillations are caused by the seasonal activity of land plants in the northern hemisphere, the steady increase because of the use of fossil energy and deforestation (after KEELING & WHORF 1994)

The steady increase of $CO_2$ content from about 280 ppm (in pre-industrial times) to the present 360 ppm $CO_2$ is proven worldwide. What has so far only been acknowledged to a lesser degree are the effects on other global metabolic cycles, which are changed because of growing anthropogenic activities. Figure 296 demonstrates natural N-binding compared to that caused by man, which has reached at least a similar magnitude to the $CO_2$ increase.

In temperate latitudes under climatic conditions without extremes, sustainable use of forests has been possible for centuries, but even in central Europe there are problems of forest damage. In regions with variable rainfall

**Fig. 296.** The global N balance is characterised by the almost constant natural N fixation (biological N fixation in terrestrial ecosystems and fixation of N from electrical discharge) and the heavily increased anthropogenic N fixation (industrial manufacture of fertilisers, for example Haber-Bosch process), N fixation in burning of fossil energy and N fixation because of legumes in agriculture (after VITOUSEK 1994)

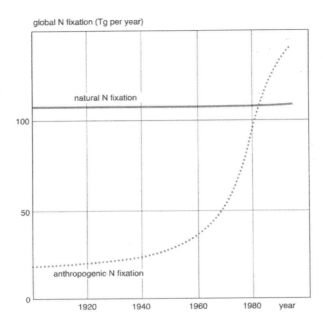

global N fixation (Tg per year)

natural N fixation

anthropogenic N fixation

the rate of erosion is a great problem in areas where trees have been felled. After trees have been cut, soils are washed away, making re-forestation very difficult. In tropical rain forests the use of wood according to the European traditional forestry is economic nonsense (see pp. 156).

Almost two thirds of the original forests of the earth have been lost for ever. Of the $8.08 \times 10^9$ ha covered by forest about 8000 years ago, only $3.04 \times 10^9$ ha are left. This frightening result was given in a recent documentation of the World Wide Life Fund (WWF) on the global condition of forests. Maintaining what is left is by no means secure. Today, about $17 \times 10^6$ ha of original forest is destroyed annually by large-scale deforestation, industrial wood cutting, road building and other human interventions or replaced by species for man-made forests with little ecological value. The worrying fact is that during recent years destruction has not been reduced, but accelerated. The WWF therefore proposes to create a global net of protected areas comprising 10% in each area of tropical and subtropical forests, as well as those of temperate and boreal forests. In Europe alone, 100 forest regions are being proposed for this project.

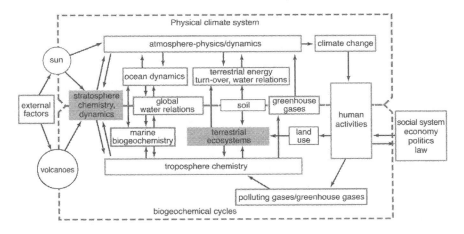

**Fig. 297.** Attempts are made to gain a better understanding of the system of the earth by using global models within the framework of the IGBP (International Geosphere-Biosphere Programme) by linking, for example, the physical climate system, the biogeochemical cycles and human activity. From this a prognosis on future developments may be made (after IGBP 1993)

## Large Programs and Global Programs

The existence of a global environmental problem is shown by the steadily increasing treaties and conferences between states and groups of states. Agreements of increased co-operation to curb global warming, to protect the ozone layer, against acid rain, for clean air, water and soil as well as in the waste industries and the re-cycling of raw material are steadily rising. Effects on the basic problem and actual improvements have so far only been in certain places (for example the water quality of the river Rhine). Less money for meetings, congresses, political conferences and committees, and instead clearer political decisions and more money for research in pure science (see Chap. 1), would perhaps be more successful. The wealth of organisations, of large projects and large research centres addressing these problems has created a flood of new acronyms and leads more and more to the development of its own language. Some acronyms are mentioned in Figs. 297–299 insofar as they can claim to hint at ecological research. It is surely sensible to watch these activities critically and add new ideas (often "common sense" would be helpful). Many of these large projects and network research facilities are trying to understand the physical-chemical dynamics on a global scale (Figs. 297, 299).

**Fig. 298.** Main activities (numbered *1–5*) of the international research program Diversitas and some special objectives (*6–10*). The *top line* lists grant-giving authorities (after DIVERSITAS 1996)

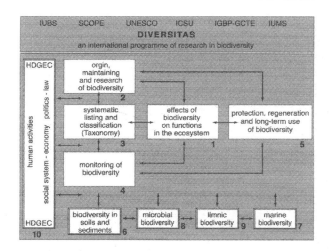

**Fig. 299.** Network of different international organisations and their programmes to research the earth holistically (after IGBP 1993)

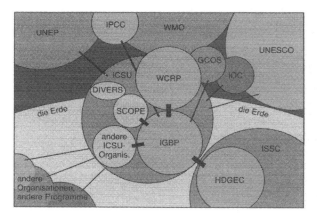

| DIVERS | Diversitas: Biodiversity Programme |
|---|---|
| GCOS | Global Climate Observing System |
| HDGEC | Human Dimensions of Global Environmental Change |
| ICSU | International Council of Scientific Unions |
| IGBP | International Geosphere-Biosphere Programme |
| IPCC | Intergovernmental Panel on Climate Change |
| IOC | Intergovernmental Oceanographic Commission |
| ISSC | International Social Science Council |
| SCOPE | Scientific Committee on Problems of the Environment |
| UNEP | United Nations Environment Programme |
| UNESCO | United Nations Educational, Scientific and Cultural Organization |
| WCRP | World Climate Research Programme |
| WMO | World Meteorological Organization |

The world of organisms has clearly been neglected. Whether a program like DIVERSITAS (see also Fig. 298) can catch up the backlog is questionable. Many problems have been known for a long time and, although the exact logging of data is surely important it is only one side of the coin; the other side is the implementation of measures which will have an effect and can solve the problem. In this there is an ever-increasing gap in the racing technological developments of industrial countries and the "treading of water" and "healing symptoms" of almost all countries.

## Testimonies

What must be done to achieve a development for sustainable use enabling a life worth living for our children and children's children? This is not a science problem, but a question of the socio-political system. The impression is given that the mass of the population is not exactly treated as adult citizens. Long-due decisive measures are shied away from and in effect "loans" are taken, to be paid back by future generations and the state is seen as a self-service shop. Positive role models have become rare and the law courts do not make decisions which would give any direction. Change is only possible if all pull in the same direction. Indeed, change is particularly important at the level of the individual, as only at the level of the **intact family** is it possible to hand down sustainable behaviour from generation to generation.

The loss of moral concepts and traditions leads to disturbing developments. In Hungary, Taiwan and many other countries children are the precondition for a fulfilled life for about 90% of the population. Only 46% in USA and 49% in Germany of the adult population see children as giving meaning to their life. However, this so-called self-realisation of the individual, excessive liberalisation and limitless lust leads to chaos.

Neurophysiological research in the USA has shown in recent years that the human brain is particularly impressionable during the early stage of development and growing up. In earlier times it was very clear that the job of **mother** in an intact family was the best guarantee for the emotional and spiritual well-being of children and thus of human kind. This has been forgotten in many western countries during the last decades. The state, but much more so the individual, is brought to task.

In his old age WALTER formulated his thoughts which are briefly considered in the final section.

Whether clever or dim, powerful or weak, psychologically strong or labile, whether determined or unfocused, susceptible to addictions, criminality, psychological illnesses, optimistic or disheartened, whether thus, lifelong happy or unhappy, all depends on the impressions stored in the human brain during the early phase of development.

Humanity and close human contact are becoming rarer and rarer, being reduced to short and non-committal telephone calls or even short e-mail notes. Each person has become a number: an identification number, a tax number, a social security number, various customer numbers etc. A name is only used on the envelope. But for how long? What most people once regarded as a relationship to nature is unknown to today's youth in the technologically advanced countries. They have no idea what has been denied them. It is the quality of life and not the material standard of living which is most important; the inner being and not the outward appearance.

Quality of life and standard of living are not necessarily opposites. Experience has proved, however, that the more one values outward appearance, the more one loses one's inner life, which is therefore, never spoken of.

Quality of life is also expressed outwardly in a healthy and natural way of living with the sensible use of one's vitality, rejection of all addictive substances and the preference for a simple life in modesty and self-control. Those who have a true relationship with nature and who are familiar with nature's enormous variety and greatness, do not feel themselves to be the centre of creation and know that they are only a tiny, fleeting protein lump in the eternity of the universe.

WALTER says to this topic:

"Not only does one come to know the **external environment**, which has been the topic of this volume, to which we physically belong and which we can investigate by using our intelligence, but one also comes to know the other aspect of mankind, his inner consciousness, which is not subject to logic, and for which philosophers have developed many different complicated designations, and which is generally referred to as the "soul". These thoughts cannot be expressed in words, nor can they be proven. It is the free decision of each individual to acknowledge this and to regard himself as part of the whole. Without this recognition, there is no true freedom for mankind. It makes one independent of the judgement of others, providing an inner security, peace and composure, as well as an inner joy. This has nothing to do with mortality or eternity. The absolute knows no limits. It is within us and around us. This is the most important conclusion for the youth searching for the purpose of life, the result of a long life dedicated to the research of life on the entire earth. It was a life full of miraculous wonder in a time in which the wonder of miracles is not believed in, and which has

**Fig. 300.** Situation of mankind. Symbolic portrayal by Jordan Pop-Ilier (Macedonia) from UN CCD: *Comics to combat Desertification*, p. 75; UN CCD, First Conference of the Parties, Rome, 29.9.–10.10.1997

lost touch with the centre of all things. It is always necessary to swim against the dirty current before arriving at the pure source ascending from the deep." (see also WALTER: *Testimony of an ecologist*).

## Questions

1. What is sustainable, workable agriculture, what is sustainable forestry?
2. Where is the annual production of green producers greater, on land or in the ocean?
3. How does the relation of biomass of producers and the annual productivity in marine and terrestrial ecosystems differ?
4. What is a "Red List Species"?
5. Why is the term "explosion" suitable to describe the population increase of humanity?
6. How do minimum requirements of humans differ from the normally demanded standard of living?

# References

Only some basic, mostly more recent publications are listed here.
For more detailed references see particularly WALTER 1973, 1986, as well as WALTER & BRECKLE 1990, 1991, 1991a, 1994.

AHTI, T.L. & JALAS, J. 1968: Vegetation zones and their sections in Northwestern Europe. Ann. Bot. Fenn. 5, 169–211.

AHTI, L.T, & KONEN, T. 1974: A scheme of vegetation zones for Japan and adjacent regions. Ann. Bot. Fenn. 11, 59–88.

AICHINGER, E. 1933: Vegetationskunde der Karawanken. Pflanzensoz. Hd. 2, 329 pp.

ALBERT, R. 1982: Halophyten. In KINZEL, H.: Pflanzenökologie und Mineralstoffwechsel. pp 33–204. Ulmer/Stuttgart.

ALBERT, R., BRECKLE, S.-W. & CHOO, Y.-S. 1996: Südkorea – Ökologische Exkursion der Universitäten Wien und Bielefeld 1996, Wien 276 pp.

ALEKSANDROVA, V.D. 1971: On the principles of zonal subdivision of arctic vegetation. Bot. Z. 56, 3–21 (Russ.).

ANDERSON, G.D. & HERLOCKER, D.J. 1973: Soil factors affecting the distribution of the vegetation types and their utilisation by wild animals in Ngorongoro crater, Tanzania. J. Ecol. 61, 627–651

ARROYA, M.T.K., ZEDLER, P.H. & FOX, M.D. 1995: Ecology and biogeography of Mediterranean ecosystems in Chile, California, and Australia. Ecol. Stud. 108, 455 pp.

AUBREVILLE, A. 1938: La forêt équatoriale et les formations forestières tropicales africaines. Scientia (Como) 63, 157.

AXELROD, D.I. 1973: History of the Mediterranean ecosystems in California. Ecol. Stud. 7, 225–277.

BARBOUR, M.G. & MAJOR, J. 1977: Terrestrial vegetation of California. Wiley-Intersci. Publ. 1002 p.

BARTHLOTT, W., LAUER, W. & PLACKE, A. 1996: Global distribution of species diversity in vascular plants: towards a world map of phytodiversity Erdkunde **50**, 317–327.

BAZILEVICH, N.I. & RODIN, L.E. 1971: Geographical regularities in productivity and the circulation of chemical elements in the earth's main vegetation types. Soviet Geogr.-Transl. American Geogr Soc/New York.

BAZILEVICH, N.I., RODIN, L.E. & ROZOV, N.N. 1970: Untersuchungen der biologischen Produktivität in geographischer Sicht. V. Tag. Geogr. Ges. USSR, Leningrad (Russ.).

BEADLE, N.C.W. 1981: The Vegetation of Australia. Vegetationsmonographien der einzelnen Großräume Bd. IV. Fischer/Stuttgart 690 p.

BELSKY, A.J. & CANHAM, C.D. 1994: Forest gaps and isolated savanna trees. BioScience **44**, 77–84.

BLASCO, F. 1977: Outlines of ecology, botany and forestry of the mangals of Indian subcontinent. Ecosystems of the World, (ed. V. J. CHAPMAN), Vol. 1, 241–260.

BLISS, L.C. & WIELGOLASKI, F.E. (eds.) 1973: Primary production and production process, Tundra Biome. Proc. Conf. Dublin, Swedish IBP Comm., Stockholm 250 p.

BLONDEL, J. & ARONSON, J. 1995: Biodiversity and ecosystem function in the mediterranean basin: human and non-human determinants. Ecol. Stud. **109**, 43–120.

BÖCHER, T.W., HJERTING, J.P. & RAHN, K. 1972: Botanical studies in the Atuel Valley, Mendoza Province, Argentina. Dansk Botan. Ark. **22**, 195–358.

BOERBOOM, J.H.A. & WIERSUM, K.F. 1983: Human impact on tropical moist forest. In: HOLZNER, W., WERGER, M.J. A., & IKUSIMA, I. (eds.): Man's impact on vegetation. Junk/The Hague 83–106.

BOURLIÈRE, F. (ed.) 1983: Tropical savannas. Ecosystems of the World **13**, 730 p.

BOX, E.O. 1995: Factors determining distributions of tree species and plant functional types. Vegetatio **121**, 101–116.

BRANDE, A. 1973: Untersuchungen zur postglazialen Vegetationsgeschichte der Neretwa-Niederung (Dalmatien, Herzegowina). Flora **162**, 1–44.

BRAUN-BLANQUET, J. & JENNY, H. 1926: Vegetationsgliederung und Bodenbildung in der alpinen Stufe der Zentralalpen. Denkschr. Schweiz. Naturforsch. Ges. **63**, 183–349.

BRECKLE, S.-W. 1971: Vegetation in alpine regions of Afghanistan. In DAVIS, P.H. *et al.* (eds.): Plant Life of South-West Asia. Proceedings of the Symposium 1970/ Edinburgh, 107–116.

BRECKLE, S.-W. 1973: Mikroklimatische Messungen und ökologische Beobachtungen in der alpinen Stufe des afghanischen Hindukusch. Bot. Jahrb. System. **93**, 25–55.

BRECKLE, S.-W. 1974: Notes on alpine and nival flora of the Hindu Kush, East Afghanistan. Bot. Notiser (Lund) **127**, 278–284.

BRECKLE, S.-W. 1976: Zur Ökologie und zu den Mineralstoffverhältnissen absalzender und nichtabsalzender Xerohalophyten. Habil.-Schr. Bonn, 170 p., Cramer (Diss. Bot.).

BRECKLE, S.-W. 1983: Temperate Deserts and Semideserts of Afghanistan and Iran. In: WEST, N.E. (ed): Temperate Deserts and Semideserts. Ecosystems of the World (ed.: GOODALL, D.W.) **5**, 271–319 Elsevier/Amsterdam

BRECKLE, S.-W. 1986: Studies on halophytes from Iran and Afghanistan. II. Ecology of halophytes along salt gradients. Proceed. Roy. Soc. Edinburgh **89B**, 203–215.

BRECKLE, S.-W., AGACHANJANZ, O.E. & WUCHERER, W. 1998: Der Aralsee: Geoökologische Probleme. Naturwiss. Rdschau **51**, 347–355.

BROWN, K.S.J. & AB'SABER, A.N. 1978: Ice age refuges and evolution in the neotropics: correlation and paleoclimatological, geomorphological and pedological data with modern biological endemism. Paleoclimas (San Paulo) **5**, 1–30.

BRÜNING, F.F. 1973: Species richness and stand diversity in relation to site and succession. Amazonia **4**, 293–320.

BUCHER, E.H. 1982: Chaco and Caatinga – South American arid savannas, woodlands and thickets, 48–79. In: HUNTLEY, B.J. and WALKER, B.H. (eds.).

BURROWS, C.J. 1990: Processes of vegetation change. U. Hyman/London 551 p.

CABRERA, A.L. & WILLINK, A. 1973: Biogeografia de America Latina. Washington, D.C., 120 p. + map.

CALDWELL, M.M. & MANWARING, J.H. 1994: Hydraulic lift and soil nutrient heterogeneity. Israel J. Plant Sciences **42**, 321–330.

CALDWELL, M.M., RICHARDS, J.H. & BEYSCHLAG, W. 1991: Hydraulic lift: ecological implications of water efflux from roots. In: ATKINSON, D. (ed): Plant root growth – an ecological perspective. Blackwell/Oxford.

CAMPELL, B. 1985: Ökologie des Menschen. Harnack/München 232 pp.

CANNEL, M.G.R. 1982: World forest biomass and primary production dates. Academic Press, London-New York, 391 pp.

CASTRI, F. DI 1973: Climatographical comparison between Chile and the western coast of North America. Ecol. Stud. **7**, 21–36.

CASTRI, F. DI 1981: Mediterranean-type shrublands of the world. In: CASTRI, F. DI, GOODALL, D. W. & SPECHT, R. L. (eds.): Mediterranean-type shrublands. Vol. 11, 643 pp. Ecosystems of the World. Amsterdam 1–52.

CASTRI, F. DI & MOONEY, H. A. (eds.) 1973: Mediterranean type ecosystems. Ecol. Stud. 7.

CERNUSCA, A. 1976: Bestandesstruktur, Bioklima und Energiehaushalt von alpinen Zwergstrauchbeständen. Oecologia Plantarum 11, 71–102.

CHAPIN, D. M., BLISS, L. C. & BLEDSOE, L. J. 1991: Environmental regulation of nitrogen fixation in a high-arctic lowland ecosystem. Can. J. Bot. 69, 2744–2755.

CHAPMAN, V. J. 1976: Mangrove vegetation. Vaduz, 477 pp.

CHONG-DIAZ, G. 1988: The Cenozoic saline deposits of the Chilean Andes between $18°00'$ and $27°00'$ south latitude. Lecture Notes in Earth Sciences 17, 137–151.

CLEMENTS, F. F., WEAVER, J. E. & HANSON, H. C. 1929: Plant Competition. Carnegie Inst. Wash., Publ. 398.

CORNELISSEN, J. H. C. & THOMPSON, K. 1997: Functional leaf attributes predict litter decomposition rate in herbaceous plants. New Phytol. 135, 109–114.

COUTINHO, L. M. 1982: Ecological effect of fire in Brasilian Cerrado, 273–291. In: HUNTLEY, B. J. & WALKER, B. H. (eds.).

CROWLEY, G. M. 1994: Quaternary soil salinity events and Australian vegetation history. Quarternary Science Reviews 13, 15–22.

CUMMING, D. H. M. 1982: The influence of large herbivores on savanna structure in Africa, 217–245. In: HUNTLEY, B. J. and WALKER, B. H. (eds.).

DAFIS, SP. & LANDOLDT, E. (eds.) 1975: Zur Vegetation und Flora Griechenlands. Veröff. Geobot. Inst. Zürich 50, 237 pp.

DAVIS, G. W. & RICHARDSON, D. M. 1995: Mediterranean-type Ecosystems. The function of biodiversity. Ecol. Stud. 109, 366 pp.

DELCOURT, P. A. & DELCOURT, H. R. 1987: Long-term forest dynamics of the temperate zone. Ecol. Stud. 63, 439 pp.

DIERSCHKE, H. 1994: Pflanzensoziologie. Ulmer/Stuttgart, 683 pp.

DIERSSEN, K. 1996: Vegetation Nordeuropas. Ulmer/Stuttgart, 838 pp.

DINGER, B. E. & PATTEN, D. T. 1974: Carbon dioxide exchange and transpiration in species of *Echinocereus* (Cactaceae) as related to their distribution within the Piñaleno Mountains, Arizona. Oecologia 14, 389–411.

DIVERSITAS, 1996: DIVERSITAS: An international programme of biodiversity science. Operational plan. DIVERSITAS, Paris 42.

DOKUTSCHAJEV, U.V. 1898: Zur Lehre über die Naturzonen. St. Petersburg (Russ.).

DONIGIAN, A. S.j. *et al.* 1995: Modeling the impacts of agricultural management practices on soil carbon in the central U.S. Soil management and greenhouse effect. CRC Press/Boca Raton, 121–135.

DUVIGNEAUD, P. 1974: La synthèse écologique. Population, communautés, écosystèmes, biosphère, noosphère, 296 pp. Paris.

EDMONTON, R.L. (ed.) 1982: Analysis of coniferous forest ecosystems in the Western United States. US/IBP Synthesis Series 14, 419 pp.

EDWARDS, C.A., REICHLE, D.F. & CROSSLEY, D.A. jr. 1970: The role of soil invertebrates in turnover of organic matter and nutrients. Ecol. Stud. 1, 147–172.

EHLERS, W. 1996: Wasser in Boden und Pflanze. Ulmer/Stuttgart 272 pp.

EITEN, G. 1982: Brazilian "savannas", 25–79. In: HUNTLEY, B.J., and WALKER, B.H. (eds.).

ELLENBERG, H. 1975: Vegetationsstufen in perhumiden bis perariden Bereichen der tropischen Anden. Phytocoenologia 2, 368–387.

ELLENBERG, H. 1996: Vegetation Mitteleuropas mit den Alpen in ökologischer, dynamischer und historischer Sicht. 5. Aufl., Ulmer/Stuttgart 1096 pp.

ELLENBERG, H., MAYER, R. & SCHAUERMANN, J. 1986: Ökosystemforschung, Ergebnisse des Sollingprojektes 1966–1986. Ulmer/Stuttgart 507 pp.

ELLENBERG, H. & BERGEMANN, A. (eds.) 1990: Entwicklungsprobleme Costa Ricas. ASA-Studien 18, Breitenbach/Saarbrücken.

ERN, H. 1966: Die dreidimensionale Anordnung der Gebirgsvegetation auf der Iberischen Halbinsel. Bonner Geogr. Abh. 37, 136 pp.

ERNST, W. & WALKER, G.H. 1973: Studies on hydrature of trees in miombo woodland in South Central Africa. J. Ecol. 61, 667–686.

EVENARI, M., SHANAN, L. & TADMOR, N. 1982: The Negev. The challenge of a desert. 2. edit., 437 pp. Cambridge, Mass.

FRANGI, J. 1975: Sinopsis de la comunidades vegetales y el medio de las sierras de Tandil (Provincia Buenos Aires). Bol. Soc. Argentina de Botan. 16, 293–319.

FREITAG, H. 1971a: Die natürliche Vegetation des südostspanischen Trockengebiets. Bot. Jahrb. 91, 147–208.

FREITAG, 1971b: Die natürliche Vegetation Afghanistans. Vegetatio 22, 285–344.

FRENCH, N.R. (ed.) 1979: Perspectives in grassland ecology. Ecol. Stud., 32, 204 pp.

GAMS, H. 1927: Von den Follatères zur Dent de Morcles. Beitr. Geobot. Landesaufn. Schweiz 15, 760 pp.

GAUSSEN, H., MEHER-HOMJI, V.M., LEGRIS, P. et al. 1972: Notice de la feuille Rajasthan (1:1 Mill.). Travaux Sect. Sc. et Techn., Inst. Français de Pondichéry, Serie No 12.

GEIGER, R. 1928: Das Klima der bodennahen Luftschicht. Braunschweig, 460 pp.

GIGON, A. 1974: Ökosysteme. Gleichgewichte und Störungen. In: LEIBUNDGUT, H. (ed.) Landschaftsschutz und Umweltpflege. Huber, Frauenfeld, 16–39.

GLINKA, K.D. 1914: Die Typen der Bodenbildung. Berlin 215 pp.

GOLDAMMER, J.G. (ed.) 1990: Fire in the tropical biota: ecosystem processes and global challenges. Ecol. Stud. 84, 497 pp.

GOLLEY, F.B. & MEDINA, F. (eds.) 1975: Tropical ecological systems. Ecol. Stud. 11.

GORYSCHINA, T.K. (ed.) 1974: Biologische Produktion und ihre Faktoren im Eichenwald der Waldsteppe. Arb. Forstl. Versuchsst. d. Univ. Leningrad. "Wald an der Worskla" 6, 1–213 (Russ.).

GREBENSCHTSCHIKOW, O.S. 1972: Ökologisch-geographische Gesetzmäßigkeiten in der Pflanzendecke der Balkan-Halbinsel. Akad. Wiss. Ser. Geogr. Nr. 4, Moskau, with vegetation map (Russ.).

GROENENDAEL, J.M. VAN, KLIMES, L. KLEMESOVA, J., HENDRIKS, R.J.J. 1996: Comparative ecology of clonal plants. Phil. Trans. R. Soc. London, B351, 1331–1339.

GROOMBRIDGE, B. (ed.) 1992: Global biodiversity. Status of the earth's living resources. Chapman & Hall/London 585 pp.

GROVES, R.H., BEARD, J.S., DEACON, H.J. et al. 1983: The origins and characteristics of mediterranean ecosystems. In: DAY, J.A. (ed.): Mineral nutrients in mediterranean ecosystems. S. Afr. Nat. Sci. Progr. Rep. No. 71, 1–18.

GRUHL, H. 1975: Ein Planet wird geplündert. Die Schreckensbilanz unserer Politik. 376 pp., S. Fischer/Frankfurt.

HAECKEL, E. 1866: Generelle Morphologie der Organismen, Bd. II, Reimer/Berlin.

HAINES, B. 1975: Impact of leaf-cutting ants on vegetation development at Barro Colorado Island. Ecol. Stud. 11, 99–111.

HALVORSON, W.L. & PATTEN, D.T. 1974: Seasonal water potential changes in Sonoran Desert shrubs in relation to topography. Ecology 55, 173–177.

HAMILTON III, W.J. & SEELY, M.K. 1976: Fog basking by the Namib Desert beetle, Onymacris unguicularis. Nature 262, 284–285.

HENNING, I. 1975: Die La Sal Mountains, Utah. Akad. Wiss., Mainz, Math.-Naturw. Klasse Nr. 2, Wiesbaden

HENNING, I. 1994: Hydroklima und Klimavegetation der Kontinente. Münstersche Geographische Arbeiten 37, 144 pp.

HEYWOOD, V.H. (ed.) 1995: Global biodiversity assessment. UNEP, Cambridge Univ. Press, 1140 pp.

HILLEL, D. 1980: Applications of soil physics. Acad. Press/ New York.

HOFMANN, G. 1997: Mitteleuropäische Wald- und Forst-Ökosystemtypen in Wort und Bild. AFZ/Der Wald Sonderheft 3–85.

HOFFMANN, A.J. & ARMESTO, J.J. 1995: Modes of seed dispersal in the mediterranean regions in Chile, California, and Australia. Ecol. Stud. **108**, 289–310.

HOLDGATE, M.W. (ed.) 1970: Antarctic Ecology. Vol. **2**, 394 pp. New York.

HOLDING, A.J., HEAL, O.W., MACLEAN, S.F. & FLANGAGAN, P.W. (eds.) 1974: Soil organisms and decomposition in tundra. Proc. Microbiol. Meet., Fairbanks 1973. Swedish IBP Comm./Stockholm.

HOLDRIDGE, L.R., GRENKE, W.C., HATHEWAY, W.H. et al. 1971: Forest environments in tropical life zones. Pergamon Press Oxford.

HOPKINS, B. 1974: Forest and Savanna (West Africa). 2. edit., 154 pp., Ibadan/London.

HORVAT, I., GLAVAC, V. & ELLENBERG, H. 1974: Vegetation Südosteuropas. 768 pp. with 2 vegetation maps. Fischer/Stuttgart.

HUECK, K. 1966: Die Wälder Südamerikas. Vegetationsmonographien der einzelnen Großräume, Bd. II, 296 pp., Fischer/Stuttgart.

HUECK, K. & SEIBERT, P. 1972: Vegetationskarte von Südamerika. Vegetationsmonogr. d. einz. Großräume, Bd. II a, Fischer/Stuttgart.

HULTEN, E. 1932: Süd-Kamtschatka. Vegetationsbilder 23. Reihe, Heft 1/2, Jena.

HUMPHRIES, C.J., WILLIAMS, P.H. & VANE-WRIGHT, R.I. 1995: Measuring biodiversity value for conservation. Ann. Rev. Ecol. Syst. **26**, 93–111.

HUNTLEY, B.J. & MORRIS, J.W. 1978: Savanna ecosystem project. Phase I summary and phase II progress. South Africa Nat. Sc. Progr., Rep. No **29**, 52 pp.

HUNTLEY, B.J. & WALKER, B.H. (eds.) 1982: Ecology of tropical savannas. Ecol. Stud. **42**, 669 pp.

HÜPPE, J. 1993: Development of NE European heathlands – palaeoecological and historical aspects. Scripta Geobot. **21**, 141–146.

HÜTTEL, C. 1975: Root distribution and biomass in three Ivory Coast rain forest plots. Ecol. Stud. 11, 123–130.

IGBP 1993: Global Chance: reducing uncertainties. IGBP, Royal Swed. Acad./Stockholm 40 pp.

INCHAUSTI, P. 1995: Competition between perennial grasses in a neotropical savanna: the effect of fire and of hydric-nutritional stress. J. Ecol. 83, 231–243.

IWAKI, H., MONSI, M. & MIDORIKAWA, B. 1966: Dry matter production of some herb communities in Japan. 11th Pacific Science Congress/Tokyo.

JANZEN, B. H. 1978: Seeding patterns of tropical trees. In: TOMLINSON, P. B. & ZIMMERMANN, M. H.: Tropical trees as living systems. Cambridge Univ. Press, 83–128.

JELTSCH, F. 1992: Modelle zu natürlichen Waldsterbephänomenen. Dissertation Univ. Marburg, 113 pp.

JOHANSSON, D. 1974: Ecology of vascular epiphytes in West Africa rain forest. Acta Phytogeogr. Suecica 59, 1–129.

JÜRGENS, N., BURKE, A., SEELY, M. K. & JACOBSON, K. M. 1997: Deserts. In: COWLING, R. M., RICHARDSON, D. M. & PIERCE, S. M. (eds.): Vegetation of Southern Africa. Cambridge University Press, 189–214.

KAAGMAN, M. & FANTA, J. 1993: Cyclic succession in heathland under enhanced nitrogen deposition: a case study from the Netherlands. Scripta Geobot. 21, 29–38.

KÄMMER, F. 1974: Klima und Vegetation von Teneriffa besonders im Hinblick auf den Nebelniederschlag. Scripta Geobot./Göttingen 7, 78 pp.

KÄMMER, F. 1982: Flora und Fauna von Makaronesien. 179 pp., Selbstverlag/Freiburg i. Br.

KAPPELLE, M. 1995: Ecology of mature and recovering Talamancan montane *Quercus* forests, Costa Rica. Acad. Proefschrift, Amsterdam 270 pp.

KEARNEY, T. H., BRIGGS, L. J. et al. 1914: Indicator significance of vegetation in Tooele Valley, Utah. J. Agric. Res. 1, 365–417.

KEELING, C. D. & WHORF, T. P. 1994: Atmospheric $CO_2$ records from sites in the SIO air sampling network. p. 16–26. In: BODEN, T. A. et al. (eds.): Trends '93: A compendium of data on global change. ORNL/CDIAC-65. Carbon Dioxide Information Analysis Center, Oak Ridge Nat. Lab/Oak Ridge.

KIRA, T., ONO, Y. & HOSOKAWA, T. (eds.) 1978: Biological production in a warm temperature evergreen oak forest of Japan. JIBP Synthesis, vol. 18.

KLÖTZLI, F. 1975: Edellaubwälder im Bereich der südlichen Nadelwälder Schwedens. Ber. Geobot. Inst. Rübel 43, 23–53.

KLÖTZLI, F. 1987: On the global position of the evergreen broadleaved (non ombrophilous) forest in the subtropical and temperate zones. Veröff. Geobot. Inst. ETH, Stift. Rübel/Zürich **98**.

KLÖTZLI, S. 1997: Umweltzerstörung und Politik in Zentralasien – eine ökoregionale Systemuntersuchung. Europäische Hochschulschriften IV/17: Peter Lang/Bern 292 pp.

KNAPP, R. 1973: Die Vegetation von Afrika. Veget. Monogr. d. einz. Großräume, Bd. III, 626 pp., Fischer/Stuttgart.

KRUGER, F.J., MITCHELL, D.T. & JARVIS, J.U.M. 1983: Mediterranean-type ecosystems. Ecol. Stud. **43**, 552 pp.

KRUTZSCH, W. 1992: Paläobotanische Klimagliederung des Alttertiärs (Mitteleozän bis Oberoligozän) in Mitteldeutschland und das Problem der Verknüpfung mariner und kontinentaler Gliederungen (klassische Biostratigraphien – paläobotanisch-ökologische Klimastratigraphie – Evolutions-Stratigraphie der Vertebraten). N. Jb. Geol. Paläont. Abb. **186**, 137–253.

KÜCHLER, A.W. 1974: A new vegetation map of Kansas. Ecology **55**, 586–604.

KÜHNELT, W. 1975: Beiträge zur Kenntnis der Nahrungsketten in der Namib (Südwestafrika). Verh. Ges. f. Ökologie/Wien **4**, 197–210.

KUMMEROW, J. 1981: Structure of roots and root systems. Ecosystems of the World (Amsterdam), vol. **11**, 269–288.

KUNKEL, G. (ed.) 1976: Biogeography and ecology in the Canary Islands. Junk, The Hague.

KUNTZE, H., ROESCHMANN, G. & SCHWERDTFEGER, G. 1994: Bodenkunde. 5. Aufl., Ulmer/Stuttgart 424 pp.

KUTTLER, W. (ed.) 1995: Handbuch zur Ökologie. Analytica/Berlin 2. Aufl.

LAMOTTE, M. 1975: The structure and function of a tropical savanna ecosystem. Ecol. Stud. **11**, 179–222.

LAMOTTE, M. 1982: Consumption and decomposition in tropical grassland ecosystem at Lamto, Ivory Coast, 414–429. In: HUNTLEY, B.J. and WALKER, B.H. (eds.).

LARCHER, W. 1977: Ergebnisse des IPB-Projekts „Zwergstrauchheide Patscherkofel". Produktivität und Überlebensstrategien von Pflanzen und Pflanzenbeständen im Hochgebirge. Sitz.-Ber. Österr. Akad. d. Wiss., Mathemat.-naturw. Kl., Abt. I, **186**, 301–386, Wien.

LARCHER, W. 1994: Ökologie der Pflanzen. 5. Aufl., Ulmer/Stuttgart.

LAUER, W., RAFIQPOOR, M.D. & FRANKENBERG, P. 1996: Die Klimate der Erde. Erdkunde **50**, 275–300.

LAVAGNE, A. 1972: La végétation de l'Ile de Port-Cros, Marseille, 30 pp.

LAVAGNE, A. & MOUTTE, P. 1974: Bull. Carte Végétation de la Provence et des Alpes du Sud. Marseille, 129 pp.

LEICHT, H. 1996: Altbäume – Tierökologische Bedeutung für die Praxis. Ber. Bayer. Landesamt Umweltschutz **132**, 86–93.

LERCH, G. 1991: Pflanzenökologie. Akad.-Verlag/Berlin 535 pp.

LEUSCHNER, C. 1993: Forest dynamics on sandy soils in the Lüneburger Heide area, NW Germany. Scripta Geobot. **21**, 53–60.

LEVINA, F. J. 1964: Die Halbwüstenvegetation der nördlichen Kaspischen Ebene. 344 pp., Moskau-Leningrad (Russ.).

LEVITT, J. 1972: Responses of plants to environmental stresses. vol. 1+2; (1980: 2nd. edit.) Acad. Press/New York 497+606 pp.

LEWIS, J. P. & COLLANTES, M. B. 1975: La vegetación de la Provincia de Santa Fe. Bol. Soc. Argentina de Bot. **16**, 151–179.

LIETH, H. 1964: Versuch einer kartographischen Darstellung der Produktivität der Pflanzendecke auf der Erde. Geogr. Taschenbuch/Wiesbaden, 72–80.

LIETH, H. & WHITTAKER, R. H. (eds.) 1975: Primary productivity of the biosphere. Ecol. Stud. 14.

LOGAN, R. F. 1960: The Central Namib Desert, South West Africa. Publication **758**, 162 pp. Nat. Ac. Sc., Washington D. C.

LONGMAN, K. A. & JENIK, J. 1974: Tropical forest and its environment (Ghana). 196 pp., Thetford, Norfolk.

LÖTSCHERT, W. 1974: Über die Vegetation frostgeformter Böden auf Island. Ber. Forschungsst. Neori As (Island) **16**, 1–15.

MacARTHUR, R. H. 1972: Geographical ecology: patterns in the distribution of species. Harper & Row/New York.

MALTHUS, T. R. 1798: Essay on the principles of population. London.

MANI, M. S. (ed.) 1974: Ecology and biogeography in India. 773 pp., The Hague.

MANN, H. S. (ed.) 1977: The spectre of desertification. Ann. Arid Zone/Jodhpur **16**, 279–394.

MANN, H. S., LAHIRI, A. N. & PAREEK, O. P. 1976: A study on the moisture availability and other conditions of unstabilized dunes etc. Ann. Arid Zone/Jodhpur **15**, 270–286.

MARTIN, P. S. 1984: Prehistoric overkill: the global model. In: MARTIN, P. S. & KLEIN, R. G. (eds.): Quaternary extinctions: a prehistoric revolution. Tucson, Univ. of Arizona Press, 354–403.

MAYER, H. 1974: Wälder des Ostalpenraums. Fischer/Stuttgart 344 pp.

MAYER, H. 1977: Karte der natürlichen Wälder des Ostalpenraums. Cbl. Ges. Forstwesen/Wien **94**, 147–153.

MEDINA, E. 1968: Bodenatmung und Stoffproduktion verschiedener tropischer Pflanzengemeinschaften. Ber. Dtsch. Bot. Ges. **81**, 159–168.

MEDINA, E. 1974: Dark $CO_2$ fixation, habitat preference and evolution within the Bromeliaceae. Evolution **28**, 677–686.

MENAULT, J.C. & CESAR J. 1982: The structure and dynamics of a West African Savanna, p. 80–100. In: HUNTLEY, B.J. & WALKER, B.H. (eds.).

MEUSEL, H., SCHUBERT, R., et al. 1971: Beiträge zur Pflanzengeographie des Westhimalajas. Flora **160**, 137–194, 370–432, 573–606.

MILLER, P.C. (ed.) 1981: Resources by chaparral and matoral. Ecol. Stud. **39**, 455 pp.

MIYAWAKI, A. 1983: Conservation and recreation of vegetation and its importance to human existence. Look Japan **28** (323), 10.2.83.

MONTGOMERY, G.G. & SUNQUIST, M.E. 1975: Impact of sloths on neotropical forest. Energy and nutrient cycling. Ecol. Stud. **11**, 69–98.

MOONEY, H.A. & PARSONS, D.J. 1973: Structure and function of the Californian chaparral – an example from San Dimas. Ecol. Stud. **11**, 83–112.

MOSER, W., BRZOSKA, W., ZACHHUBER, K. & LARCHER, W. 1977: Ergebnisse des IBP-Projekts „Hoher Nebelkogel 3184 m". Sitz.ber. Österr. Akad. d. Wiss., Wien, Mathemat.-naturw. Kl., Abt. I, **186**, 387–419.

MÜLLER, H.J. 1991: Ökologie. 2. Aufl. UTB 318 Fischer/ Stuttgart, 415 pp.

MÜLLER-DOMBOIS, D. 1987: Natural dieback in forests. BioScience **37**, 575–583.

NETSCHAJEVA, N.T. (ed.) 1975: Die Biogeozönosen der östlichen Karakum (Charakterisierung der Hauptkomponenten). Akad. Wiss. USSR, Repetek Sandwüsten-Station, 74 pp., Aschchabad (Russ.).

NOBEL, P.S. 1976: Water relations and photosynthesis of a desert CAM plant *Agave deserti*. Plant Physiol. **58**, 576–582.

NOBEL, P.S. 1977a: Water relations of flowering *Agave deserti*. Bot. Gaz. **138**, 1–6

NOBEL, P.S. 1977b: Water relations and photosynthesis of a Barrel Cactus, *Ferocactus acanthoides* in Colorado Desert. Oecologia **27**, 117–133.

NUMATA, M., MIYAWAKI, A. & ITOW, D. 1972: Natural and seminatural vegetation in Japan. Blumea **20**, 435–481.

OBERDORFER, E. 1965: Pflanzensoziologische Studien auf Teneriffa und Gomera. Beitr. Naturk. Forsch. SW-Deutschl. **24**, 47–104.

ODUM, E. P. 1971: Fundamentals of ecology. 3. ed., 574 pp. Saunders Co/Philadelphia – London – Toronto.

OGAWA, H., YODA, K. & KIRO, T. 1961: A preliminary survey on vegetation of Thailand. Nature Life SE Asia **1**, 21–157.

OVERBECK, F. 1975: Botanisch-geologische Moorkunde. 719 pp., Neumünster.

OZENDA, P. 1975: Sur les étages de végétation dans les montagnes du bassin méditerranéen. Doc. Cartogr. Ecol./Grenoble **16**, 1–32.

OZENDA, P. 1994: Végétation du continent européen. Delachaux et Nestlé/Lausanne 271 pp.

PECCIO, A. 1981: Die Zukunft in unserer Hand. 244 pp. Verlag Fritz Molden.

PETERS, R. 1997: Beech Forests. Geobotany **24**, Kluwer Acad. Publ. 169 pp.

PETERS, R., GENTRY, A. W. & MENDELSOHN, R. O. 1989: Valuation of an Amazonian rainforest. Nature **339**, 655–656.

POPP, M. 1995: Salt resistance in herbaceous halophytes and mangroves. Progress in Botany **56**, 416–429.

POPOV, A. I. (ed.) 1971–1975: Die natürlichen Verhältnisse Westsibiriens. Lief. I-V. Verlag d. Moskauer Univ. (Russ.).

QUIDEAU, S. A., CHADWICK, O. A., GRAHAM, R. C. & WOOD, H. B. 1996: Base cation biogeochemistry and weathering under oak and pine: a controlled long-term experiment. Biogeochemistry **35**, 377–398.

QUINTANILLA, V. 1974: Les formations végétales du Chili temperé. Doc. Cartogr. Ecol./Grenoble, **14**, 33–80.

RAUH, W. 1973: Über Zonierung und Differenzierung der Vegetation Madagaskars. Akad. Wiss. Mainz, Math.-Naturwiss. Kl. 1. Wiesbaden.

RAWITSCHER, F. 1948: The water economy of the "Campos cerrados" in southern Brazil. J. Ecol. **36**, 237–268.

READ, D. J. & MITCHELL, D. T. 1983: Decomposition and mineralization processes in mediterranean-type ecosystems and in heathlands of similar structure. Ecol. Stud. **43**, 208–232.

REHDER, H. 1966: Klimadiagrammkarte der Alpen. Flora B, **156**, 78–93.

REICHLE, D. E. 1970: Analysis of temperate forest ecosystems. Ecol. Stud. **1**, 304 p.

REICHHOLF, J. H. 1990: Der unersetzbare Dschungel. Leben, Gefährdung und Rettung des tropischen Regenwaldes. BLV/München 207 pp.

RODIN, L. E., BAZILEVICH, N. I., GRADUSOV, B. P. & YARILO-VA, E. A. 1977: Trockensavanne von Rajputan (Wüste Thar). Aridnye pochvy, ikh genesis, geokhimia, ispol'novaniye, 195–225, Moskva (Russ.)

RUCKDESCHEL, W. 1996: Landbewirtschaftung als ökologische Schlüsselfunktion. Ber. Bayer. Landesamt Umweltschutz 132, 61–72.

RUNDEL, P. W. 1982: The matorral zone of central Chile. Ecosystems of the world 11, 175–201.

RUTHERFORD, M. C. 1982: Woody plant biomass distribution in *Burkea africana* savannas, 120–141. In: HUNTLEY, B. J. & WALKER, B. H. (eds.).

RUTHSATZ, B. 1977: Pflanzengesellschaften und ihre Lebensbedingungen in den andinen Halbwüsten Nordwest-Argentiniens. 168 pp. Diss. Bot., 39, J. Cramer/Vaduz.

SARMIENTO, G. 1996 Biodiversity and water relations in tropical savannas. Ecol. Stud. 121, 61–75.

SCHALLER, F. 1993: Was heißt und zu welchem Ende betreibt man Tropenökologie? - Schrift. Verein z. Verbreit. Naturwiss. Kenntnisse in Wien. 132, 73–88.

SCHARFETTER, R. 1938: Das Pflanzenleben der Ostalpen. Wien 419 pp.

SCHMID, E. 1961: Vegetationskarte der Schweiz. Geobot. Landesaufnahme Schweiz 39, 52 pp.

SCHMITTHÜSEN, J. 1956: Die räumliche Ordnung der chilenischen Vegetation. Bonner Geogr. Abh. 17, 86 pp.

SCHÖNWIESE, C.-D. 1994: Klimatologie. UTB 1793, Ulmer/Stuttgart, 436 pp.

SCHREIBER, K. F. 1977: Wärmegliederung der Schweiz aufgrund von phänologischen Geländeaufnahmen in den Jahren 1969–1973 (4 Kartenblätter 1:200000). Eidgen. Drucks. Zentr./Benn.

SCHROETER, C. 1926: Das Pflanzenleben der Alpen. 2. Aufl., Zürich, 1288 pp.

SCHULZE, E.-D. 1970: Der $CO_2$-Gaswechsel der Buche *(Fagus silvatica* L.) in Abhängigkeit von den Klimafaktoren im Freiland. Flora 159, 177–232.

SCHULTZ, J. 1995: Die Ökozonen der Erde. 2. Aufl. UTB 1514, Ulmer/Stuttgart, 535 pp.

SCHEFFER, P. & SCHACHTSCHNABEL, P. 1992: Lehrbuch der Bodenkunde. Enke/Stuttgart, 491 pp.

SEELY, M. K. 1978: Grassland productivity. S. Afric. J. of Sci 74, 295–297.

SEELY, M. K. & HAMILTON III, W. J. 1976: Fog catchment sand trenches by Tenebrionid beetles, *Lepidochora,* from the Namib Desert. Science 193, No. 4252.

SEIBERT, P. 1968: Übersichtskarte der natürlichen Vegetationsgebiete von Bayern 1:500 000 mit Erläuterungen. Schriftenr. f. Vegetationskunde/Bad Godesberg **13**.

SIMMONS, I. G. 1996: Changing the face of the earth. Blackwell Public/Oxford.

SIMPSON, B. B. & HAFFER, J. 1978: Speciation patterns in the Amazonian forest biota. Ann. Rev. Ecol. Syst. **9**, 497–518.

SKOGLAND, T. 1983: Wild reindeer foraging-niche organization. Holarct. Ecol. **7**, 345–379.

SOLBRIG, O. T., MEDINA, E. & SILVA, J. F. (eds.) 1996: Biodiversity and savanna ecosystem processes. Ecol. Stud. **121**, 233 pp.

SPECHT, R. L. 1973: Structure and functional response of ecosystems in the mediterranean climate of Australia. Ecol. Stud. **7**, 113–120.

SPRENGER, A. & BRECKLE, S.-W. 1997: Ecological studies in a submontane rainforest in Costa Rica. Bielefelder Ökologische Beiträge **11** (Contributions to tropical ecology research in Costa Rica), 77–88.

SPRUGEL, D. G. 1976: Dynamic structure of wave-generated *Abies balsamea* forest in the northeastern United States. J. Ecol. **64**, 889–912.

STANJUKOVITSCH, K. V. 1973: The mountains of the USSR. 412 S. Duschanbe (Russ.).

SUNDING, P. 1972: The vegetation of Gran Canaria. Norske Vid.-Akad. Oslo, I Math.-Naturv. Klasse Ny Serie No **29**, 186 pp.

SUNDING, P. 1973: A botanical bibliography of the Canary Islands. 2. ed., Bot. Garden, Univ. of Oslo.

TAYLOR, A. R. 1973: Ecological aspects of lightning in forests. Ann. Proc. Tall Timber Fire Ecol. /Tallahassee **13**, 455–482.

TERBORGH, J. 1991: Lebensraum Regenwald, Zentrum biologischer Vielfalt. Spektrum Akad. Verl./Heidelberg, 253 pp.

THOMAS, M. F. 1974: Tropical geomorphology. London.

TIELBÖRGER, K. 1997: The vegetation of linear desert dunes in the north-western Negev, Israel. Flora **192**, 261–278.

TINLEY, K. L. 1982: The influence of soil moisture balance on ecosystem patterns in Southern Africa. 175–192. In: HUNTLEY, B. J. & WALKER, B. H. (eds.).

TISCHLER, W. 1984: Einführung in die Ökologie. 3. Aufl. Fischer/Stuttgart, 437 pp.

TOMLINSON, P. B. & ZIMMERMANN, M. H. 1976: Tropical trees as living systems. Cambridge Univ. Press.

TROLL, C. 1960: Die Physiognomik der Gewächse als Ausdruck der ökologischen Lebensbedingungen. Verhdl. Dt. Geographentag **32**, 97–122.

TROLL, C. 1967: Die klimatische und vegetationsgeographische Gliederung des Himalaya-Systems. Ergeb. Forsch. Untern. Nepal Himalaya **1**, 353–388.

VARESCHI, V. 1980: Vegetationsökologie der Tropen. 253 pp., Ulmer/Stuttgart.

VITOUSEK, P.M. 1994: Beyond global warming: ecology and global change. Ecology **75**, 1861–1876.

VOGGENREITER, F. 1974: Geobotanische Untersuchungen an der natürlichen Vegetation der Kanareninsel Tenerife. Diss. Bot. **26**, 718 pp., Cramer/Lehre.

WAGNER, H. 1971: Karte der natürlichen Vegetation (1:1 000 000). Österreich-Atlas.

WALKER, J. & GILLISON, A.N. 1982: Australian Savannas. 5–24. In: HUNTLEY, B.J. & WALKER, B.H.

WALKER, L.C. 1998: The North American Forests. 398 p., CRC Press/Boca Raton.

WALTER, H. 1960: Standortslehre. 2. Aufl., 566 p. Ulmer/Stuttgart.

WALTER, H. 1967: Die physiologischen Voraussetzungen für den Übergang der autotrophen Pflanzen vom Leben im Wasser zum Landleben. Z. f. Pflanzenphys. **56**, 170–185.

WALTER, H. 1968: Die Vegetation der Erde, Bd. II: Gemäßigte und arktische Zonen. 1001 pp., Fischer/Jena-Stuttgart.

WALTER, H. 1973: Die Vegetation der Erde, Bd. I: Tropische und subtropische Zonen. 3. Aufl., 743 pp., Fischer/Jena-Stuttgart.

WALTER, H. 1973a: Ökologische Betrachtungen der Vegetationsverhältnisse im Ebrobecken (NO-Spanien). Acta Bot. Acad. Sc. Hungaricae **19**, 193–402.

WALTER, H. 1974: Die Vegetation Osteuropas, Nord- und Zentralasiens. 452 pp., Vegetationsmonographien, Fischer/Stuttgart.

WALTER, H. 1975: Betrachtungen zur Höhenstufenfolge im Mediterrangebiet (insbesondere in Griechenland) in Verbindung mit dem Wettbewerbsfaktor. Veröff. Geobot. Inst. Zürich **55**, 72–83.

WALTER, H. 1975a: Über ökologische Beziehungen zwischen Steppenpflanzen und alpinen Elementen. Flora **164**, 339–346.

WALTER, H. 1976: Die ökologischen Systeme der Kontinente (Biogeosphäre). Prinzipien ihrer Gliederung mit Beispielen, 131 pp. Fischer/Stuttgart.

WALTER, H. 1977: The oligotrophic peatlands of Western Siberia – The largest peino-helobiom in the world. Vegetatio **34**, 167–178.

WALTER, H. 1981: Höchstwerte der Produktion von natürlicher Riesen-Staudenvegetation in Ostasien. Vegetatio **44**, 37–41.

WALTER, H. 1986: Allgemeine Geobotanik (UTB 284). 3. Aufl., 279 pp., Ulmer/Stuttgart.

WALTER, H. 1989: Bekenntnisse eines Ökologen. Erlebtes in acht Jahrzehnten und auf Forschungsreisen in allen Erdteilen. 6. Aufl., Fischer/Stuttgart, 353 pp.

WALTER, H. 1990: Vegetationszonen und Klima. 6. Aufl., Ulmer/Stuttgart, 382 pp.

WALTER, H. & ALECHIN, W.W. 1936: Grundlagen der Pflanzengeographie (Russ.). Moskau-Leningrad.

WALTER, H. & BOX, E.O. 1983: Overview of Eurasian continental deserts and semideserts, 3–269. In: Ecosystems of the World Vol. **V**. Amsterdam.

WALTER, H. & BRECKLE, S.-W. 1985: Ecological Systems of the Geobiosphere. Vol. 1: Ecological Principles in Global Perspective. 242 pp. Springer/Berlin

WALTER, H. & BRECKLE, S.-W. 1986: Ecological Systems of the Geobiosphere. Vol. 2: Tropical and Subtropical Zonobiomes. 465 pp. Springer/Berlin

WALTER, H. & BRECKLE, S.-W. 1989: Ecological Systems of the Geobiosphere. Vol. 3: Temperate and Polar Zonobiomes of Northern Eurasia. 581 pp. Springer/Berlin

WALTER, H. & BRECKLE, S.-W. 1990: Ökologie der Erde, Bd. 1: Ökologische Grundlagen in globaler Sicht, 238 pp., UTB Große Reihe, 2. Aufl. Fischer/Stuttgart.

WALTER, H. & BRECKLE, S.-W. 1991: Ökologie der Erde, Bd. 2: Spezielle Ökologie der Tropischen und Subtropischen Zonen, 461 pp. UTB Große Reihe, 2. Aufl., Fischer/Stuttgart.

WALTER, H. & BRECKLE, S.-W. 1991a: Ökologie der Erde, Bd. 4: Spezielle Ökologie der Gemäßigten und Arktischen Zonen außerhalb Euro-Nordasiens, 586 pp. UTB Große Reihe, Fischer/Stuttgart.

WALTER, H. & BRECKLE, S.-W. 1994: Ökologie der Erde, Bd. 3: Spezielle Ökologie der Gemäßigten und Arktischen Zonen Euro-Nordasiens, 726 pp. UTB Große Reihe, 2. Aufl., Fischer/Stuttgart.

WALTER, H., HARNICKELL, F. & MÜLLER-DOMBOIS, D. 1975: Klimadiagramm-Karten der einzelnen Kontinente und ökologische Klimagliederung der Erde. 36 pp. Fischer/Stuttgart.

WALTER, H. & KREEB, K. 1970: Die Hydratation und Hydratur des Protoplasmas der Pflanzen. Protoplasmatologia, Bd. II C 6, 306 pp., Wien.

WALTER, H. & LIETH, H. 1967: Klimadiagramm-Weltatlas. Fischer/Jena.

WALTER, H. & MEDINA, E. 1971: Caracterizacion climatica de Venezuela sobre la base de climadiagramas de estaciones particulares. Bol. Socied. Venez. de Cienc. Natur. **29**, 211–240.

WALTER, H. & STEINER, M. 1936: Die Ökologie der ostafrikanischen Mangroven. Ztschr. f. Bot. **30**, 63–193.

WALTER, H. & STRAKA, H. 1970: Arealkunde, Floristisch-Historische Geobotanik. 2. Aufl., 473 pp., Stuttgart.

WATTENBERG, I. & BRECKLE, S.-W. 1995: Tree species diversity of a premontane rain forest in the Cordillera de Tilaran, Costa Rica. Ecotropica **1**, 21–30.

WEISCHET, W. 1980: Die ökologische Benachteiligung der Tropen. 2. Aufl., Teubner/Stuttgart.

WHITE, I.D., MOTTERSHEAD, D.N. & HARRISON, S.J. 1992: Environmental Systems. Chapman & Hall/London, 616 pp.

WHYTE, R.O. 1974: Tropical grazing lands. 222 pp., The Hague.

WICKENS, G.E. 1993: Vegetation and ethnobotany of the Atacama desert and adjacent Andes in northern Chile. Opera Botanica **121**, 291–307.

WILLERT, J. VON, ELLER, B.M., BRINCKMANN, E. & BAASCH, R. 1982: $CO_2$ gas exchange and transpiration of *Welwitschia mirabilis* Hook fil. in the Central Namib Desert. Oecologia **55**, 1 21–29.

WILLERT, J. VON, ELLER, B.M., WERGER, M.J.A. & BRINCKMANN, E. 1990: Desert succulents and their life strategies. Vegetatio **90**, 133–143.

WILSON, E.O. (ed.) 1992: Ende der biologischen Vielfalt? Der Verlust an Arten, Genen und Lebensräumen und die Chancen für eine Umkehr. Spektrum Akad. Verlag/Heidelberg, 557 pp.

WISSMANN, H. VON, 1961: Stufen und Gürtel der Vegetation und des Klimas in Hochasien und seinen Randgebieten (Teil B). Erdkunde **15**, 19–44.

YURTSEV, B.A. 1981: Relikte von Steppenkomplexen in Nordostasien. 168 pp. „Nauka", Novosibirsk (Russ.).

ZINKE, P.J. 1973: Analogy between the soil and vegetation types of Italy, Greece and California. Ecol. Stud. **7**, 61–82.

ZUKRIGL, K., ECKHARDT, G. & NATHER, J. 1963: Standortskundliche und waldbauliche Untersuchungen in Urwaldresten der niederösterreichischen Kalkalpen. Mitt. Forstl. Bundesversuchsanst./Mariabrunn **62**, 244 pp.

# Explanations of Foreign Technical Words Used

(g = Greek, l = Latin, r = Russian, gen. = genitive)

| | |
|---|---|
| a-, an- (g) | = negating prefix, without |
| ad- (l) | = on, to |
| aequus (l) | = equal |
| aér, gen. aéros (g) | = air |
| aggregátio (l) | = aggregate, accumulation |
| alkali (arab.) | = potash |
| allélos (g) | = mutual, reciprocal |
| ámorphos (g) | = amorphous, shapeless |
| amphi (g) | = (a)round, surrounded by |
| análysis (g) | = analysis, explanation |
| ánemos (g) | = wind |
| ánnuus (l) | = annual |
| antagónisma (g) | = antagonism, contest, competition |
| ánthos (g) | = flower |
| ánthropos (g) | = human |
| áper, from apértus (l) | = open, uncovered |
| arché (g) | = beginning, origin |
| áridus (l) | = dry |
| assimilátio (l) | = assimilation, adjustment, modification |
| atmós (g) | = vapour |
| autós (g) | = self |
| bactéria (g) | = rod |
| básis (g) | = base, foundation |
| biénnis (l) | = biennial |
| bíos (g) | = life |
| boréas (g) | = northern wind |
| burozem, from buryi (r) | = brown and zemljá (r) = soil |
| capíllus (l) | = hair |
| carpos, see karpos | |
| causa (l) | = reason |
| chamaí (g) | = low |
| chernózem, from chórnyi (r) | = black and zemljà (r) = earth |
| chlorós (g) | = green |
| chorein (g) | = wander |
| círcum (l) | = around |

| | |
|---|---|
| – cóla (l) | = living |
| con, cum (l) | = together |
| cútis (l) | = skin |
| de- (l) | = from, away |
| déndron (g) | = tree |
| día (g) | = through the middle |
| dichotomein (g) | = divide into two |
| diffúsio (l) | = spreading, expansion |
| dis- (l) | = apart, un- |
| dissimilátio (l) | = to make dissimilar, unlike |
| dissociáre (l) | = separate |
| domináre (l) | = rule, dominate |
| dynamis (g) | = power |
| eidos, oides (g) | = appearing |
| endémos (g) | = native, endemic |
| ephémeros (g) | = lasting one day |
| epí (g) | = on |
| epígaios (g) | = on earth |
| eu (g) | = good, beautiful |
| eurys (g) | = wide |
| evaporátio (l) | = to lose moisture, evaporation |
| e, ex (l) | = from, out of |
| exó (g) | = outside |
| extensívus (l) | = extensive, covering a large area |
| factor (l) | = influence, power |
| flos, gen. floris (l) | = flower |
| flúctuus (l) | = waving, swaying |
| fóssilis (l) | = dug out |
| frúctus (l) | = fruit |
| fúngos (l) | = fungus |
| gámos (g) | = marriage |
| ge, gaia (g) | = earth |
| génesis (g) | = origin |
| graphein (g) | = to write |
| gútta (l) | = drop |
| gymnós (g) | = naked |
| hals, gen. halos (g) | = salt |
| hápax (g) | = once |
| harmonia (g) | = correct relation |
| hélos, gen. héleos (g) | = swamp |
| hélios (g) | = sun |
| hémi- (g) | = half |
| hérba (l) | = grass |
| héteros (g) | = dissimilar, the other |
| hólos- (g) | = complete, whole |
| homoios (g) | = similar, equal |
| húmidus (l) | = moist, humid |
| húmus (l) | = soil |
| hydor, gen. hydatos (g) | = water |

| | |
|---|---|
| hygrós (g) | = moist |
| hypér (g) | = above |
| hypó (g) | = under |
| jedom, from jeda (r) | = food (ice loess, which is "eaten up" by the warmth of summer) |
| incrustáre (l) | = to cover with a shell |
| indicáre (l) | = to show, indicate |
| indivíduum (l) | = indivisable |
| inténsus (l) | = fierce, intense |
| ínter (l) | = between |
| ión (g) | = wandering |
| ísos (g) | = similar, the same |
| karpós (g) | = fruit |
| kínesis (g) | = change, movement |
| klimax, Gen. klimatos (g) | = ladder |
| koinós (g) | = together |
| kormós (g) | = shoot |
| kryos (g) | = frost |
| lábilis (l) | = transitory |
| lamélla (l) | = small leaf |
| laurus (l) | = victory, triumph, laurel |
| leptós (g) | = tender |
| letális (l) | = deadly |
| lígnum (l) | = wood |
| límne (g) | = sea |
| líthos (g) | = stone |
| lítus, gen. litoris (l) | = bank |
| lógos (g) | = teaching, science |
| lysis (g) | = solution |
| makrós (g) | = large, big |
| malakós (g) | = soft, limp |
| máximum (l) | = biggest |
| mesós (g) | = moderate |
| métron (g) | = measure |
| mikrós (g) | = small |
| mínimum (l) | = smallest, least |
| molécula (l) | = little lump |
| mónos (g) | = single, alone |
| morphé (g) | = form |
| mors, gen. mórtis (l) | = death |
| mykes (g) | = fungus |
| myrmex (g) | = ant |
| nekrós (g) | = dead |
| nemorális (l) | = grove, deciduous forest |
| nómos (g) | = law |
| ob- (l) | = inversed, opposite |
| oikos, oikia (g) | = house, household |
| olígos (g) | = little |

| | |
|---|---|
| ópsis (g) | = appearance |
| óptimus (l) | = best |
| órganon (g) | = tool |
| órnis, gen. órnithos (g) | = bird |
| óros (g) | = high ground, mountain |
| ósis (g) | = pressure |
| osmós (g) | = press out |
| oxys (g) | = acid, sharp |
| palaiós (g) | = old |
| parasítos (g) | = eating with someone |
| páthos (g) | = suffering, illness |
| pédon (g) | = soil |
| peine (g) | = hunger, lack of |
| perénnis (l) | = persisting, lasting |
| perí (g) | = around |
| períodos (g) | = range, extend |
| permeábilis (l) | = permeable |
| phagein (g) | = to eat |
| phainein (g) | = make visible |
| phainomai (g) | = to appear |
| phanerós (g) | = obvious, clear |
| phásis (g) | = appearance |
| phílos (g) | = loving |
| phobein (g) | = to flee |
| phóbos (g) | = fear |
| phos, gen. photós (g) | = light |
| phykos (g) | = seaweed |
| phyllon (g) | = leaf |
| physis, Gen. physeos (g) | = nature, life |
| phytón (g) | = plant |
| planktós (g) | = roaming, sway |
| plasma (g) | = thing, formation |
| plus, gen. plúris (l) | = more |
| podsól, from pod (r) | = below, underneath and solá (r) = ash |
| póros (g) | = opening, through passage |
| pótens (g) | = able, efficient |
| pous, gen. podós (g) | = foot |
| prímus (l) | = first |
| pro (g) | = before, previously |
| próblema (g) | = debatable point |
| proveníre (l) | = emerge, come out |
| psammos (g) | = sand |
| pseudés (g) | = wrong |
| psilós (g) | = naked |
| pterídion (g) | = little feather |
| pyr, gen. pyrós (g) | = fire |
| quótiens (l) | = how many times |

| | |
|---|---|
| reáctio (l) | = counter effect |
| redúctio (l) | = return, restore |
| regeneráre (l) | = to recreate |
| reprodúcere (l) | = bring forth, produce |
| resorbére (l) | = swallow, engulf |
| rhíza (g) | = root |
| rúdus, gen. rúderis (l) | = rubble |
| saprós (g) | = rotten |
| secrétio (l) | = separation, secretion |
| sémi- (l) | = half |
| sílva (l) | = forest |
| símilis (l) | = at the same time |
| sklerós (g) | = hard |
| sócius (l) | = together |
| sólum (l) | = soil |
| spéctrum (l) | = scheme, face |
| spérma, | = seed |
| gen. spérmatos (g) | |
| sphaira (g) | = globe, ball |
| spiráre (l) | = breath |
| spontáneus (l) | = voluntarily |
| spóros, sporá (g) | = seed |
| stádium (l) | = state, condition |
| stenós (g) | = narrow |
| stérilis (l) | = sterile |
| structúra (l) | = building |
| sub- (l) | = below |
| substrátum (l) | = basis, foundation |
| súbtilis (l) | = fine |
| succuléntus (l) | = juicy |
| sym-, syn- (g) | = together |
| syrozém, | = soil |
| from syrój (r) raw | |
| and zemljá (r) | |
| systema (g) | = putting together |
| télos (g) | = goal |
| temperáre (l) | = moderate |
| terra (l) | = earth |
| thállos (g) | = green shoot |
| theoría (g) | = look at, consider |
| thermós (g) | = warm |
| théros (g) | = summer |
| tomé (g) | = division, cut |
| tónos (g) | = tension |
| tópos (g) | = position, place |
| tótus (l) | = complete |
| trans (l) | = away, over |
| trophé (g) | = nourishment |
| túber (l) | = bulb |

| | |
|---|---|
| túbus (l) | = tube |
| túrgor (l) | = swelling |
| typos (g) | = characteristic, type |
| ulígo, gen. uliginis (l) | = swamp |
| últra (l) | = beyond |
| unifórmis (l) | = regular, even, uniform |
| usíon (g) | = processing |
| vácuum (l) | = empty space |
| várians (l) | = changeable |
| vegetáre (l) | = give life |
| vicárius (l) | = in place of, substitute |
| vírus (l) | = salt, poison |
| víta (l) | = life |
| volúmen (l) | = content |
| voráre (I) | = consume |
| xerós (g) | = dry |
| xylon (g) | = wood |
| zön see koinós | |
| zóne (g) | = girdle, belt |
| zóon (g) | = animal |

# Subject Index

Made in the USA
San Bernardino, CA
29 August 2015